STUDIES OF *SARGASSUM* AND THE *SARGASSUM* COMMUNITY

James N. Butler
Byron F. Morris
Jill Cadwallader
Allan W. Stoner

With contributions by:
Barbara M. Baum
Laura Brown
Rosamond H. Butler
Kevin Cain
Ellen Deason
Paul Deslauriers
A. Barr Dolan
Allan Douglas
Michael Fine
Tom Flanagan
Jane Frick
Susan Fritz
Joy Geiselman
Holly S. Greening
Emily Jo Hoffman
Lisa M. Horrocks
Anthony H. Knap
Deborah D. Mogelberg
Gretchen Rupp
Thomas D. Sleeter
Karen Stockbridge
Joe Swierzbinski
Dorset W. Trapnell
Barbara Trimingham
Jan Tuttleman

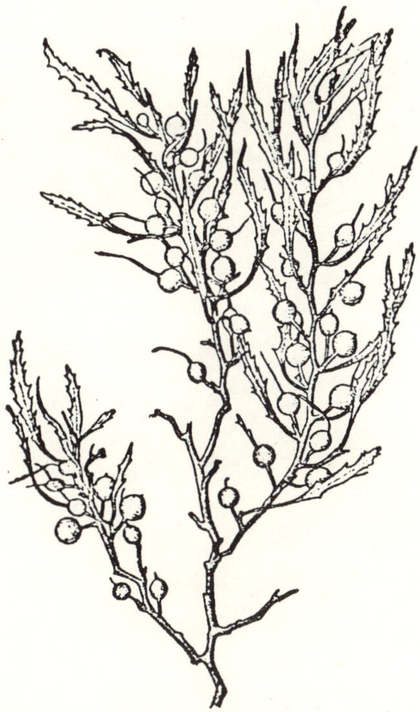

Fig. 3. *Sargassum I.* Typical specimen. Nat. size.

Bermuda Biological Station
Special Publication No. 22
1983

Author's addresses at the time of publication:

Prof. James N. Butler
Division of Applied Sciences
Harvard University
Pierce Hall, 29 Oxford St.
Cambridge, MA 02138
(617) 495-2845

Dr. Byron F. Morris
National Marine Fisheries Service
Box 43, 701 C St.
Anchorage, AK 99513
(907) 271-5006

Ms. Jill Cadwallader
Bermuda Biological Station for Research
Ferry Reach 1-15
Bermuda
(809) 297-1880

Dr. Allan W. Stoner
Sea Education Association
P.O. Box 6
Woods Hole, MA 02543
(617) 540-3954

Illustrations: Front cover: Photo by James N. Butler. Back cover: Krummel (1891). Title page: Winge (1923).

ISBN 0-917642-22-8

ABSTRACT

The quantity and distribution of *Sargassum* in the Sargasso Sea, as estimated by various investigators, is reviewed. There has apparently been no significant change in the biomass of *Sargassum* from 1933 to 1981, except for an area northeast of the Antilles (20-25°N, 62-68°W) where measurements made in November 1977 and November 1980 were about 0.1% of values measured in February and March 1933. Because of the lack of change in the Bermuda region, the Bahamas region, or the Gulf Stream region, this effect does not appear to be due to pollution or broad climatic changes; it is most likely due to a seasonal or long-term shift of the currents defining the southwestern boundary of the Sargasso Sea.

Motile macrofauna on *Sargassum* were identified and counted for 244 samples taken at Station 'S' near Bermuda, and for 155 samples taken from other parts of the western North Atlantic and Caribbean. The nature of this community and its seasonal variations are discussed. Preliminary studies of the food web, as well as of the microfauna (copepods) and sessile fauna are reported. Although petroleum hydrocarbon residues are found in all samples of *Sargassum* and its larger associated fauna, the variability of individual samples is so extreme that no correlation of community structure with hydrocarbon pollution could be demonstrated.

PREFACE

This book was begun so that the uncompleted results of faunistic studies of the *Sargassum* community, carried out at the Bermuda Biological Station from 1971 to 1976 under the direction of Byron F. Morris, could be presented in an accessible format.

In 1976, Morris became Assistant Director of the Biological Station and Project Leader of the Bermuda Inshore Waters Investigation; in 1977 he moved to Alaska and is currently the Outer Continental Shelf Coordinator for the National Marine Fisheries Service in Anchorage. The partially completed *Sargassum* studies were never published (except for some food web data included in the 1976 book "Marine Pollutant Transfer" edited by Windom and Duce - see Appendix VI).

Even though the grant "Transfer of Persistent Pollutants in Sargassum Communities" (NSF-IDOE GX 32883) expired in 1978, we felt it important to make these unpublished field data available to future workers on this problem, and therefore compiled and edited what seemed to be valuable.

The quantitative surveys of *Sargassum* made from the R/V Westward during the period 1977 to 1981 by Allan Stoner have been included because of their obvious relevance to our work. We are able to present these data in greater detail than available elsewhere, and we have made some previously unpublished comparisons using them. Stoner's work was supported by the Sea Education Association of Woods Hole, Massachusetts.

We want to emphasize that this report is not intended to be a final scientific statement. We want to make data and observations, many of which cannot be repeated, available to other researchers.

Christine Lawton patiently typed the many drafts of this text, drew some of the illustrations, and never once complained about revisions. This last phase of the work was supported in part by a grant from the Zemurray Foundation.

James N. Butler
Bermuda, August 1982

CONTENTS

1. THE QUANTITY OF SARGASSUM IN THE SARGASSO SEA

1.1 HISTORY

Beginning with Columbus, drifting *Sargassum* found in the central North Atlantic was believed to have been detached from littoral regions or from a submerged shelf area in the vicinity of the Sargasso Sea (Humboldt's "Fucusbank"). Meyen (1838) was the first to suggest that *Sargassum* was truly pelagic and propagated by the partitioning of older plants. Krummel (1891) gave a historical survey of information on the Sargasso Sea and Winge (1923) gave a good summary updating Krummel's survey. Although the argument is still not completely resolved -- as recently as 1968 Weis repeated the belief in a benthic origin of pelagic *Sargassum* species -- the studies of Winge (1923) and Parr (1939) have resulted in general acceptance that the *Sargassum* species found in the Sargasso Sea (*S. natans* and *S. fluitans*) are almost entirely holopelagic.

Many early investigators attempted to define the boundaries of the Sargasso Sea on the basis of the abundance of *Sargassum*, but as a result of temporal and spatial inhomogeneities in its distribution, a considerable variety of boundaries have been proposed for the Sargasso Sea. (Figs. 1.1-1.5).

The Sargasso Sea is now defined in terms of physical boundaries -- the Gulf Stream to the west, the North Atlantic Current to the north, the Canaries Current to the east, and the North Equatorial Current to the south (Fig. 1.6; Dietrich, 1965). *Sargassum* may be found anywhere within these boundaries, and although it may be more consistently abundant in certain sectors, abundance at any single location is quite unpredictable. As Parr (1939) recognized, it is probably impossible to predict *Sargassum* quantities accurately on a local scale.

The boundaries of *Sargassum* distribution deduced by both Krummel (1891) (Fig. 1.4) and Winge (1923) (Fig. 1.5) are quite similar, however, and approximate the accepted physical boundaries sufficiently to suggest that within any given region of the gyre, one can expect to find some *Sargassum*. The question then becomes not one of presence or absence, but of the quantities that could be expected to occur within various regions and at different times. Three studies have addressed this question.

Krummel (1891), using ships' logs, simply noted the frequency of occurrence for each 1° square (that is, the percentage of time at which *Sargassum* was observed within each square, disregarding the quantities which were present at any given time. Winge obtained his samples from plankton tows and graded them qualitatively from "absent" to "luxuriantly abundant", but as Parr noted, there is no way to translate such

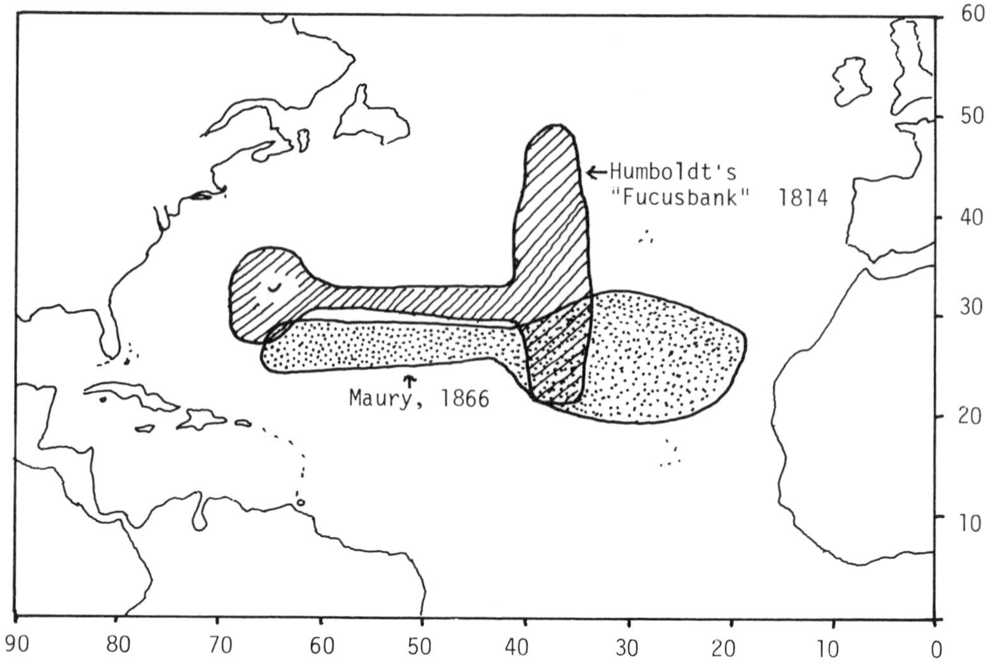

Fig. 1.1. Boundaries of the Sargasso Sea as deduced by observations of the limits of *Sargassum*, according to Humboldt (1814) and Maury (1866).

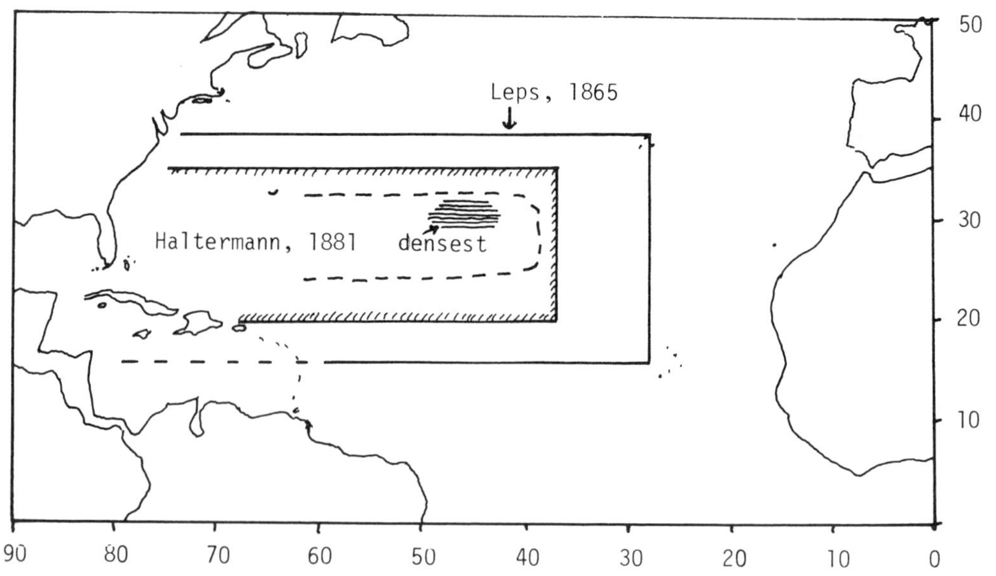

Fig. 1.2. Boundaries of the Sargasso Sea according to Leps (1865) and Haltermann (1881).

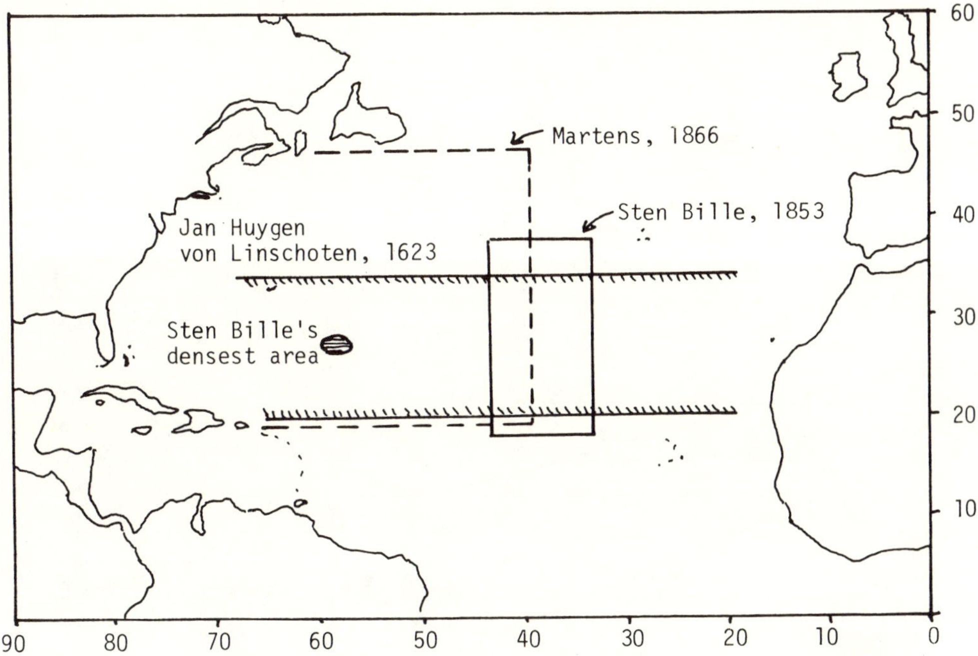

Fig. 1.3. Boundaries of the Sargasso Sea according to Jan Huygen von Linschoten (1623), Sten Bille (1853) and Martens (1866).

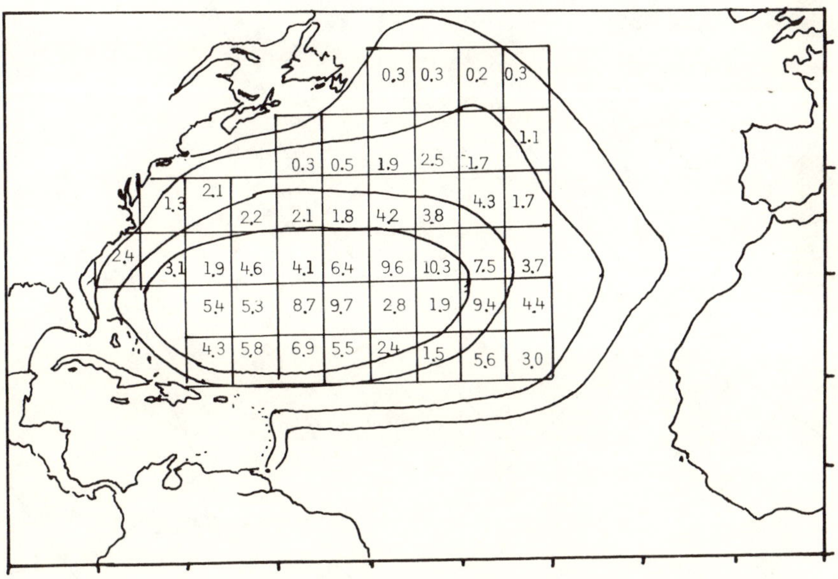

Fig. 1.4. The isophykodes of Krummel (1891) showing Sargassum frequencies of respectively 10, 5, 1, and 0.3 percentage. In the 5° sections are printed the frequencies for the entire year. (Winge, 1923)

Fig. 1.5. Observations of Sargassum (black spots) and observations of its absence (open circles). Greater quantities of Sargassum are represented by larger spots. (Winge, 1923)

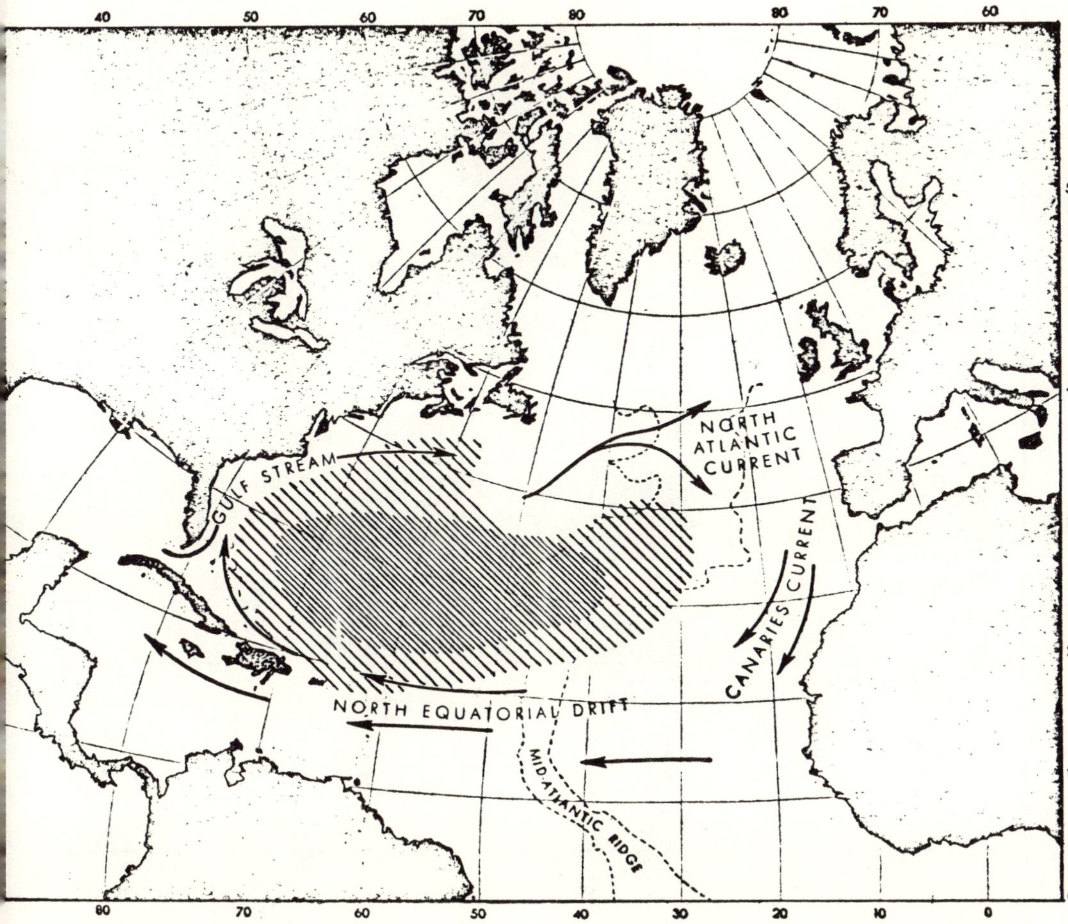

Fig. 1.6 Circular system of currents in the North Atlantic defines the Sargasso Sea. The darker shading indicates Krümmel's boundary and the lighter shading indicates Wingë's boundary. (Ryther, 1956) From "The Sargasso Sea" by John Ryther.

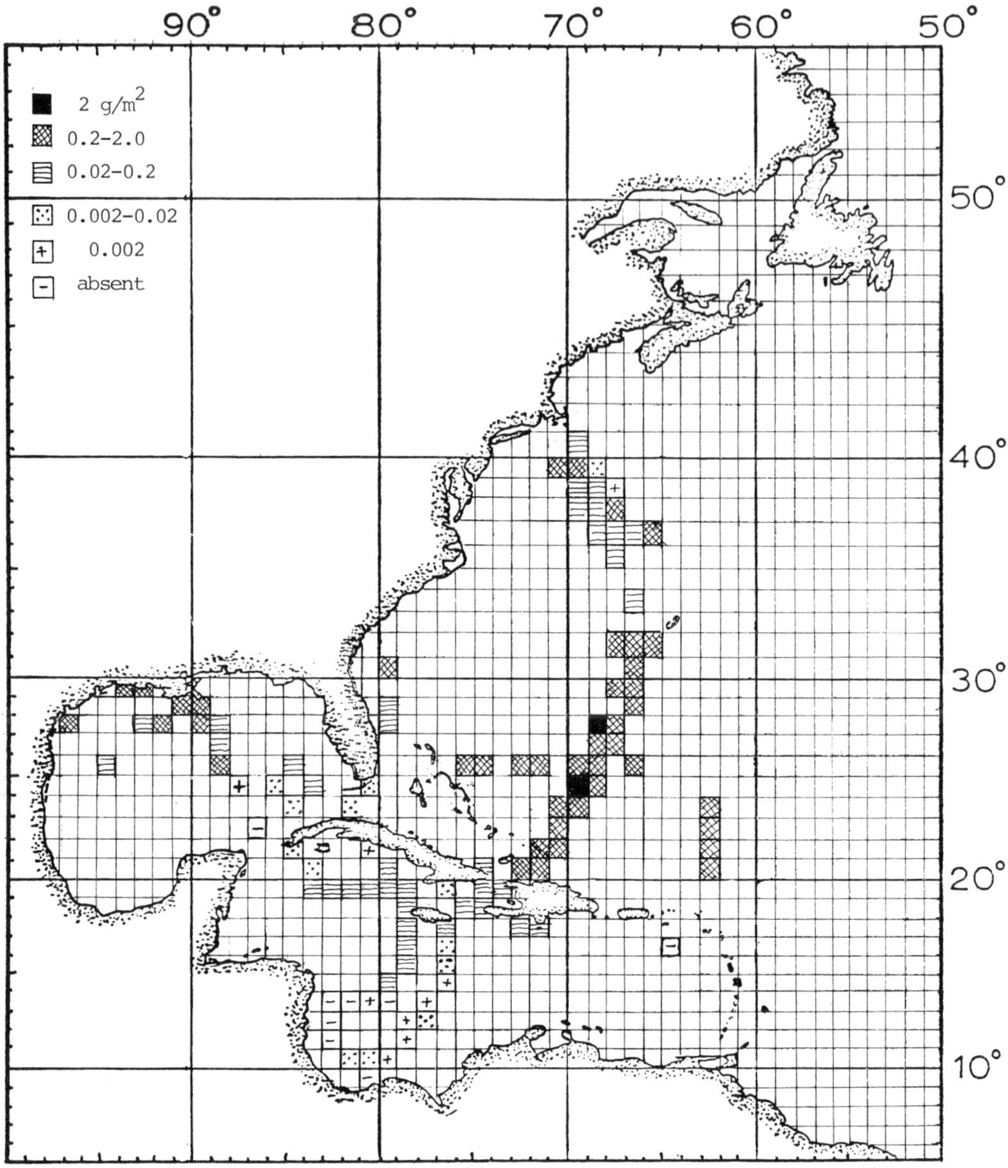

Fig. 1.7. Distribution of Sargassum as determined by Parr (1939). When more than one tow was taken within a one-degree square, the total Sargassum collected was divided by the total area skimmed to obtain a weighted mean value.

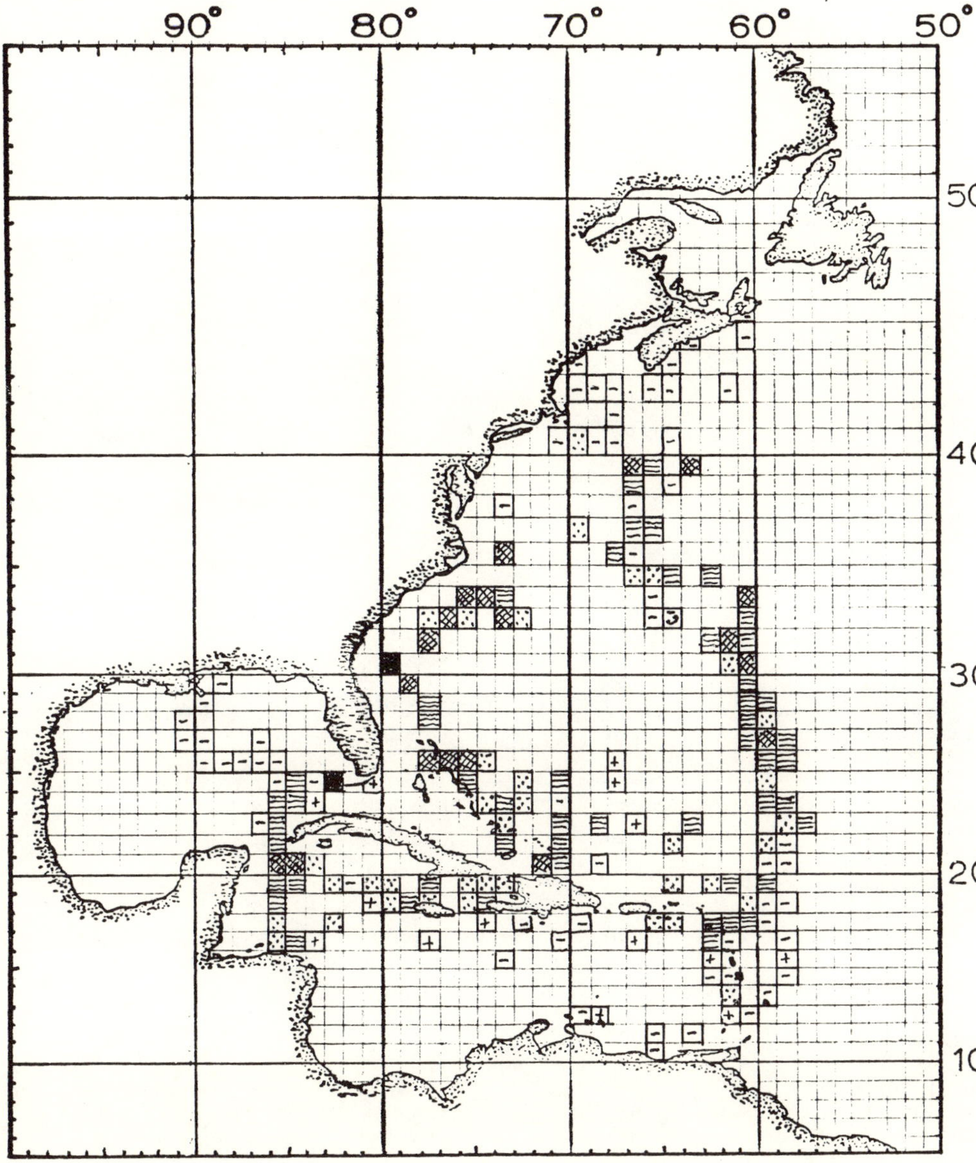

Fig. 1.8. Stations at which neuston tows were made by Stoner during 1977-81 (Stoner, 1982)

Table 1.1. Abundance of Pelagic *Sargassum* in various regions of the north Atlantic Ocean and Caribbean Sea. Values for 1933-35 are derived from data of Parr (1939) and those for 1977-81 from data of Stoner (1982).

Geographic Region	Sargassum biomass (g/m^2)	Range of biomass (g/m^2)	number of tows	% of tows yielding Sargassum
Sargasso Sea*				
1933-35	1.24	0.022-19.58	86	100
1977-81	0.072	0-1.12	76	76
Bahama Islands*				
1933-35	0.59	0.053-1.62	7	100
1977-81	0.165	0.002-1.73	22	100
Gulf of Mexico*				
1933-35	0.26	0.001-0.66	22	100
1977-81	0	0	15	0
Gulf Stream				
1933-35	0.21	0.042-0.57	4	100
1977-81	0.28	0-1.84	13	69
Florida Straits				
1933-35	0.047	0.007-0.052	3	100
1977-81	0.637	0-6.02	10	60
Caribbean Sea				
1933-35	0.028	0-0.15	62	84
1977-81	0.018	0-0.195	71	68
Yucatan Strait*				
1933-35	0.013	0-0.036	6	83
1977-81	0.140	0-1.03	13	85
Shelf of North America				
1933-35	no data			
1977-81	0.0034	0-14	14	7
Slope of North America*				
1933-35	0.008	0.001-0.015	2	100
1977-81	0	0	7	0

*mean *Sargassum* biomass values for the two time periods were different, as determined by the Mann-Whitney U test. $p \leq 0.01$ except for Bahamas, where $p \leq 0.05$ Station locations and boundaries of geographic regions are shown in the map of Appendix I.

descriptions into absolute quantities.

Parr (1939) did the first and one of the few quantitative studies of Sargassum, and calculated that the average amount of Sargassum within the Sargasso Sea was about 0.5 to 1.5 g/m^2 for the period 1933-35. His full data set is reproduced in Appendix I. In Sections 1.2 and 1.3, Parr's results will be compared to more recent quantitative measurements.

Unfortunately, Parr's study was limited to the region west of Bermuda, and nothing is known about the quantitative abundance of Sargassum in the eastern regions of the Sargasso Sea.

Figure 1.7 presents a semi-quantitative distribution of Sargassum for each 1° square as reported by Parr. It can readily be seen that the region of highest abundance in the areas he studied was in the southwestern sector of the Sargasso Sea, where quantities ranged from 0.2 to 2.0 g/m^2, with greater than 2 g/m^2 at 26°N, 68°W. The region to the north of Bermuda had about an order of magnitude less (0.02-0.2 g/m^2). In the Caribbean Sea, distribution was patchy, and quantities ranged from 0 to 0.2 g/m^2.

The Gulf of Mexico was also patchy, but a secondary maximum was found in the northern region where quantities approached 1.0 g/m^2. Maximum and minimum quantities within each square generally did not differ by more than a factor of ten from the mean quantity.

Morris (1971; see also Butler, Morris and Sass, 1973) sampled Sargassum from the R/V Panulirus II near Bermuda during 1971 and 1972. His results are presented and discussed in Section 1.3.

Stoner (1982) surveyed the western Sargasso Sea, the Caribbean, and the Gulf of Mexico on board the R/V Westward, during the period 1977-81. The area covered and number of observations were comparable to Parr's 1933-35 study (Figure 1.8). However, Stoner's tows were much shorter (1 nautical mile) than Parr's (typically 10-40 miles, but occasionally as much as 174 miles). The patchy distribution of Sargassum results in more frequent zero values for the shorter tows (Section 1.3). Stoner's full data set is given in Appendix I, and an abbreviated comparison with Parr's data is given in Table 1.1.

Highest biomass was found in the Florida Straits (mean 0.64 g/m^2, range 0.6.02), Gulf Stream (mean 0.28 g/m^2, range 0-1.84), and near the Bahamas (mean 0.165 g/m^2, range 0.002-1.73). Lowest biomass was found in the Sargasso Sea (mean 0.072 g/m^2, range 0-1.12), the Gulf of Mexico (zero for 15 tows). The values for the continental shelf and slope were also mostly zero, as would be expected. Stoner's results are compared in detail with Parr's results in Sections 1.2 and 1.3.

TABLE 1.2 SARGASSUM COLLECTED AT STATION 'S', 1971-72

Sta.	Date	m^3	naut. miles	Tar g	Tar mg/m^3	Sargassum g	Sargassum mg/m^3	Species
341	12 May 71	?	?	?	0.1	0.	0	
342	25 May 71	?	?	?	11.	0.13?	4.8	not noted
343	7 June 71	45.4	0.82	4.61	102.	2.08	46.	"
344	21 June 71	36.4	0.66	2.54	70.	26.0	712.	"
345	5 July 71	32.2	0.58	0.474	14.7	0	0	
346	6 July 71	29.5	0.53	0.102	3.44	2.04	69.	"
347	19 July 71	26.7	0.48	0.618	23.2	11.4	427.	"
348	5 Aug 71	45.	0.81	6.65	148.5	26.9	602.	"
349	18 Aug 71	36.	0.65	7.95	221.	35.1	974.	"
350	30 Aug 71	26.	0.47	0.338	13.0	24.0	925.	"
351	14 Sept 71	23.	0.41	1.29	56.0	14.2	617.*	"
352	1 Oct 71	73.	1.31	1.07	14.6	72.1	986.	"
353	13 Oct 71	70.	1.26	27.55	394.	165.1	2358.	"
354	(no samples)							
355	24 Nov 71	33.	0.59	13.2	400.	220.	6666.	"
356	15 Feb 72							predominantly
	1600h	27.	0.48	0.50	18.5	4.25	158.	S. natans
	2100h				5.8		710.	
	2200h				19.		156.	
	2300h				77.		675.	
	2400h				10.		529.	
	0100h				10.		4848.	
	0200h				50.		3006.	
	0300h				103.		1609.	
	0400h				294.		512.	
	0500h				102.		3879.	
	0700h				100.		1040.	
357	28 Feb 72	60.	1.08	2.00	33.3	3.6	60.	S. natans
358	5 Apr 72	45.	0.81	3.70	82.	64.5	1430.	"
359	19 Apr 72	70.	1.26	1.30	18.6	0	0.	
360	29 Apr 72	36.	0.65	3.80	105.6	6.60	183.	S. natans
361	12 May 72	20.	0.36	1.40	70.	217.	10870.	"
362	7 June 72	42.	0.76	1.90	45.2	1.00	24.	"
363	28 June 72	25.	0.45	1.00	40	0.	0.	
364	13 July 72	29.	0.52	5.30	182.8	4.60	159.	S. natans
365	8 Aug 72							
	1200h	27?	0.5?	3.1?	114.	.84?	31.	"
	1500h				77.		450.	"
							50.	S. fluitans
	1700h				140.		28.	S. natans
	1800h				158.		266.	"
	1900h				40.		158.	"
	2000h				100.		2963.	"
	2100h				114.		1803.	"
	2200h				25.		82.	"
	2400h				79.		129.	"
	0200h				38.		13.	"
	0300h				120.		67.	"
	0500h				11.		0.	

Sta.	Date	m^3	naut. miles	Tar g	Tar mg/m^3	Sargassum g	Sargassum mg/m^3	Species
365	0600h	27?	0.5?		13.		100.	S. natans
	0700h				117.		892.	"
	0800h				25.		100.	"
	0900h				36.		950.	"
366	31 Aug 72	27.	0.49	0.10	3.7	2.19	81.	"
367	13 Sept 72	29.	0.52	4.70	162.	1.80	62.	"
						2.20	76.	S. fluitans
368	26 Sept 72	18.	0.32	3.80	211.	0.59	33.	S. natans
						0.20	11.	S. fluitans
						0.11	6.	S. hystrix
369	10 Oct 72	31.	0.56	4.00	129.	6.60	213.	S. natans
370	23 Oct 72	22.	0.40	1.10	50.	0.18	8.3	"
						8.29	377.	unidentified
371	3 Nov 72	20.	0.36	7.50	375.	463.6	23180.	S. fluitans
372	28 Nov 72	34.	0.61	0.30	8.8	33.3	979.	S. natans
373	11 Dec 72	21.	0.38	0.20	9.5	45.5	2167.	"
						6.0	286.	S. polyceratum
374	27 Dec 72	35.	0.63	4.50	129.	4.20	120.	S. natans
						8.80	251.	S. fluitans

NOTES: The neuston net had 0.3 mm mesh (Morris, 1971) and sampled surface surface water approximately 10 cm deep and 30 cm wide. The water flow was measured in m^3. To convert m^3 to nautical miles, multiply by 1.80×10^{-2}; to convert m^3 to km towed, divide by 30. To convert mg/m^3 to mg/m^2, divide by 10.

*Sargassum from Station 351 was erroneously reported as 1420 mg/m^2 (14200 mg/m^3) by Butler, Morris and Sass (1973).

The species of Sargassum collected were recorded in detail by Parr (1939) (See Appendix I), and in less detail by Morris (Section 1.3, Table 1.2) and Stoner (1982). In the area near Bermuda (Table 1.3), Parr usually found both *S. natans* (I) and *S. fluitans* (III) -- at stations 1 to 3, 30 to 38, 129 to 132, and 139 to 145, together with other species or subspecies at some stations (37, 141 to 145). At only four stations (119 to 121, 138) did he find *S. natans* only. Morris (Table 1.2) found *S. natans* at all stations but three (359 and 363 were both zero; 371 was a very large haul of *S. fluitans*).

Even though Morris found *S. fluitans* at fewer stations than did Parr, the relative amounts by weight are similar. Summing all the weights for Parr's collections in the Bermuda region

Table 1.3. Comparison of selected data on the abundance of *Sargassum* near Bermuda (28° to 36°N, 60° to 68°W) (from Parr, 1939, pp. 86-90).

NO.	DATE	LOCATION	NAUT. Miles	WET WEIGHT (g/m^2)	SPECIES
1	13 Feb 33	31°25.5'N, 66°45'W	3.0	.067 .268	I = *S. natans* III = *S. fluitar*
2	13 Feb 33	31°24'N, 66°52.5'W	8.0	.138 .490	I III
3	13 Feb 33	31°22'N, 67°3'W	10.0	.151 .542	I III
30	22 Jan 34	36°21'N, 66°39'W	5.9	.048 .163	I III, trace V
31	22 Jan 34	36°18'N, 66°21'W	36.8	.005 .120	I III
32	23 Jan 34	31°57'N, 65°52'W	14.5	.595 .346	I III, trace X
33	23 Jan 34	31°44'N, 66°2'W	14.0	.187 .151	I III
34	24 Jan 34	30°9'N, 66°51'W	8.0	.929 .050	I III
35	24 Jan 34	29°56'N, 66°53'W	19.7	.612 .102	I III, trace II
36	24 Jan 34	29°38'N, 67°00'W	14.8	1.262 .271	I III, trace II
37	24 Jan 34	29°26'N, 67°9'W	16.8	.795 .072 .024	I III II
38	24 Jan 34	29°11'N, 67°17'W	15.2	1.348 .132	I III, trace II, X
119	20 July 34	36°58'N, 66°42'W	38.8 }	.095	I, trace III
120	20-21 July 34	36°21'N, 65°58'W	62.1 }		
121	22-23 July 34	36°03'N, 65°52'W	53.5	.188	I
129	22 Aug 34	36°34'N, 67°29'W	61.3	.161 .015	I III
130	22-23 Aug 34	35°56'N, 67°21'W	15.5	.130 .006	I III
131	26-27 Aug 34	36°04'N, 68°12'W	92.3	.058 .003	I III
132	27 Aug 34	36°53'N, 68°58'W	29.1	.297 .059	I III
138	22 Jan 35	35°53'N, 67°17'W	28.8	.024	III
139	23 Jan 35	33°42'N, 65°31'W	30.9	.023 .085	I III

.40	23 Jan 35	33°7.5'N, 65°2.5'W	49.9	.038	I
				.117	III, trace II
.41	6 Feb 35	29°56'N, 65°55'W	17.65	.313	I
				.165	III
				.023	II, traces VIII, X
.42	6 Feb 35	29°38'N, 66°04'W	21.5	.411	I
				.206	III
				.075	II-VIII
				.019	V, trace X
.43	6 Feb 35	29°06'N, 66°23'W	44.8	.108	I
				.099	III
				.013	II
				.018	X, trace VIII
.44	7 Feb 35	28°30'N, 66°41'W	25.0	.265	I
				1.401	III
				.024	X, traces II, VIII
.45	7 Feb 35	28°03'N, 66°56'W	28.6	.253	I
				.140	III
				.021	X
				.021	II, traces VIII, X

Note: Parr's net was 2 ft wide, coarse (0.7-2 cm mesh), towed at approximately 10 knots (5 m/sec). To convert Parr's pounds per mile to g/m^2, multiply by 0.402. Multiply nautical miles by 1.852 to get km. Full data are given in Appendix I.

(Table 1.3), one obtains 63% _S. natans_, 35% _S. fluitans_ and 2% other. Morris's data (Table 1.2), covering a much smaller area of ocean surface (5.0 x 10 m^2, compared to Parr's 8.66 x 10 m^2) gives 44% _S. natans_, 35% _S. fluitans_ and 2% other.

Stoner reports that his collections appeared to yield mostly _S. natans_, but quantitative data are not available, since the samples were only occasionally sorted.

1.2 SPATIAL DISTRIBUTION OF _SARGASSUM_

The irregular and patchy spatial distribution of _Sargassum_ is well documented. Tradational tales tell of islands of weed thick enough to stop ships, but more reliable reports (e.g., Winge, 1923 and Parr, 1939) emphasize that the normal distribution of _Sargassum_ is in clumps 10-50 cm diameter which tend to line up in the direction of the wind in rows spaced 20-50 m apart. Occasionally, "periods of relatively calm weather will

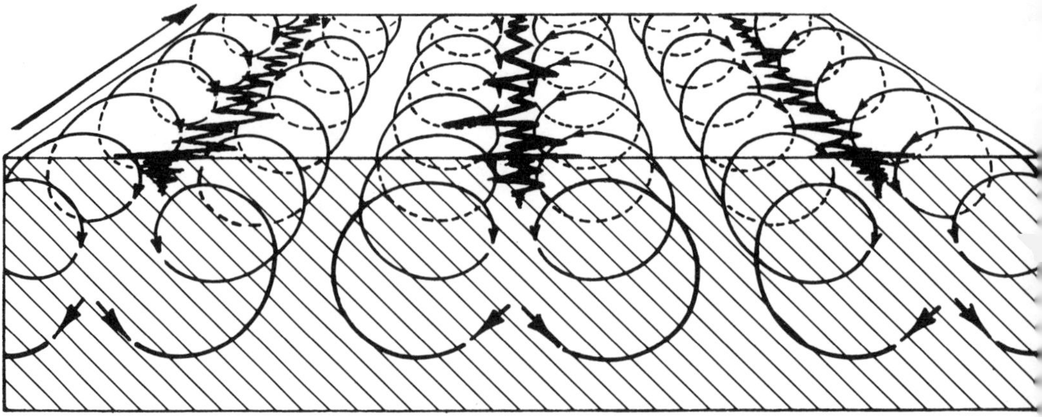

Fig. 1.9. Lines (windrows) of Sargassum can be stabilized by the Langmuir circulation. The wind (arrow at upper left) sets up counter-rotating horizontal eddies. Sargassum and other flotsam collects at the downwelling (dark zigzags). From "The Sargasso Sea" by John Ryther. Copyright © 1956 by Scientific American, Inc. All rights reserved.

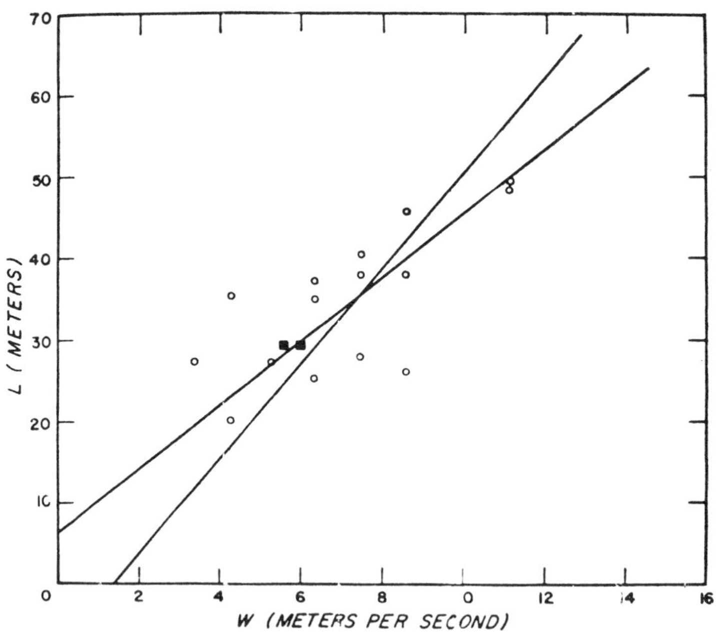

Fig. 1.10. Spacing of windrows (L) as a function of wind speed W. Open circles are observations by Woodcock, solid squares by J. Beers. Lines are regression estimates. (From Faller and Woodcock, 1964).

favor the concentration of drifting weeds into larger and larger aggregations, reducing the frequency of the rafts as their average size increases" (Parr, 1939). As will be noted below, these rafts of Sargassum accumulate at stable downwelling convergences. While they are by no means thick enough to stop a ship, they may be as much as 50 m across and give the illusion of solidity.

Winge (1923) mentioned several times that Sargassum often collects "in long stripes parallel with the direction of the wind". Langmuir (1938) also observed this phenomenon and did experiments to prove that it is the result of wind-induced helical vortices in the surface waters (Figure 1.9). Both a steady wind and a crossed-wave pattern appear to be necessary to establish these vortices (Craik and Leibovich, 1976; Faller, 1978; Leibovich and Paolucci, 1980). Faller and Woodcock (1964) measured the distance between adjacent "windrows" of Sargassum and demonstrated that the spacing increases with increased wind speed (Figure 1.10).

The abundance of Sargassum in windrows sometimes seems to increase with increasing wind speed, but this has not been demonstrated quantitatively; it is likely to be a result of increased visibility as the Sargassum is packed more densely by stronger surface currents.

The wind-driven Langmuir circulation could conceivably introduce bias in samples taken by neuston net, if the direction of the tow was in the direction of the wind. Parr towed while the R/V Atlantis was traveling between hydrographic stations, apparently without regard for wind direction. His long tows probably prevented any bias. Morris always towed in a circle about 1 km diameter to avoid sampling bias. Stoner towed from the port side of the R/V Westward while the ship was partially hove to on a port tack; hence his tows were all essentially perpendicular to any windrows which may have been present.

The duration of windrows after the winds have abated has not been studied, but from Langmuir's (1938) observation of a shift in direction by 90 degrees over a period of 20 minutes, they probably have a relaxation time of less than one hour.

The distribution of Sargassum is also governed by the mesoscale circulation of the Sargasso Sea, a system of eddies with a spatial scale of about 150 to 300 km and a time scale of months, analogous to atomspheric weather systems (MODE Group 1978; Robinson, 1982). Perhaps the best-studied systems in this size range are the Gulf Stream rings (Parker, 1971; Backus *et al.*, 1981), which are circulating bodies of water about 100 km in diameter. On the Sargasso Sea side of the Gulf Stream, the interior of the rings has a temperature and salinity more characteristic of the cold continental slope water than of the

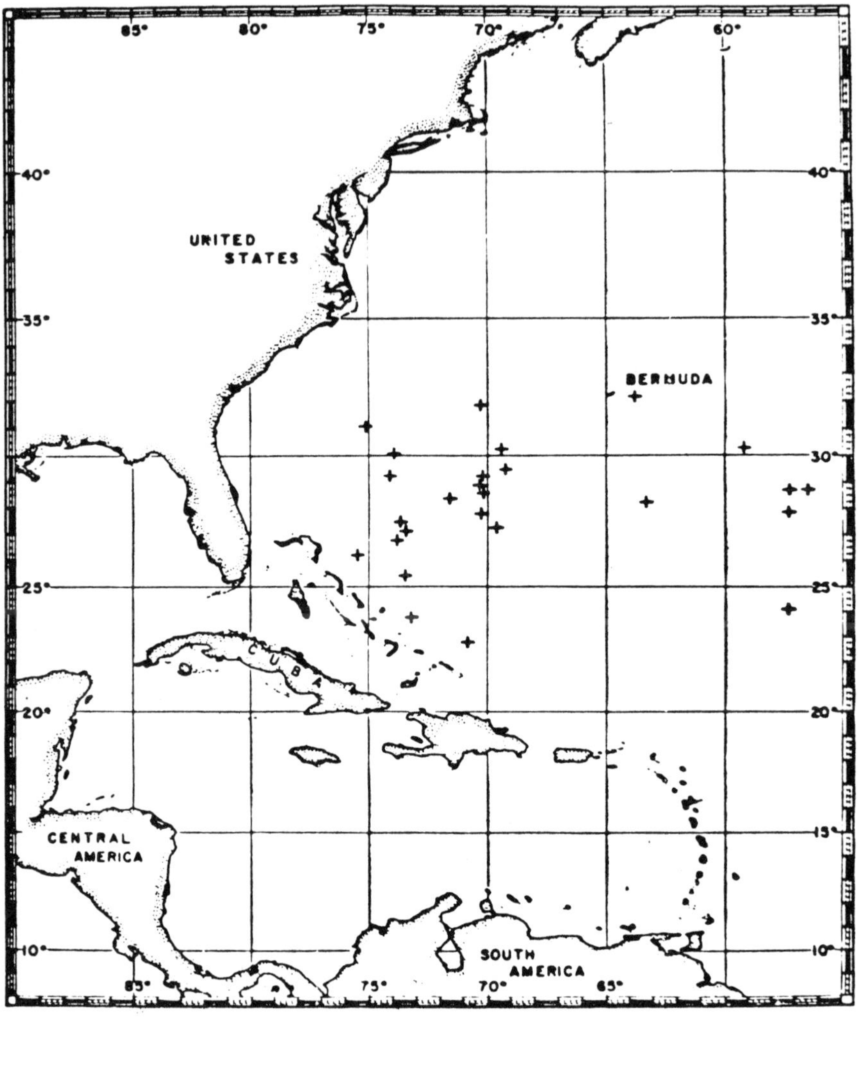

Fig. 1.11. Thermal front observations made during the years 1958-63. (Voorhis, 1969)

surrounding region (Hagan _et al._, 1978). Their counterparts, warm-core rings, carry Sargasso Sea water into the region between the Gulf Stream and the East Coast of North America (Leetmaa, 1977; Watts and Olson, 1978). Stoner, for example, observed large rafts of _Sargassum_ in a warm-core ring to the north of the Gulf Stream, where _Sargassum_ is only rarely observed.

Satellite tracking methods and new dynamic theories have stimulated research into the mesoscale circulation of all the oceans (Wunsch, 1977; Robinson, 1982). Although the details are complicated, the basic motions are well-established and certainly must be one of the principal factors governing the unpredictable and patchy distribution of _Sargassum_. Wherever there is a downwelling convergence, there one might expect to find _Sargassum_, together with tar, plastics, wood, and any other floating material.

Butler and Morris (1974) showed that the Fourier analysis of the 2-year time series at Station 'S' (Table 1.2, discussed in Section 1.3) gave a clear annual cycle of _Sargassum_ (but not of tar) and a 7 to 10 week cycle of both. This latter cycle was identified as being possibly associated with mesoscale circulation.

Another occasion when _Sargassum_ tends to be abundant is through concentrating forces at the edge of thermal fronts. Voorhis and Hersey (1964) and Voorhis (1969) observed 26 thermal fronts in the southern Sargasso Sea between 1958 and 1963. These fronts have been found between 22°N and 32°N, (Fig. 1.11) extending in an east-west direction for several hundred kilometers. As these fronts migrate northward, generally in the winter and spring, they concentrate _Sargassum_ on their leading edge. Their location is unpredictable and their persistence and stability is unknown; but it seems plausible that sharp increases in _Sargassum_ off Bermuda might be explained by the passage of such fronts.

More survey work would be necessary to establish regions of highest _Sargassum_ abundance in the Sargasso Sea. Parr's (1939) survey (Figure 1.7) suggests that peak abundance may be in the region of 25-35°N and 65-75°W, that is, southwest of Bermuda. But Stoner (1981) found the most _Sargassum_ in the Gulf Stream and little in the southwestern Sargasso Sea (See Table 1.1, Figure 1.8, and Section 1.3). However, both Parr's (1939) and Stoner's (1982) observations were restricted to the western region of the Sargasso Sea the Caribbean and Gulf of Mexico. Although no quantitative observations have been made to the east of about 60°W, both Haltermann (as quoted in Kuntze, 1881) and Winge (1923) report that the densest _Sargassum_ is found to the east of Bermuda. Conover and Sieburth (1964), on a transit between Rhode Island and the Canal Zone in the Fall of 1962, found the densest concentration of _Sargassum_ north and east of Bermuda.

Parr (1939) thought that there were no areas of continual high abundance, but rather that the distribution is continually changing. This is consistent with recent studies of mesoscale circulation mentioned above. He realized that a complete picture of Sargassum distribution and abundance would require observations repeated many times over a large number of routes, a difficult and expensive task indeed.

Since windrows and patches are observable from the bridge of a ship, perhaps aerial or high resolution satellite surveillance could locate and photographically quantify areas of abundance to which ships could be dispatched for ground truth measurements. However, to our knowledge such a study has not yet been made.

1.3 HAS THE AMOUNT OF SARGASSUM CHANGED IN THE LAST 50 YEARS?

Quantitative surveys of pelagic organisms are time-consuming and expensive, but are essential if the biological effects of human activity on the open oceans are to be observed directly. When this project was first conceived, one of the central questions we asked was whether the enormous increase in petroleum pollution of the open oceans during the period 1950-1970 had affected their ecology. Anecdotal evidence (for example, Cousteau, 1970) supported polemical pronouncements about the "death of the oceans", but no quantitative evidence seemed to be available.

Sargassum is an obvious choice for such a study, since it is confined to the ocean surface in a region with no continental land mass, and would be influenced by one major type of pollution: pelagic tar. Direct evidence of petroleum hydrocarbons in Sargassum and some of its larger associated fauna was published during the last decade (Burns and Teal, 1973; Morris, Cadwallader, Geiselman and Butler, 1976 - see Appendix VI). Because the petroleum hydrocarbon content of Sargassum itself varies by more than a factor of 100 (Morris et al., 1976) one might expect to see changes in the populations of associated fauna correlated with the presence of petroleum. However, we were not able to see such effects because of the extremely large natural variability of faunal populations on the small, relatively isolated clumps of Sargassum (See Section 2.9).

Sargassum and some of its associated fauna have been shown to accumulate arsenic, mercury, and germanium (Johnson and Braman, 1975). The distribution of As(III), As(V) and alkyl As were the same as in surface Sargasso Sea water. Barnacles and shrimp showed higher Ge and Hg content than did other members of the community. PCB, DDT, and other chlorinated hydrocarbons are more concentrated in the sea surface microlayer at convergences such as the Langmuir windrows (Bidleman and Olney, 1974; Bidleman et al., 1976). However, no reliable analyses of these compounds

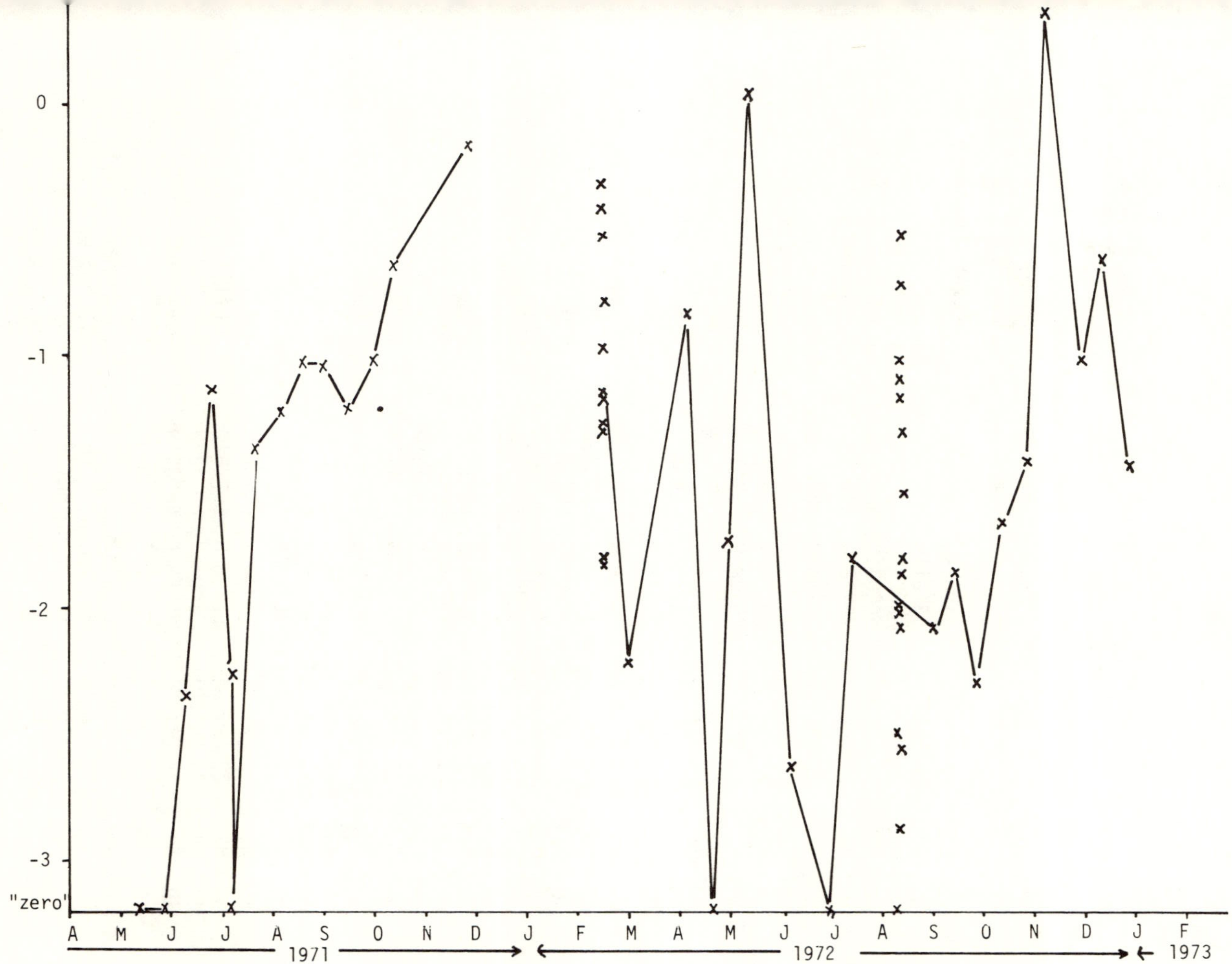

Fig. 1.12. <u>Sargassum</u> collected at Station 'S': Time series 1971-72.

Table 1.4 Selected data on the abundance of *Sargassum* near Bermuda (28° to 36°N, 60° to 68°W) from Stoner, 1982.

STATION	DATE	LOCATION	WET WEIGHT (g/m^2)
W36-N1	19 Oct 77	34°01'N, 62°03'W	0.028
W36-N2	20 Oct 77	33°21'N, 60°14'W	0.220
W36-N3	21 Oct 77	32°29'N, 60°14'W	0.021
W36-N4	22 Oct 77	30°07'N, 60°25'W	0.012
W36-N5	23 Oct 77	28°12'N, 60°33'W	0.190
W42-N2	22 Oct 78	31°06'N, 62°31'W	0.100
W42-N3	22 Oct 78	31°09'N, 61°52'W	0.264
W42-N4	22 Oct 78	30°25'N, 61°28'W	0.018
W42-N5	23 Oct 78	29°20'N, 60°27'W	0.113
W48-N1	16 Oct 79	36°24'N, 65°17'W	0.027
W48-N2	17 Oct 79	34°44'N, 65°04'W	0.018
W48-N3	18 Oct 79	31°31'N, 61°04'W	0.160
W52-N8	13 June 80	36°35'N, 66°48'W	0.159
W52-N9	15 June 80	35°39'N, 66°09'W	0.0
W52-N10	17 June 80	34°58'N, 65°08'W	0.0
W52-N11	18 June 80	34°29'N, 64°38'W	0.0
W52-N12	19 June 80	32°51'N, 64°33'W	0.0
W52-N13	20 June 80	32°26'N, 64°14'W	0.0
W52-N14	23 June 80	33°08'N, 65°17'W	0.0
W52-N15	24 June 80	33°24'N, 65°24'W	0.0
W52-N16	25 June 80	33°32'N, 65°44'W	0.0
W52-N17	26 June 80	34°23'N, 66°11'W	0.0
W58-N32	24 June 81	35°26'N, 67°07'W	0.044
W58-N34	25 June 81	34°06'N, 66°12'W	0.011
W58-N36	1 July 81	34°26'N, 64°58'W	0.051
W58-N37	2 July 81	36°47'N, 65°07'W	0.036
W60-N1	20 Oct 81	36°39'N, 63°05'W	1.381
W60-N2	22 Oct 81	32°18'N, 60°28'W	0.071
W60-N4	24 Oct 81	31°03'N, 60°43'W	0.113
W60-N5	25 Oct 81	30°28'N, 60°34'W	0.845

Note: All tows were made with a 1.0 m wide, 0.55 mm mesh net for 1 nautical mile (1.852 km) at 1 to 2 knots (0.5-1 m/sec). See Appendix I for full original data.

in the Sargassum community have come to our attention (see, for example, Jonas and Pfaender, 1976; Risebrough et al., 1976).

Surface-skimming tows with a neuston net (designed by David, 1965) were taken by B.F. Morris regularly between March 1971 and December 1972, at hydrographic station 'S' 20 miles southeast of Bermuda ($32^{o}10'$ N, $64^{o}30'$ W). The quantities of Sargassum (wet weight) and pelagic tar collected over an approximately one kilometer tow path are given in Table 1.2 and Figure 1.12. (These results were also presented by Butler, Morris and Sass, 1973.) The range of abundance found was 0 to 2.3 g/m^2, similar to that found by Parr (1939) and Stoner (1982). A more detailed comparison will be made below.

The occurrence of Sargassum was sporadic. Although periods of abundance were unpredictable, there appeared to be some seasonality, with maximum quantities occurring in the late autumn. This may reflect the movement past Bermuda of oceanic downwellings and thermal fronts (discussed in Section 1.2 above) as well as the effects of summer growth.

On two occasions, repeated tows were made at approximately hourly intervals to assess the short-term variability of neuston sampling. It is clear from Table 1.2 and Figure 1.12 that the variability within a 24 hour period at one location is two orders of magnitude - nearly as great as the variability over a 2-year period. Fourier transformer time series analysis (Butler and Morris, 1974) showed that on the hourly time scale there was only Markov-type frequency dependence and no distinctive frequency of recurrence; but the two-year series showed a clear seasonal cycle as well as a 7-10 week cycle probably associated with mesoscale circulation.

Tables 1.3 and 1.4 give a selection of data from the Bermuda region as given by Parr and Stoner, respectively. Comparison of Stoner's subset (30-40°N, 60-70°W) with Morris's results is made in Figure 1.13. Parr's data for this region is displayed in the distribution of Figure 1.14. The three distributions overlap substantially and some of the apparent differences result from the different length of tow used.

The height of the histogram bars in Figures 1.13 and 1.14 are proportional to the length of tow used to take the sample. Since both Morris and Stoner used short tows, the sum of all their Sargassum weights divided by the total area of ocean skimmed (weighted mean) would be the best number (Hess, 1981) to compare with Parr's long tows. The weighted means for Morris's and Stoner's data sets are 0.12 g/m^2 and 0.10 g/m^2 respectively. These values are very close to the weighted mean of Parr's data (0.188 g/m^2) and on Figure 1.14 they have been plotted as if they were additional tows of Parr's data set. It is clear from their central position in Parr's distribution that there is no

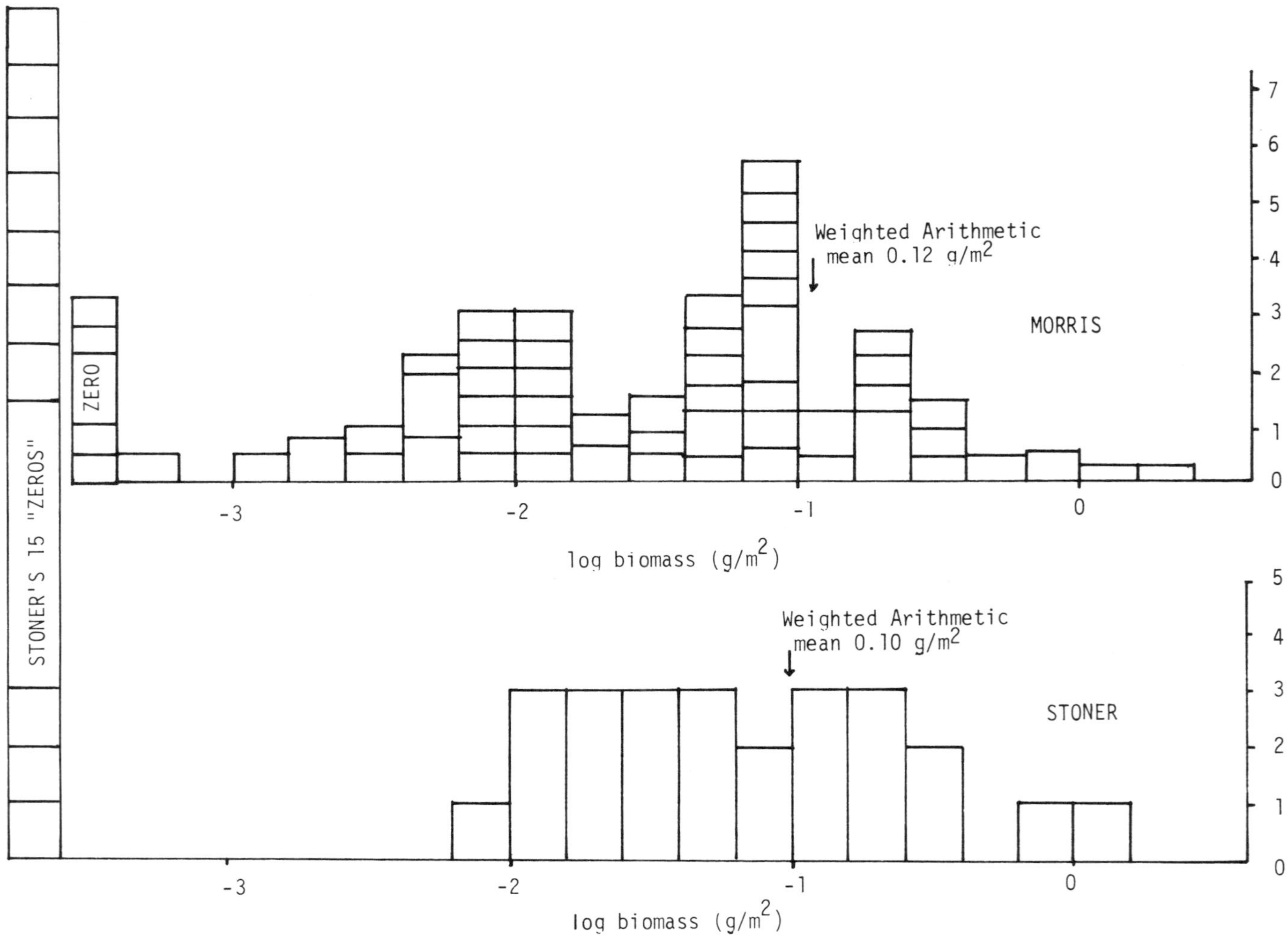

Fig. 1.13. Comparison of Stoner data for northwest Sargasso Sea (30-40°N, 60-70°W), 1977-81
... collected at Station 'S' by Morris, 1971-72

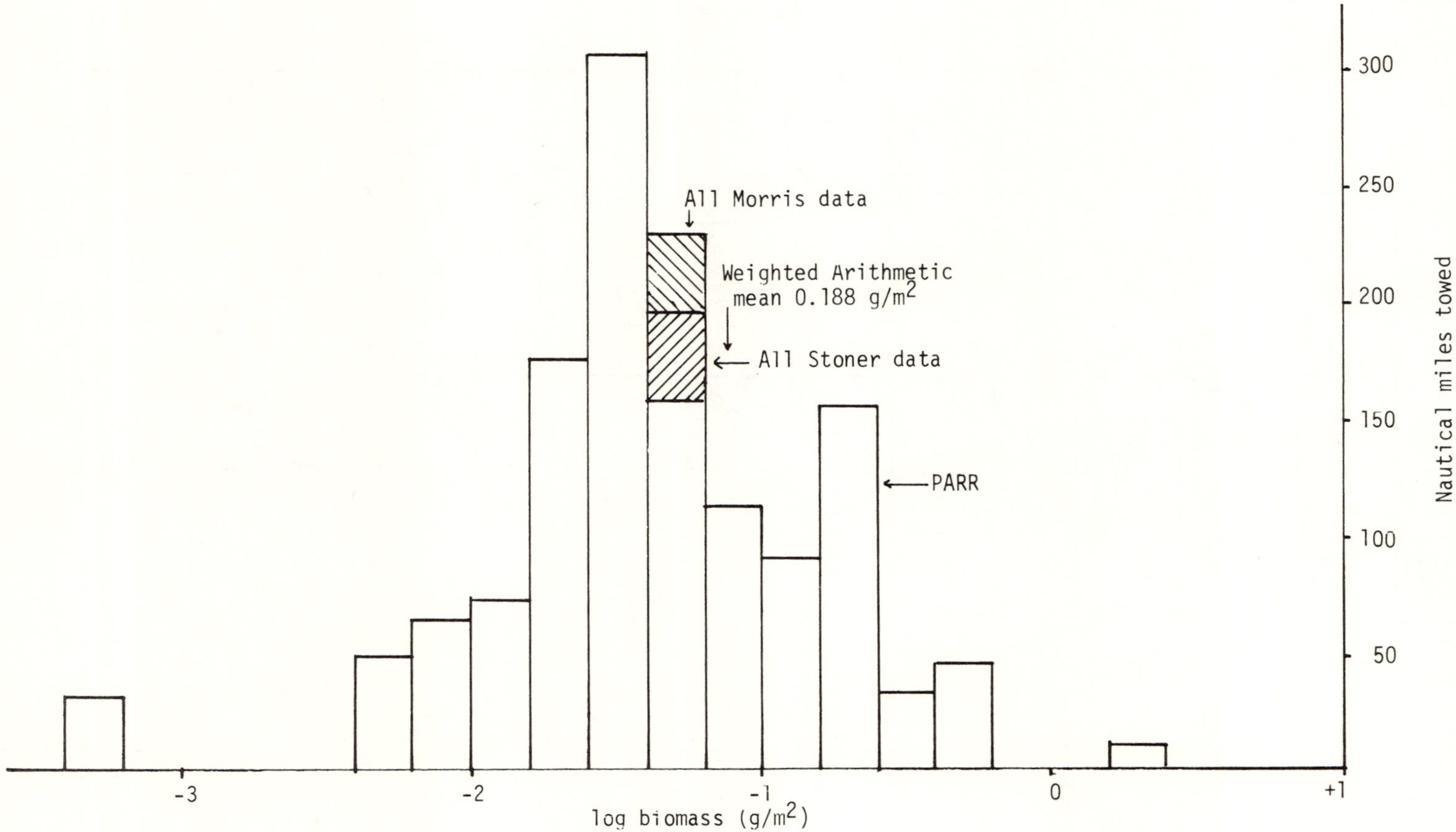

Fig. 1.14. Comparison of Parr, Stoner and Morris data for 30-40°N, 60-70°W (northwest Sargasso Sea)

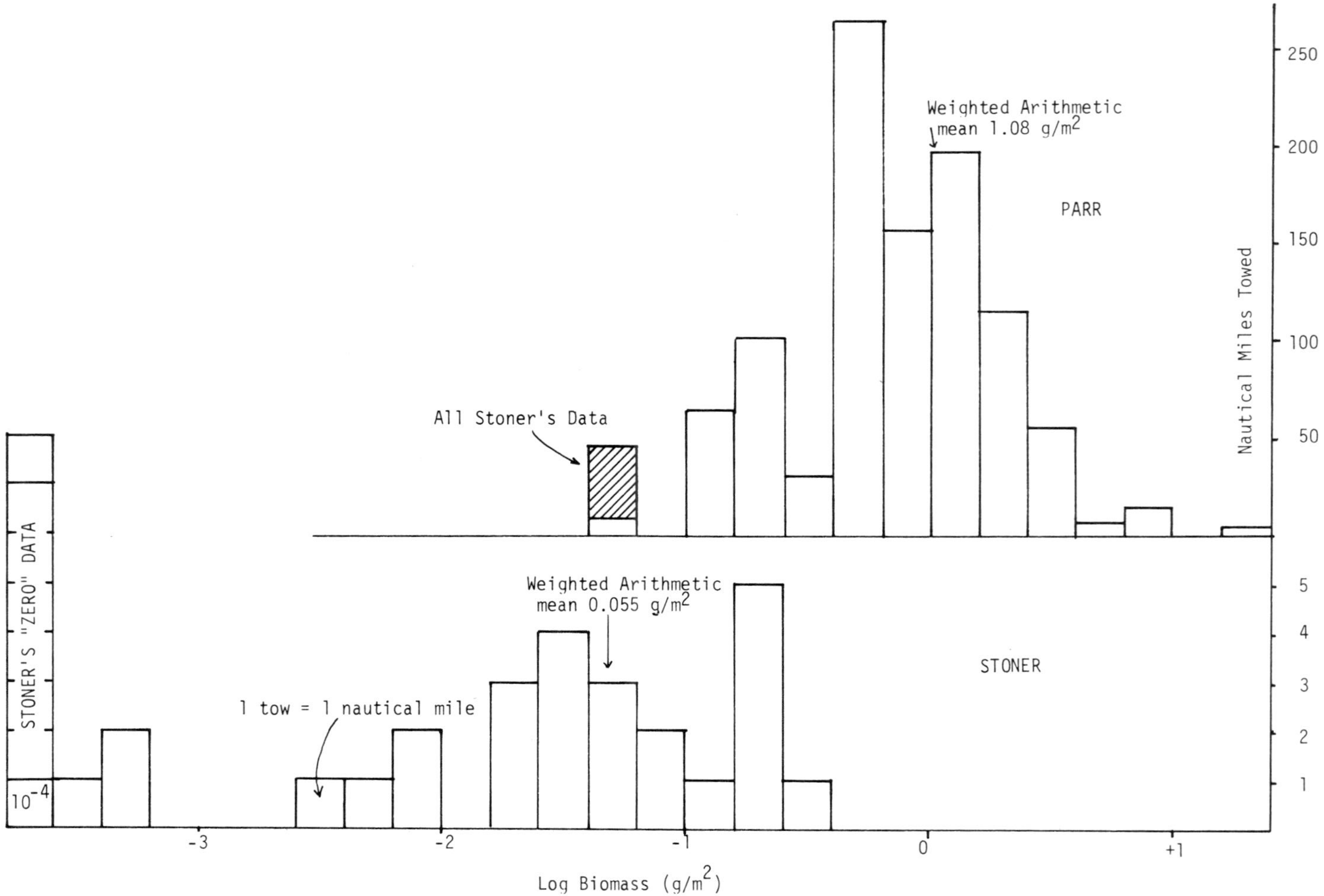

Fig. 1.15. Comparison of Parr's and Stoner's data for the southwestern Sargasso Sea (20-30°N, 57-70°W)

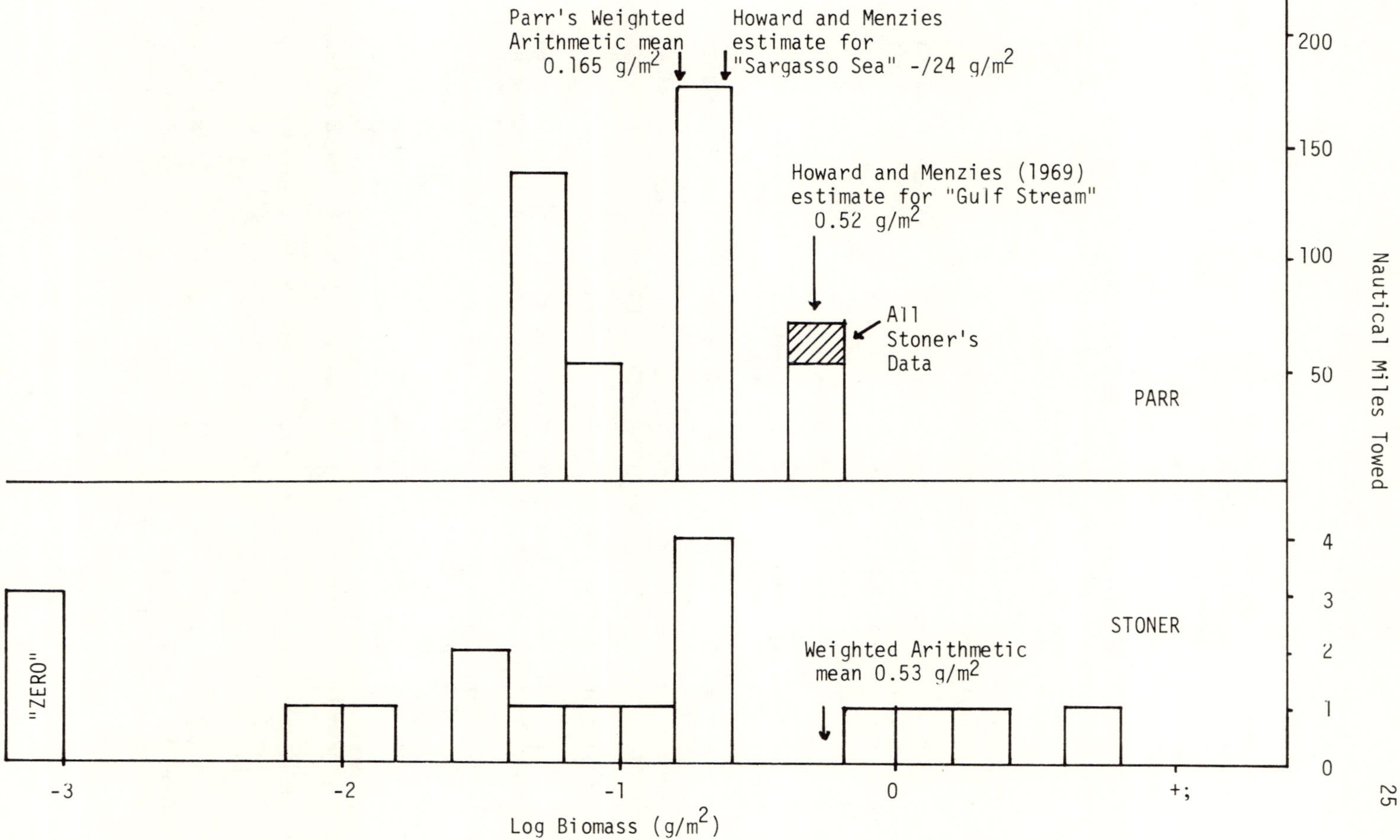

Fig. 1.16. Comparison of data for the Gulf Stream region (30-40°N, 70-80°W), obtained by Parr, Howard and Menzies, and Stoner.

significant difference between the three data sets, and hence that there has probably been no significant change in biomass of Sargassum in the Bermuda region since 1933.

Howard and Menzies (1969) estimated the amount of Sargassum by visual observations during 1966 from the bridge of the R/V Eastward, and give a standing crop of 0.52 g/m^2 for the "Gulf Stream" and 0.24 g/m^2 for the "Sargasso Sea" -- actually all these observations were made in the region 31-35°N, 72-79°W, which we have called the "Gulf Stream Region" (30-40°N, 70-80°W) for comparison purposes (Figure 1.15). In this region, the weighted average of Stoner's data was higher than Parr's four tows from that region, 0.54 g/m^2 vs 0.165 g/m^2. The two estimates by Howard and Menzies fall between, but because of the variability in all the data (Figure 1.15, lower half), it seems likely that there has been no significant increase in the Gulf Stream region from 1933 to 1981.

The Bahamas region (20-30°N, 70-80°W -- Figure 1.16) likewise shows no significant difference between Parr's and Stoner's observations. Weighted arithmetic means were 0.55 g/m^2 and 0.33 g/m^2 respectively; and the total of Stoner's data fell in the middle of Parr's distribution.

There is, however, a highly significant difference between Parr's and Stoner's results in the southern Sargasso Sea (20-30°N, 57-70°W -- see Table 1.1 and Figure 1.17). The weighted arithmetic means were 1.08 g/m^2 for Parr's 1933-35 survey and 0.055 g/m^2 for Stoner's 1977-81 survey -- an apparent decrease by a factor of 20. If all Stoner's data are summed and treated as a single tow from Parr's set, they fall at the lowest end of Parr's distribution (Figure 1.17), and the chance that the averages are the same is only a few percent.

Thus the only significant differences in Sargassum biomass appear to have been in the southwestern Sargasso Sea. Agreement among the four studies in other parts of the western North Atlantic appears to be good, and implies that most of the obvious causes of sampling bias were not a major factor. (For example, Stoner used shorter tows, finer net mesh, and slower speed of towing than did Parr.) Furthermore, effects of pelagic tar (discussed above and in Section 2.9), nutrient depletion, or climatic variation (Parr and others have proposed that the northern boundary of the Sargasso Sea is defined by death of the alga in winter) might be expected to be greater in the more highly stressed pelagic community of the northern Sargasso Sea, but in that region these effects appear to be small compared to variability in sampling (Figure 1.14).

Although most of Parr's stations were further to the west (65-70°W) than Stoner's (58-61°W), there is some overlap, as indicated in Figure 1.18. Near 25°N, 67°W, Stoner obtained

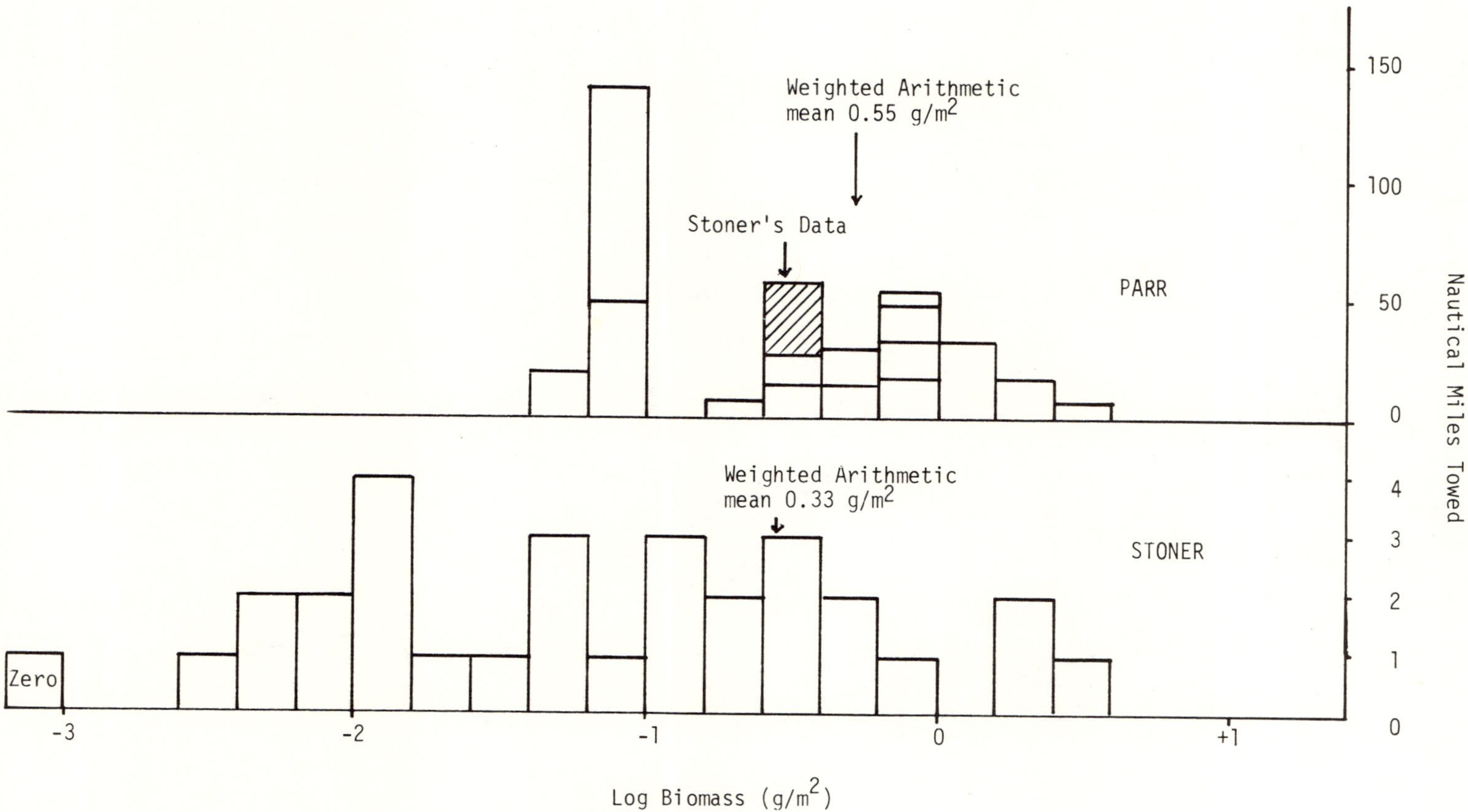

Fig. 1.17. Comparison of data from the Bahamas region (20-30°N, 70-80°W), obtained by Parr and Stoner.

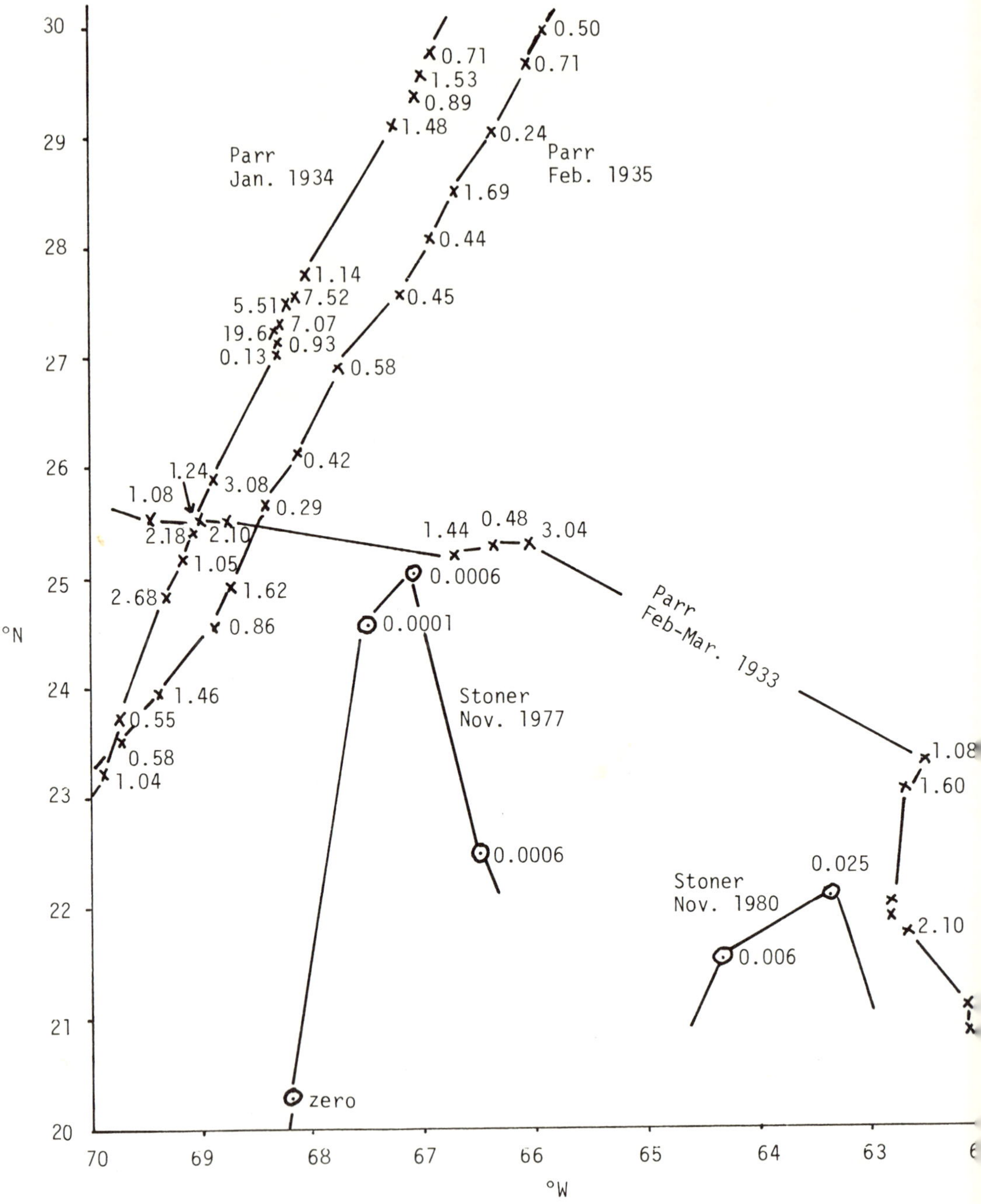

Fig. 1.18. Comparison of Parr's and Stoner's cruise tracks. Additional tows were made by Stoner east of 61°W (See Fig. 1.8 and App.

Sargassum amounts less than 0.001 g/m^2 in a place where Parr had found 0.48 to 3.04 g/m^2. This difference far exceeds any year-to-year variability observed by either investigator.

What, then, is the explanation of the large differences observed in the southwestern Sargasso Sea? There is a difference in season when the observations were made: Parr's cruises were made in January through March, and Stoner's were all in October and November. The area in which surface convergences were most intense might shift with the season. Mesoscale eddies travel long distances. For example, McDowell and Rossby (1978) describe a mesoscale eddy at 24°N, 70°W which apparently originated near the Mediterranean outflow in the eastern North Atlantic.

It is not possible to say, without further field work, whether there are large seasonal shifts of *Sargassum* abundance in this region, although Morris's time series from Bermuda (Figure 1.12) implies there might be. Whether there is much less *Sargassum* there now than fifty years ago, and what the cause of the change might be, is a matter only for speculation at this time.

1.4 *SARGASSUM* ON BERMUDA BEACHES

Parr (1939) quoted observations by the guard at Fort St. Catherine beach, that "much less weed drifted into the beaches of Bermuda in 1935 than in 1934, as judged by the amount of work required to keep the beaches clean during these two years". This agreed with a decreased average amount in his tows: 0.58 g/m^2 for 1935 *vs* 1.56 g/m^2 for 1934. He also noted that there was about a factor of five less *S. natans* in 1935 compared to 1934, due partly to decreased total biomass and partly to the increased amount of *S. fluitans* and other species.

The notion of being able to sample the deep ocean by beachcombing is an attractive one, and so weekly 1-meter transects of six representative Bermuda beaches were taken by Morris and others from November 1971 to January 1973. The approximate dimensions of the beached *Sargassum* bed were used to estimate the volume in m^3. As with the neuston samples, considerable seasonality and large fluctuations were observed. Within wide limits, the *Sargassum* quantities were generally unpredictable. The highest quantities found were 1.7 m^3 per meter of shoreline and the lowest were essentially zero. Fig. 1.19 displays the weekly quantities of *Sargassum* for the 6 beaches. A summary of these results is presented in Table 1.5. The most obvious feature is the high values for the winter of 1971-72 and low values for the winter of 1972-73, with a few exceptions.

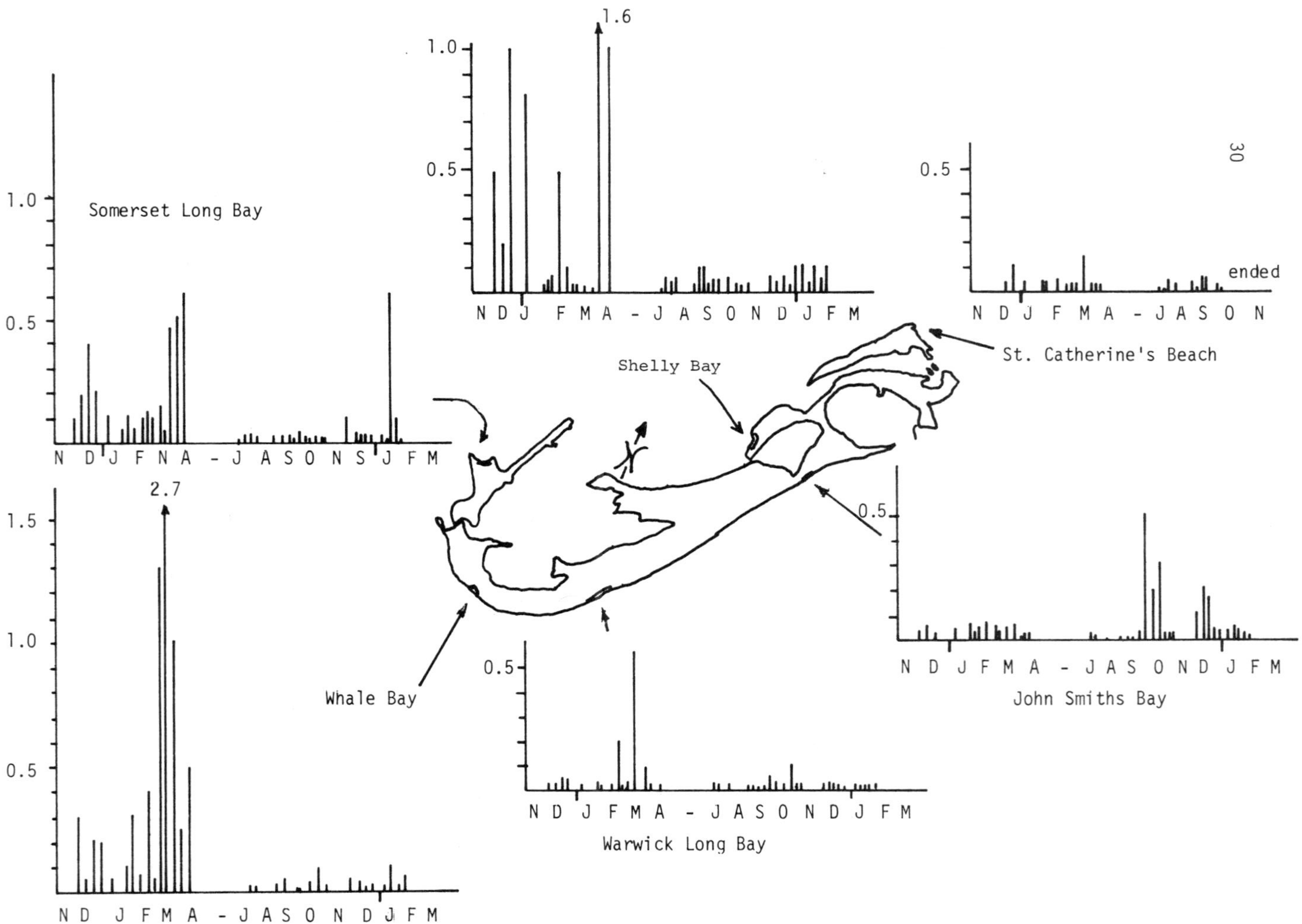

Fig. 1.19 Sargassum quantities in m^3 per m of beach, for 1971-72 (Morris, unpublished)

Table 1.5 QUANTITATIVE ABUNDANCE OF *SARGASSUM* FOUND ON SELECTED BERMUDA BEACHES (assuming 1 cu. meter = 200 kg wet weight)

Beach	Approx. length of beach km		N 1971	D	J 1972	F	M	A	M-J	J	A	S	O	N	D	J 1973	TOTAL
Somerset	0.3	m^3	150	240	660	780	210	330		0	60	30	60	30	6	222	2778
Long Bay		tons	30	48	132	156	42	66	?	0	12	6	12	6	1	44	556
Whale Bay	0.1		20	50	45	55	500	75		0	5	8	0	2	1	18	779
			4	10	9	11	100	15	?	0	1	2	0	0	0	4	156
Warwick	1.0		50	30	40	210	100	0		0	0	80	100	20	30	60	720
Long Bay			10	6	8	102	20	0	?	0	0	16	20	4	6	12	144
John Smith's	0.2		20	70	10	25	20	0		0	0	100	100	20	75	30	470
Bay			4	14	2	5	4	0	?	0	0	20	20	4	15	6	94
Shelly Bay	0.3		30	540	240	320	300	750		10	10	45	15	5	40	117	2422
			6	108	48	64	60	150	?	2	2	9	3	1	8	23	474
St. Catherine	0.2		2	20	25	10	25	2		0	0	2	0	-	-	-	86
Beach			0	4	5	2	5	0	?	0	0	0	0	-	-	-	17
TOTAL/MONTH	2.1		272	950	1020	1400	1155	1157	?	10	75	265	275	77	152	447	7255
			60	196	204	280	131	131	?	2	15	53	55	15	30	89	1451

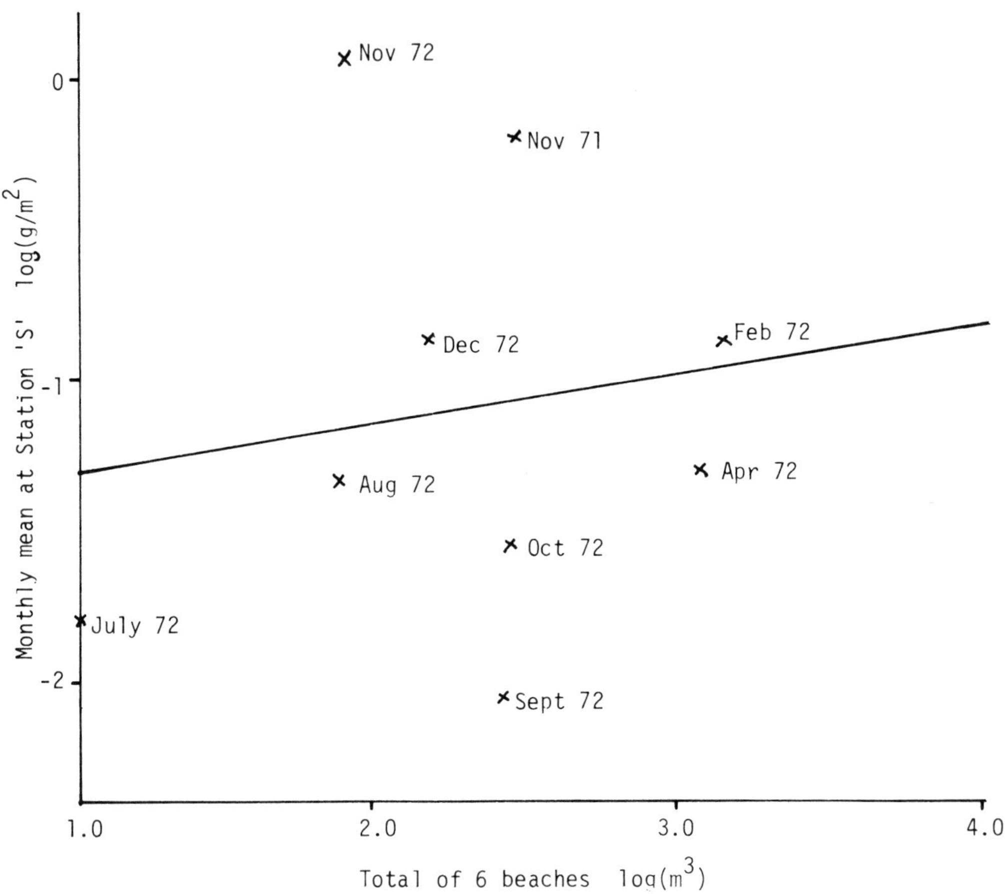

Fig. 1.20. Correlation between *Sargassum* on Bermuda Beaches (Table 1.5) and at Station 'S' (Table 1.2). R = 0.14, which is not significant.

Morris assumed that the Sargassum collected one week is different from that present the previous week and so multiplied the amount observed per meter of shoreline on the day of the harvest by the approximate length of the beach, then summed his weekly observations for each month to derive the amount of Sargassum for each beach. If Sargassum persists on the beach for less than one week, this method will underestimate the desired quantity; if it persists for more than a week, this method overestimates the quantity. Regular beach cleaning by the Dept. of Agriculture and Fisheries made the latter possibility unlikely.

One cubic meter of Sargassum (under normal compaction as found on the beaches) corresponds to 150-200 kg wet weight. Such an assumption (1 m^3 = 200 kg) was used in Table 1.5 to convert m to tons. The total for 13 months is 7225 m^3 or 1451 tons for the 2.1 km of beaches surveyed.

Knap, Iliffe and Butler (1980) found that the ratio of weekly beach tar collections to daily beach tar collection is about 3, implying an average residence time of the order of a few days for tar lumps, and presumably also for Sargassum that is not stranded above the high water mark. Hence the total input of Sargassum to the beaches surveyed by Morris might be as much as twice his estimate.

The amount of Sargassum deposited on the beaches might be expected to reflect the amount in surrounding waters, with some additional effects from prevailing winds and currents which would transport it to shore, but no significant correlation was found in the data available (Figure 1.20). Only the most general seasonal trend -- high in winter, low in summer -- seems to be apparent.

If seasonality were a dominant factor, the high values recorded on the beaches in November and December of 1971 (Table 1.5) would have been expected to recur in the winter of 1972. Although such a recurrence appears to have taken place in nearby waters (Table 1.2), it did not make a strong impact on the beaches.

1.5 IS HARVESTING SARGASSUM COMMERCIALLY FEASIBLE?

Because of the popular image of the Sargasso Sea as a dense mat of seaweed, commercial exploitation of Sargassum has often been suggested. However, methods developed for harvesting Sargassum at sea would probably have to be restricted to areas of concentration, such as windrows, convergences, and coastal embayments. Conditions conducive to these efforts are not predictable, and are often so transitory that a successful harvesting operation would have to be highly mobile.

Windrows of *Sargassum*, which could be efficient to harvest at sea, tend to be accompanied by increased sea surface roughness. When winds are light and variable, *Sargassum* disperses over the surface, making the yield from the ocean surface quite low, less than 200 mg/m^2 or 0.2 ton/km^2 (See tables 1.1 and 1.2). With harvesting restricted to windrows, lines of convergence, and to large patches when present, perhaps 10 tons/km^2 (of area cleared) could be realized.

If one gathered the mats of floating *Sargassum* trapped in enclosed bays and harbors, one could perhaps obtain a supply of several thousand tons of *Sargassum* from near the shores of Bermuda. This resource would not always be available or abundant, however, and the operation would require periods of intensive harvesting followed by long inactive periods.

Collection from Bermuda beaches would require less effort, but again quantities of *Sargassum* are unpredictable. Some beaches received over 200 kg/m of *Sargassum* during certain months of 1971-72 (Fig. 1.9). At other times less than 1 kg/m is present.

From the beaches listed in Table 1.5, one could expect approximately 1500 tons of *Sargassum* per year. Including other beaches (at least 10 km are accessible by vehicles) and restricting collection to times of greatest *Sargassum* abundance, several thousand tons of *Sargassum* might be efficiently collected in a year.

Beach-cleaning machines, currently used to remove tar from the important public beaches, could be employed as *Sargassum* harvesters. They consist of a conveyor belt which carries the upper few inches of beach sand into a hopper. The sand sifts through a screen in the bottom of the hopper. The tar and other large particles are retained. Although small amounts of *Sargassum* are also removed in this manner, large amounts are now manually raked clear of the path of the tractor to keep it from entering and clogging the hopper. Work crews also manually clear beaches of *Sargassum* on occasion, loading it on to trucks, and taking it to the municipal dump, since there is only a small demand for *Sargassum* as garden fertilizer by some Bermudians.

The most important factor which complicates the potential for commercial harvesting of *Sargassum* from the ocean surface or from the shoreline is its association, mentioned in Section 1.3, with polluting petroleum residues (pelagic tar). The quantities and characteristics of these tar lumps have been the subject of study since 1970 (Butler, Morris and Sass, 1973; Knap, Iliffe and Butler, 1978).

Because pelagic tar lumps are buoyant, they are dispersed over the ocean surface during periods of calm or of light and variable winds, and they tend to line up in windrows with the *Sargassum* during windy periods. Like *Sargassum*, they are concentrated in downwelling convergences. Quantitative measurements (Morris and Butler, 1973; See also Table 1.2) have shown that the mean quantity of tar in the ocean off Bermuda is about 10 mg/m^2 or about 4% of the amount of *Sargassum*. As expected, they are weakly correlated (Morris and Butler 1973; Butler and Morris, 1974). Tar associated with beached *Sargassum* has amounted to about 1% by weight at periods of high *Sargassum* abundance. The presence of tar in harvested *Sargassum* could require an extra step of separation and cleaning or could make it unsuitable for some purposes because of the presence in the alga of hydrocarbon residues.

In addition to its irregular occurrence and contamination by petroleum residues, the amount of available *Sargassum* would tend to be limited by the rate at which it could regenerate. The harvest would remove nutrients from the already nutrient-poor surface waters of the Sargasso Sea, and unless these were replaced at an equivalent rate, the standing crop of *Sargassum* would decrease with time until it would no longer be feasible to collect it in quantity.

2. THE SARGASSUM COMMUNITY

The animals living in association with *Sargassum* form a unique biotic community amenable to ecological investigations without several of the problems encountered in studies of other aquatic or terrestrial communities; i.e., delineation of boundaries, overlap of adjacent communities, excessive species complexity or excessive diversity.

As described in Section 1.1, *Sargassum* found in the Sargasso Sea maintains its population entirely by vegetative growth. The two common species of pelagic *Sargassum*, *S. natans* and *S. fluitans*, are holopelagic species that are thought to be derived from an ancestral littoral stock. Parr (1939) showed that the alga is restricted to the ocean surface where it maintains itself by gas-filled bladder floats. Growth and propagation result from the seasonal fragmentation (generally during winter) of larger and older clumps.

The animals inhabiting *Sargassum* have been characterized by Hedgpeth (1957) as a "displaced benthos" (Figure 2.1) since the majority of the species are derivative species or expatriate populations of littoral genera and species. The fauna and flora together form a discrete biotic community with the *Sargassum* forming a "pseudo-benthic" habitat surrounded by the pelagic ecosystem.

It should be noted that some work has been done on the ecology of communities associated with *Sargassum muticum* (Fletcher and Fletcher, 1975; Knight-Jones *et al.*, 1975; Withers *et al.*, 1975; Boalch and Potts, 1977; Norton, 1977; Chamberlain *et al.*, 1979; Gorham, 1979; Kane and Chamberlain, 1979; and Fletcher, 1980) as well as other benthic *Sargassum* species (Mukai, 1969, 1971; Taniguchi, 1975; Joshi and Gowda, 1975; Cordero, 1980; Prince, 1980; and Tararam and Wakabara, 1981) but we have not attempted to integrate these studies with those of the pelagic *Sargassum*.

2.1 COMMUNITY STRUCTURE

Attention has been given to the *Sargassum* fauna by several previous researchers. Most of these studies were descriptive faunistic surveys (Adams, 1960; Burkenroad, 1939; Dooley, 1972; Hentschel, 1922; Prat, 1935, 1940; Timmermann, 1932; Weis, 1968; Winge, 1923) and dealt only superficially with the ecology of the community. Considerable attention has been paid to the life cycles of green turtles (Fletemeyer, 1978; Carr and Meylan, 1980) and eels (Kracht and Tesch, 1981) which are partly spent in association with *Sargassum*.

Fig. 2.1. Invertebrate animals that live on floating *Sargassum*. Although probably derived from West Indian ancestors, many of these species have become especially adapted for an epiphytic life on floating *Sargassum*. The figure at top shows a large piece of *Sargassum* from which the figure on the opposite page has been taken. From: *Invertebrate Zoology* (4th Ed.), R.D. Barnes, 1980, pp. 894-895. Reproduced by permission.

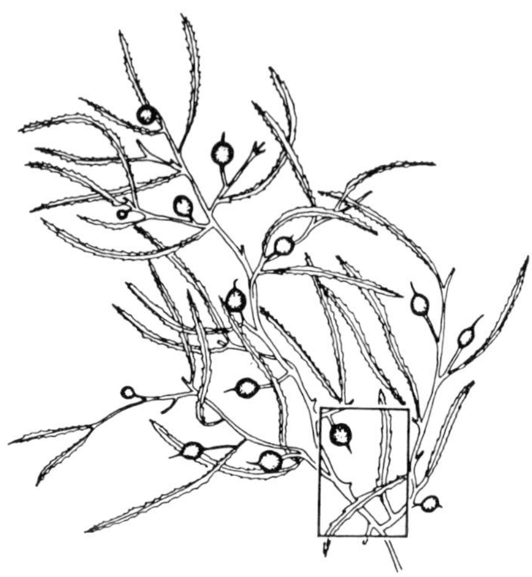

A. *Litiopa* (gastropod mollusk)
B. *Amphithoë* (amphipod crustacean) -
C. *Luconacia* (caprellid amphipod)
D. *Anemonia sargassensis* (sea anemone)
E. *Platynereis* (nereid polychaete)
F. *Sertularia* (hydroid cnidarian)
G. *Clytia* (hydroid cnidarian)
H. *Scyllaea* (nudibranch gastropod)
I. *Spirorbis* (serpulid polychaete)
J. Copepod crustacean -
K. *Zanclea* (hydroid cnidarian)
L. *Ceramium* (an epiphytic red alga)
M. *Membranipora* (bryozoan)
N. *Doto* (nudibranch gastropod)
O. *Gnescioceros* (polyclad flatworm)
P. *Obelia* (hydroid cnidarian)
Q. *Anoplodactylus* (pycnogonid) feeding on a nudibranch
R. *Fiona* (nudibranch gastropod)

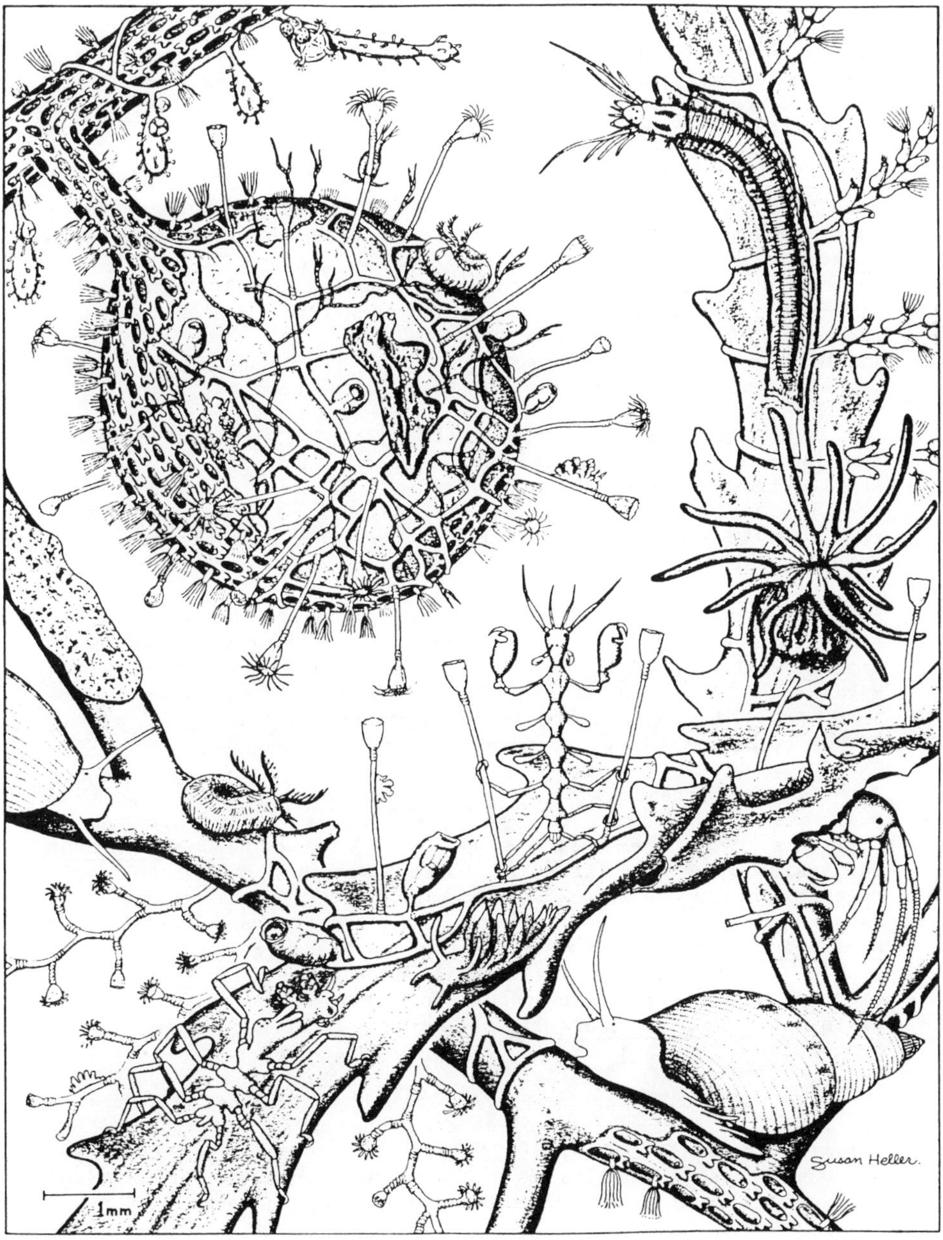
1mm
Susan Heller.

Fine (1970) counted the larger motile animals on 34 samples of *Sargassum* from the western North Atlantic, determined their diversity and some aspects of their seasonal and geographic variability. To date, however, no researchers have identified and counted all the organisms on even a single *Sargassum* clump, and the total organization of the *Sargassum* community is relatively unknown.

Morris and Mogelberg (1973) prepared an identification manual to the known *Sargassum* fauna. One hundred and one species were listed. As many as 54 species of fish are associated with *Sargassum* at some time in their life cycles (Dooley, 1972; Bortone *et al*., 1977) but only two species spend their entire life cycle on *Sargassum*. The actual number of species common to the community is considerably less, since 43 of the species listed were hydroids and most of these plus many of the other organisms are either uncommon or rare to the community.

Table 2.1 gives a phylogenetic list of the *Sargassum* fauna and indicates which species are common, uncommon, or rare in samples from the Sargasso Sea waters off Bermuda. As can be seen from the list, the basic community consists of less than 35% of the reported species. Even this is greater than the number of species that will actually be found on any individual clump.

A few species commonly occur on only one or the other of the two *Sargassum* species. (For example, *C. noliformis* grows on *S. natans* while *Halecium nanum* prefers *S. fluitans* (Morris and Mogelberg, 1973; See also Table 2.3 below). Some species are seasonal, while others are found only intermittently and may be either opportunistic or accidental strays.

The polychaete worm, *Platynereis dumerilli*, was found in most samples (227 out of 244 - see Table 2.2). Second in frequency of occurrence was the flatworm *Gnescioceros sargassicola*, while first in total abundance was the isopod *Bagatus minutus*. Only eight species of motile macrofauna were found on more than half of the samples. Fine (1970) found *Hemiaegina minuta* to be most numerous on his Sargasso Sea samples (Table 2.3, stations D3, D4 and D5) and *Bagatus minutus* (= *Janira minuta*) to be second in population, but *P. dumerillii* ranked sixth and *G. sargassicola* ranked seventh. Further comparison is made below.

The attached organisms are primarily hydroids (ten common species), an anemone, a polychaete, one barnacle, two or more foraminiferans (Spindler, 1980) and a bryozoan. The attachment of other sessile species is probably accidental and of limited occurrence, or at least of restricted geographic range.

The motile meiofauna consist primarily of cyclopoid and harpacticoid copepods, but have been poorly investigated. Two new species and one new record of cyclopoid copepods have been recently reported (Morris, 1973) as well as a new record for a tanaidacean (Greve-Jensen, 1974). Nematodes of unknown species have also been observed in the samples (Morris, unpublished).

The flora of the community is dominated by the Sargassum itself, but epiphytic green and blue-green algae are found (Carpenter, 1972; Carpenter and Cox, 1974) as well as diatoms (Carpenter, 1970) and an uninvestigated bacterial film (Conover and Sieburth, 1964). Heterotrophic bacteria and protozoa are probably also important (Caron et al., 1982).

The abundance of each species within the community may be related to season, habitat size, and geographic region. However, the most important factor governing the species composition of the Sargassum fauna appears to be historical accident. Whatever species are present on a portion of a Sargassum clump, when it is broken loose to form a new clump, will be the ancestors of the organisms present on the new clump as it enlarges and eventually fragements. This extreme form of "island biogeography" causes extreme variability in the numbers of species and individuals on each clump of Sargassum.
An inspection of Appendices III and IV will show that samples collected from the same station at nearly the same time are highly variable in their species composition.

The individuality of each Sargassum clump has largely gone unstudied, primarily because other workers (Adams, 1960; Weis, 1968; Fine, 1970) did not restrict their analyses to individual plants but worked largely with pooled samples.

As discussed in Section 1.4, Sargassum normally occurs in scattered individual clumps, which usually range in size from 10 cm to 50 cm in diameter. The clumps may be spaced one to several hundred meters apart, positioned by downwelling surface currents. This scattered distribution restricts the movement of individual animals from clump to clump and reinforces the natural individuality of each clump. Migration is probably important only when a number of colonies are swept together to form a "mat" at a major downwelling convergence.

In Morris's surface zooplankton sampling over two years, animals from the Sargassum community were never found in neuston net tows unless the net also contained some Sargassum. Simple field observation of the fidelity of the animals to their algal habitat therefore implies that Sargassum animals do not often migrate from clump to clump unless such clumps are actually in contact.

A majority of the animals living on Sargassum are dependent upon this habitat. Many are endemic to Sargassum but may occasionally be found on other flotsam. Some species are less obligate to this habitat and also occur in near-shore environments (Morris and Mogelberg, 1973).

Other factors controlling the species and populations of the

Table 2.1. The Sargassum Fauna

Group	Common (> 10%)	Uncommon (< 10%)	Rare (< 1%)
FORAMINIFERA			
	Planorbulina acervalis		
	Rosalina sp		
HYDROIDS			
	Zanclea costata	*Syncoryne mirabilis*	*Cladocoryne pelagica*
	Clytia johnstoni	*Clytia cylindrica*	*Campanularia volubilis*
	C. noliformis	*Zanclea gemmosa*	*Clytia raridentata*
	Obelia dichotoma	*Clytia longicyatha*	*Gonothyraea gracilis*
	O. hyalina	*Eucopella sargassicola*	*G. integra*
	Halecium nanum	*Obelia bicuspidata*	*Aglaeophenia perpusilla*
	Aglaeophenia latecarinata	*O. geniculata*	*A. rigida*
	Plumularia setaceoides	*Hebella calcarata*	*Aglaeophenoides mammillata*
	P. strictocarpa	*Scandia mutabilis*	*Plumularia corrugata*
	Sertularia cornicina	*Plumularia margaretta*	*P. megalocephala*
		Antenella secundaria	*Sertularia amplectens*
		Plumularia catharina	*S. exigua*
		P. diaphana	*S. turbinata*
		P. floridana	
		Dynamena quadridentata	
		Sertularia inflata	
		S. cornicina	
		S. flowersi	
		S. gracilis	
		S. mayeri	
		S. stookeyi	
SEA ANEMONE			
	Anemonia sargassensis		
FLATWORMS			
	Acerotisa notulata	*Amphiscolops sargassi*	
	Gnesioceros sargassicola		
	Hoploplana grubei		
TARDIGRADE			
			Styraconyx sargassi
POLYCHAETE WORMS			
	Platynereis dumerillii		*Platynereis coccinea*
	Spirorbis corrugatus		*Harmothoe dearborni*
			Amphinome rostrata
COPEPODS			
	Macrochiron sargassi	*Paradactylopodia oculata*	*Macrochiron avirostrum*
	Amonardia phyllopus		*Macrochiron hudsoni*
	Harpacticus gurneyi		*Pseudomacrochiron parvum*
	Paralaophonte congenera		
	Dactylopodia tisboides		
	Scutellidium longicauda		

Table 2.1. The Sargassum Fauna (cont.)

Group	Common (> 10%)	Uncommon (< 10%)	Rare (< 1%)
BARNACLES			
	Lepas pectinata	*Conchoderma virgatum* *Lepas anatifera* *L. anserifera* *L. fascicularis* *L. hilli*	
ISOPODS			
	Bagatus minutus	*Anatanais normani*	*Idotea metallica* *I. baltica* *Paradynamene benjamensis*
AMPHIPODS			
	Sunampithoe pelagica *Biancolina brassicaecephala* *Hemiaegina minuta*		*Caprella danilevskii* *Ampithoe longimana* *Luconacia incerta*
SHRIMPS			
	Latreutes fucorum *Leander tenuicornis*	*Hippolyte coerulescens*	*Hippolyte zostericola* *Tozeuma carolinense*
CRABS			
	Planes minutus *Portunus sayi*		*Portunus spinimanus*
PYCNOGONIDS			
	Anoplodactylus petiolatus	*Endeis spinosa*	*Tanystylum orbiculaire*
SHELLED MOLLUSCS			
	Litiopa melanostoma		
NUDIBRANCH MOLLUSCS			
	Doto pygmaea *Scyllaea pelagica* *Spurilla neapolitana*	*Corambella depressa* *Fiona pinnata*	*Doridella obscura*
BRYOZOANS			
	Membranipora tuberculata	*Aetea anguina* *Thalamoporella falcifera*	
TUNICATE			
		Diplosoma gelatinosum	
FISH			
		Histrio histrio *Syngnathus pelagicus*	

Sargassum community include the size and age of the colony; the timing and history of reproduction and larval settling from the plankton; the seasonal patterns of succession in the community, and perhaps the elimination of species by competition or predation.

Since many of the animals of the Sargassum community spend their entire lives on the alga, it is possible to find, for many species, all life stages from newly hatched young to mature adults within the community. In contrast with related species elsewhere, several of the flatworms, polychaetes, and pycnogonids lack planktonic larvae as do the amphipods and isopods. Although most other species have planktonic larvae, newly settled juveniles are seasonally abundant. The mechanisms by which those species with direct development maintain themselves between annual cycles is not known.

2.2 COLLECTION PROCEDURE

We have collected 244 samples of Sargassum communities from Station 'S' off Bermuda and 155 samples from cruises to other regions of the North Atlantic and Caribbean Sea (Figure 2.2). All have been sorted for the motile macrofauna, but only a few studies have been made of the microfauna and sessile fauna. These data are included in Appendices II (microfauna), III (macrofauna from Station 'S'), IV (macrofauna from various cruises), and V (sessile fauna). Some of the results are summarized in Tables 2.2 and 2.4. They will be discussed below.

Sargassum clumps were regularly collected at Station 'S' southeast of Bermuda, in conjunction with the regular biweekly hydrostation (Schroeder and Stommel, 1969; Morris and Schroeder, 1973). Long-handled dip nets 30 to 50 cm diameter with 2-3 mm mesh were used from the bow of the ship while on station. The samples were packed in plastic bags and fixed in formalin-seawater on board the Panulirus II. Other samples were similarly collected from cruises of opportunity (Fig. 2.2), and preserved for shipment to the laboratory.

In the laboratory, the Sargassum samples were individually washed several times in a 2 to 4 liter beaker to remove the motile fauna. The washings were strained through a Millipore vacuum filter and the samples were then preserved in bottles labeled with the station number, date and sample number.

Each washed Sargassum plant was placed back in its plastic bag, preserved and labeled to correspond to the sample bottles. Several twigs of Sargassum were taken from each sample, one with a large amount of sessile animal growth, and one to three with little or no such growth.

Later, the macrofauna samples were sorted from the washings,

Fig. 2.2 Cruise tracks

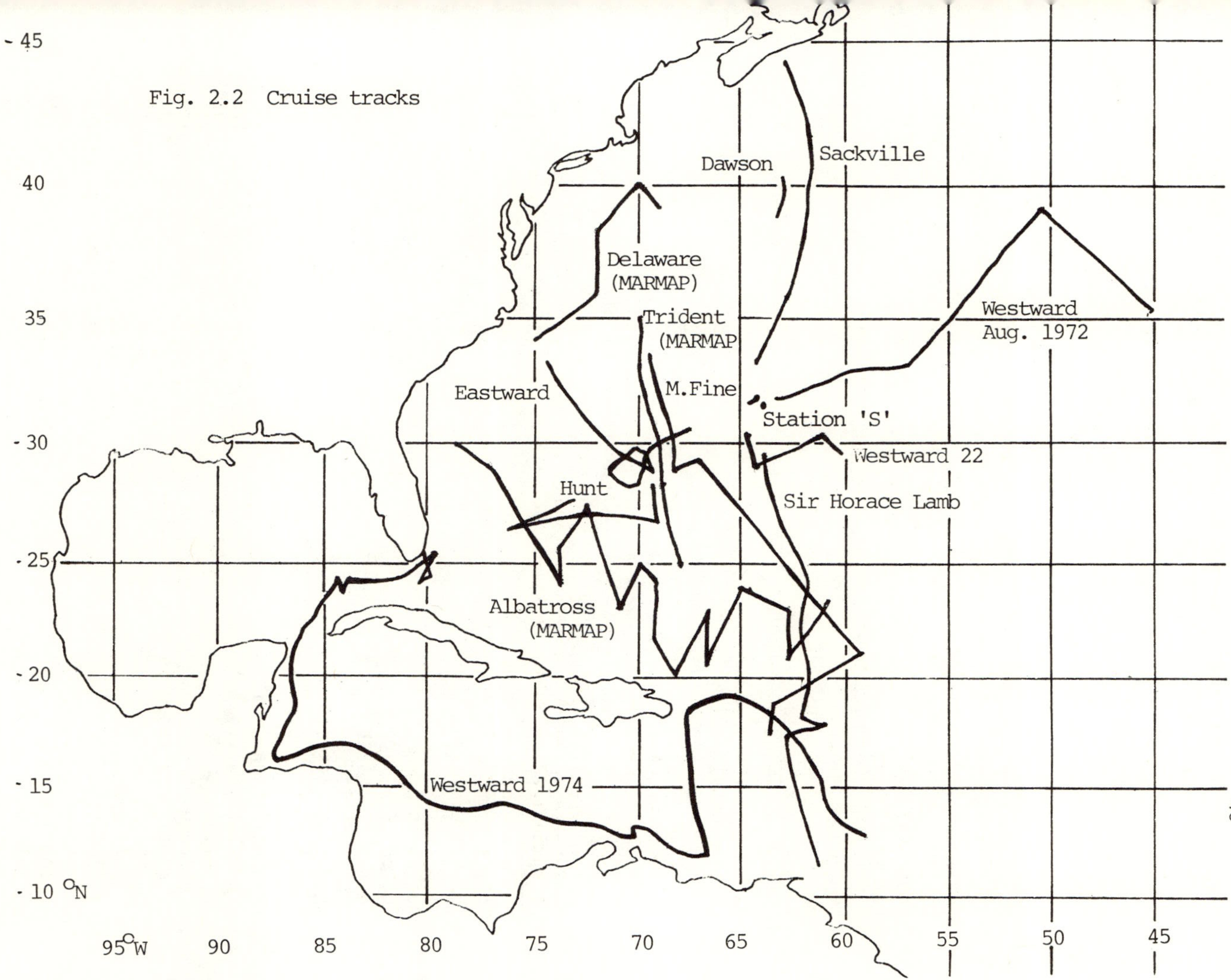

TABLE 2.2. ABUNDANCE OF MOTILE MACROFAUNA SPECIES ON 243 SAMPLES TAKEN AT STATION 'S'. See Appendix III for details.

Species	number of samples in which species occurs	total individuals found
Gnescioceros sargassicola	227	5856
Playtnereis dumerillii	208	5220
Bagatus minutus	188	7275
Anoplodactylus petiolatus	178	3269
Litiopa melanostoma	162	2472
Latreutes fucorum	154	1078
Planes minutus	146	507
Sunampithoë pelagica	144	3716
Hemiaegina minuta	106	2521
Anemonia sargassensis	97	5566
Acerotisa notulata	95	434
Hoploplana grubei	94	314
Biancolina sp.	89	240
Leander tenuicornis	87	218
Doto pygmaea (incl. unid. nudibranch "A")	83	1612
Scyllaea pelagica	73	224
Spurilla neapolitana (incl. unid. nudibranch"C")	65	435
Portunus sayi	21	21
Probopyrinella latreuticola (and other parasitic isopods)	16	26
Corambella depressa	12	38
Histrio histrio	12	14
Hippolyte coerulescens	10	13
Endeis spinosa	10	10
Syngnathus pelagicus	6	10
Amphiscolops sargassi	5	31
Anatanais normani	1	7
Idotea metallica	2	2
Janthina janthina	1	1
Lepas anatifera	2	31
Abudefduf saxtilis	1	1
larval brittle star	1	2
unidentified animals:		
nudibranch "D"	39	178
nudibranch "B"	11	26
isopods	20	148
animal described in notes to 388-1	10	61
shrimps	9	12
polychaetes	2	2
flatworms	2	2
nematodes	1	1
anemone	1	5
ostracod	1	2
others	10	62

Table 2.3. Faunal frequency (Fine, 1970)*

Series D:	1	1a	2	3	4	5	Total
Janira minuta**	32	81	120	301	206	348	1,088
Hemiaegina minuta	40	177	234	4	339	827	1,621
Sunampithoë pelagica	14	49	61	165	109	78	476
Gnescioceros sargassicola	48	74	32	5	71	39	269
Anoplodactylus petiolatus	175	42	1		80	168	466
Platynereis dumerillii	6	34	31	36	36	44	187
Litiopa melanostoma	11	8	7	24	35	132	217
Latreutes fucorum	26	14	1	1	3	2	47
Planes minutus	1	3	2	16	6	6	34
Hoploplana grubei		1	1		14	40	56
Anemonia sargassensis					53		53
Cuthona sp.	1		7		4	17	29
Spurilla neapolitana				6			6
Biancolina sp.	5	14	4	1	1	4	29
Acerotisa sp.	1	2	1		13		17
Leander tenuicornis	2	5		1			8
Histrio histrio		1	2		1	1	5
Lepas pectinata					5		5
Portunus sayi	1	3					4
Scyllaea pelagica				2		2	4
Probopyrus latreuticola	1	2					3
Polyclad sp.		2	1				3
Doto sp.						1	1
Endeis spinosa		1					1

*Station locations are
D1: 34.4°N 75.6°W
D2: 33.9°N 74.5°W
D3: 33.5°N 72.7°W
D4: 33.4°N 71.9°W
D5: 33.2°N 71.0°W

**Janira minuta is called Bagatus minutus everywhere else in this book.

Table 2.4. Frequency of occurrence of some common hydroids on Sargassum natans and S. fluitans

Hydroid species	S. natans (n=15)	S. fluitans (n=6)	n:f ratio
Clytia noliformis	12	1	12
Zanclea costata	11	5	2.2
Plumularia setaceoides	10	6	1.7
Halecium nanum	1	4	0.25
Aglaeophenia latecarinata	0	6	0

counted, and placed in separate vials. The remainder of the sample (microfauna and debris) was placed in a vial also, and the entire sample (plus 3 arbitrarily chosen twigs) placed in their corresponding labeled sample bottles.

Twenty-one clumps of *Sargassum* (15 *natans* and 6 *fluitans*) were examined for their sessile fauna (Appendix V). In these samples *Clytia noliformis* showed a strong preference for *S. natans* (Table 2.4) while *Halecium nanum* and *Aglaeophenia latecarinata* were most strongly associated with *S. fluitans*.

2.3 RELATIONSHIP BETWEEN DISPLACEMENT VOLUME, WET AND DRY WEIGHT OF *SARGASSUM*

In quantifying the amount of *Sargassum* collected, three methods were used. The simplest was to measure the volume of water displaced in a large graduated cylinder (displacement volume), and this was done with most samples. Since it is somewhat more precise, wet weight is often reported in the literature. Dry weight, which is the most rigorous basis for chemical analysis, is the most difficult to obtain, and involves at least partially destroying the specimen by drying in the sun or in an oven at 100°C.

To determine the relation between displacement volume and wet weight, 21 samples (range 4.2 to 322 g) were weighed to $\pm$ 0.1 G. Each was placed in a one-liter graduated cylinder partly full of water, and the changes in volume measured (Fig. 2.3). The mean ratio, expressed as density, is 0.962 g/ml. The t-test predicts 95% confidence limits of $\pm$ 0.18 g/ml; i.e., that 20 out of 21 values should fall between 0.78 and 1.14. Actually (Fig. 2.3) all 21 values fall between 0.81 and 1.09, indicating that the distribution is more sharply peaked than normal. There appears to be little correlation between sample size and density, but samples of intermediate size may have a slightly larger density (about 1.00). The specific gravity of seawater at 20°C and 36.0°/oo is 1.027 g/ml, and five of the values fall slightly above this; these must reflect measurement errors or the *Sargassum* would not have been floating.

To determine the relation between wet and dry weight, two samples were dried. Sample A (1009 g wet weight) was dried in the sun, and after 7 days reached an approximately constant weight of 230 g. Sample B (1017 g wet weight) was dried in a 100°C oven, and after 3 days weighed 200 g, remaining between 215 and 220 g for the next 4 days. When cooled and exposed to air (approximately 50 to 70% humidity) the weight increased to about 230 to 235 g, in agreement with the sun-dried sample. The details are shown in Figure 2.4.

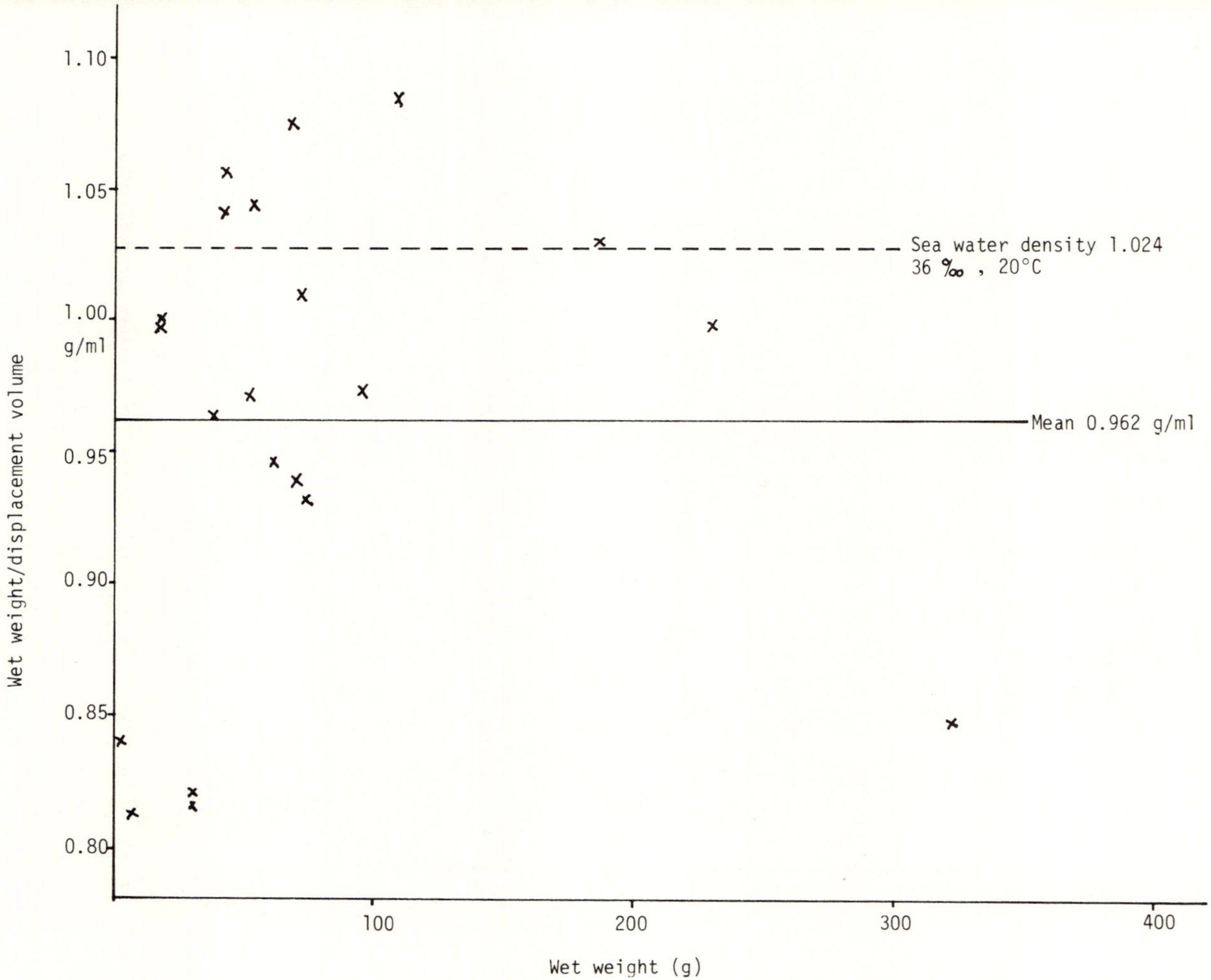

Fig. 2.3. Ratio of wet weight to displacement volume for Sargassum samples

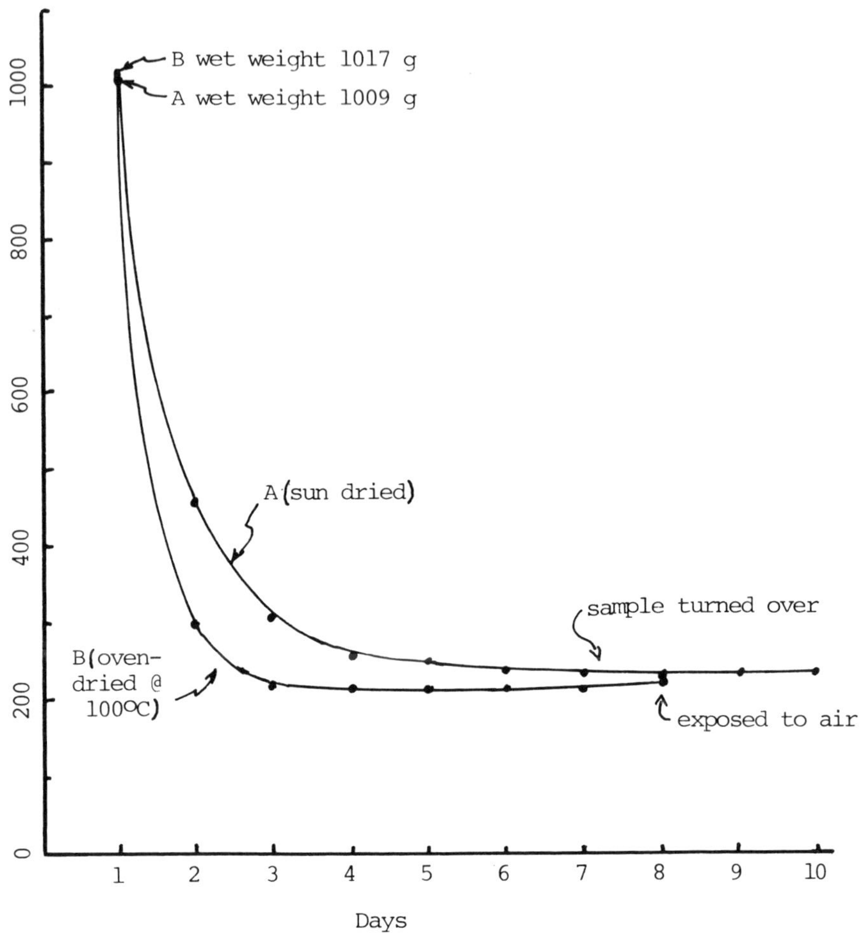

Fig. 2.4. Comparison of drying process for Sargassum: sun vs. oven.

2.4 WET AND DRY WEIGHTS OF SARGASSUM FAUNA

As part of the studies of hydrocarbons in the Sargassum community (Morris, Cadwallader, Geiselman and Butler, 1976), the relation between wet and dry weights for a number of the Sargassum animals was determined. Samples were dried in an oven at 60°C for 24 hrs, or until constant weight was achieved. Samples were also combusted in a muffle furnace at 500°C for 2 hrs, to obtain ash weights. The difference between ash and dry weights was taken to be the weight of organic material. The results are given in Table 2.5.

It should be noted that other samples may show considerable variation in wet weight per animal, but that the wet/dry weight ratios are expected to be much more consistent; 4.5 to 6.5 for all animals in this list except the isopod Biancolina, the shell mollusk Litiopa, the nudibranch mollusks Corambella and Scyllaea, and the pipefish Syngnathus. Similarly, the organic content is a consistent fraction of the dry weight (70 - 90%) except for the crabs Planes and Portunus and the shelled mollusk Litiopa.

2.5 EFFECT OF COLONY SIZE

If Sargassum clumps grow slowly compared to most of the fauna, one might expect that the total number of animals would be proportional to colony size, and that the larger colonies would show a higher diversity.

The relation between colony size (as measured by displacement volume), and the total number of motile organisms found on a colony was tested by simple linear regression (Table 2.6). All the samples from Station 'S' (Appendix III) were lumped into one group, which showed a significant correlation (r = 0.636) with a slope of 1.395 as displayed in Figure 2.5. Note that a perfectly proportional relationship (slope 1.00) would predict fewer animals on larger clumps than were actually observed.

For the various cruises (Appendix IV), the correlation coefficient was between 0.226 and 0.814 (Table 2.6), and when 139 stations from all cruises other than Station 'S' were lumped into a single sample, the correlation coefficient was 0.636, identical to that for station 'S'. The slope (Fig. 2.6) was also approximately the same (1.442).

The Shannon-Weiner function (H, defined in Table 2.6), which has been used to express the diversity of a community structure as "information content" (MacArthur and MacArthur, 1961; Watt, 1964) was calculated for each of the colonies based on the motile macrofauna populations.

Table 2.5. Wet and dry weights.

Species	no. of animals	wet weight mg	dry weight mg	wet/dry	ash weight	organic as % of dryweight
44 Anemonia sargassensis	267	366.3	81.0	4.52	10.2	87.4
46 Acerotisa notulata	99	16.4	3.3	4.97	0.8	76.
47 Hoploplana grubei	9	136.3	22.8	5.98	3.2	86.
48 Gnescioceros sargassicola	76	133.1	20.7	6.43	5.2	75.
51 Platynereis dumerillii	47	187.4	38.1	4.92	5.5	85.6
70 Bagatus minutus	131	79.9	13.9	5.75	2.8	80.0
73 Sunampithoë pelagica	30	66.8	14.1	4.74	1.8	87.2
74 Biancolina sp.	37	21.8	2.2	9.91	0.5	77.
76 Hemiaegina minuta	258	56.3	10.2	5.52	1.1	89.
78 Hippolyte coerulescens	4	198.6	39.6	5.02	6.2	84.3
80 Latreutes fucorum	22	372.1	66.4	5.60	12.2	81.6
82 Leander tenuicornis	10	1044.2	238.1	4.39	44.0	81.5
83 Planes minutus	10	3685.6	864.6	4.26	332.8	61.5
84 Portunus sayi	7	838.1	148.7	5.64	49.8	66.4
87 Anoplodactylus petiolatus	247	97.1	17.3	5.61	4.3	75.1
89 Litiopa melanostoma	27	169.3	88.9	1.90	79.5	10.6
90 Spurilla neapolitana	56	217.9	32.1	6.79	5.3	83.5
91 Corambella depressa	22	19.4	6.8	2.85	1.9	72.
93 Doto pygmaea	175	142.9	27.5	5.20	4.6	83.3
95 Scyllaea pelagica	26	173.2	119.0	1.45	38.9	67.3
100 Histrio histrio	6	811.2	174.5	4.65	36.1	79.4
101 Syngnathus pelagicus	4	893.1	353.7	2.53	69.2	80.4

Table 2.5. Wet and dry weights (cont.).

Notes and sources of organisms used for wet and dry weights. Details of collection are given in Appendices III and IV. Species numbers are those of Morris and Mogelberg, 1973.

44 Anemonia sargassensis: 369-3 (267 mixed sizes)

46 Acerotisa notulata: 348-5 (21), 358-2 (5), 358-4 (6), 358-6 (7), SHL 3-2 (17), SHL 3-3 (22), SHL 5-1 (21), (All smaller than 1 mm)

47 Hoploplana grubei: 358-2 (3), 358-6 (4), SHL 5-7 (2), 6 large, 3 small

48 Gnescioceros sargassicola: 373-2 (10), 372-5 (11), 373-2 (10), 373-4 (38), 374-2 (7)

51 Platynereis dumerillii: 276-5 (9 adults 6-12 mm, 38 juv $<$ 5 mm)

70 Bagatus minutus: 373-7 (32 mixed), 374-1 (137 mixed), 378-3 (20 adults, 44 juv), 380-5 (7 adults, 3 juv)

73 Sunampithoë pelagica: 375-3 (3), 376-7 (14 adults, 3 juv), 377-7 (5), 380-2 (5) (22 $\geq$ 3 mm, 8 $\leq$ 2 mm)

74 Biancolina sp.: 361-2 (5), 366-5 (4 adults, 8 juv), 366-7 (3 adults, 1 juv), 367-3 (5), 367-5 (7), 369-2 (2), 369-3 (2) (11 small, 26 large)

76 Hemiaegina minuta: 368-3 (8 adults, 76 juv), 373-5 (100), 375-1 (54), 375-6 (20)

78 Hippolyte coerulescens: 382-3 (2), 382-4 (1 ♀), 382-6 (1 ♀, 1 juv) (2 $\sim$ 5 mm, 2 ♀ w/eggs $\sim$ 10 mm)

80 Latreutes fucorum: 372-1 (22 adults, mostly large)

82 Leander tenuicornis: 372-1 (2 large, 1 small); second series: 371-3 (1 juv), 371-4 (3), 367-5 (2), 367-6 (1 juv) (6 large, 4 small all together)

83 Planes minutus: 376-3 (1 large); second series: 358-4 (1), 358-5 (1), 360-3 (1 juv), 360-4 (1 juv), 361-3 (4) (2 large adults, 5 smaller, 3 juv)

84 Portunus sayi: 353-2 (1), 353-5 (1), 368-1 (1), 368-5 (1), 369-1 (1), SHL 9-3 (1 10-mm adult, 1 4-mm juv) (all others 4 mm juv; over half of legs & chelopods missing on all specimens). Second series: 361-3 (1 juv), 364A-1 (1 juv), 369-3 (1 juv), 373-6 (1 adult) (2 large - 10 mm, 6 small - 4 mm)

87 Anoplodactylus petiolatus: 373-1 (70, incl 17 w/eggs), 372-1 (110 mixed), 373-1 (67, incl 12 w/eggs) (29 of 274 w/eggs)

89 Litiopa melanostoma: 372-1 (18), 380-1 (6), 380-2 (7), 380-3 (6), (19 $\geq$ 3 mm, 7 $\leq$ 1 mm)

Table 2.5. Wet and dry weights (cont.)

90 *Spurilla neapolitana*: 372-5 (10 adults, 9 juv), 376-2 (9), (10 $\geq$ 5 mm, 9 $\sim$ 3 mm, 9 $\leq$ 1 mm)

91 *Corambella depressa*: 358-6 (6), 360-2 (2), 360-4 (12), 372-4 (2), (7 $\geq$ 2 mm, 12 $\leq$ 2 mm)

93 *Doto pygmaea* (Unidentified nudibranch A): 375-2 (51), 380-2 (35), 380-3 (89), (mixed sizes)

95 *Scyllaea pelagica*: 380-2 (21 juv, 1-6 mm), 380-3 (3 juv), 382-4 (2 adults, 12 mm & 15 mm)

100 *Histrio histrio*: 369-3 (1, 1.5 cm); second series: 358-7 (1), 361-2 (1), 361-5 (1), 365-3 (2)

101 *Syngnathus pelagicus*: 369-3 (1, 8 cm); second series: 4 from Joy Geiselman, 4.5 to 9.5 cm. Ave. length 7.5 cm

---

This function shows highly variable correlation with displacement volume (-0.327 to +0.710), and a slope consistently close to zero (-0.005 to +0.013), clearly demonstrating that the diversity of the *Sargassum* community is not necessarily increased for larger rafts of *Sargassum*, even though the total number of animals increases with increasing colony size. This may be a result of the set of species being determined primarily by the initial fragmentation of a populated clump, together with the absence of immigration or emigration as the clump grows larger.

2.6 FUNCTION

Although many members of this community are plankton feeders and the community is therefore not perfectly independent, many important community processes probably occur almost entirely within the confines of the *Sargassum* habitat. The distinct physical boundaries of the habitat, the compactness of each clump, the limited number of species involved and the widespread distribution of *Sargassum* suggest that this is a valuable opportunity to study processes important to animal communities in general.

For example, what governs the number of species and abundance of individuals? Is it accident? What are the effects of habitat size, season, or age of the community? What are the associations and interrelationships between the component species? How is the species composition and diversity affected by these factors, and how do they change geographically? Does the *Sargassum* community show a definite seasonal succession? How does succession proceed? Sections 2.7 and 2.8 present preliminary attempts to answer some of these questions.

Table 2.6 TOTAL NUMBER OF MOTILE ORGANISMS AND DIVERSITY CORRELATED TO COLONY SIZE

	n	r_N	b_N	m_N	r_H	b_H	m_H
Station 'S'	244	0.636*	65.82	1.395	0.132	2.203	0.001
Sir Horace Lamb	30	0.226	16.37	0.633	0.051	1.866	0.001
Halifax	9	0.557	39.81	1.730	-0.327	2.450	-0.005
Westward 2	12	0.331	33.36	1.14	0.710*	1.225	0.013
Eastward	43	0.814*	6.788	1.564	0.200	1.949	0.002
Hunt	18	0.708*	6.151	2.247	0.014	2.131	0.000
Westward Carib.	27	0.479*	26.63	0.666	0.407	1.513	0.006
All cruises	139	0.636*	9.964	1.442	0.263*	1.812	0.003

n = number of samples in group.
N = total number of organisms counted in sample
V = displacement volume (ml)
$H = -\sum_{i=1}^{S} p_i \log_2 p_i$
S = total number of species in sample
p_i = proportion of total number of organisms in species i

correlations: $N = b_N + m_N V$ (r_N = correlation coefficient between N and V

$H = b_H + m_H V$ (r_H = correlation coefficient between H and V

Correlation coefficients marked with asterisk are significant at the 1% ($\alpha < 0.01$) level.

Fig. 2.5. Correlation of total number of motile macrofauna with Sargassum clump size. (Samples from Station 'S' detailed in Appendix III.)

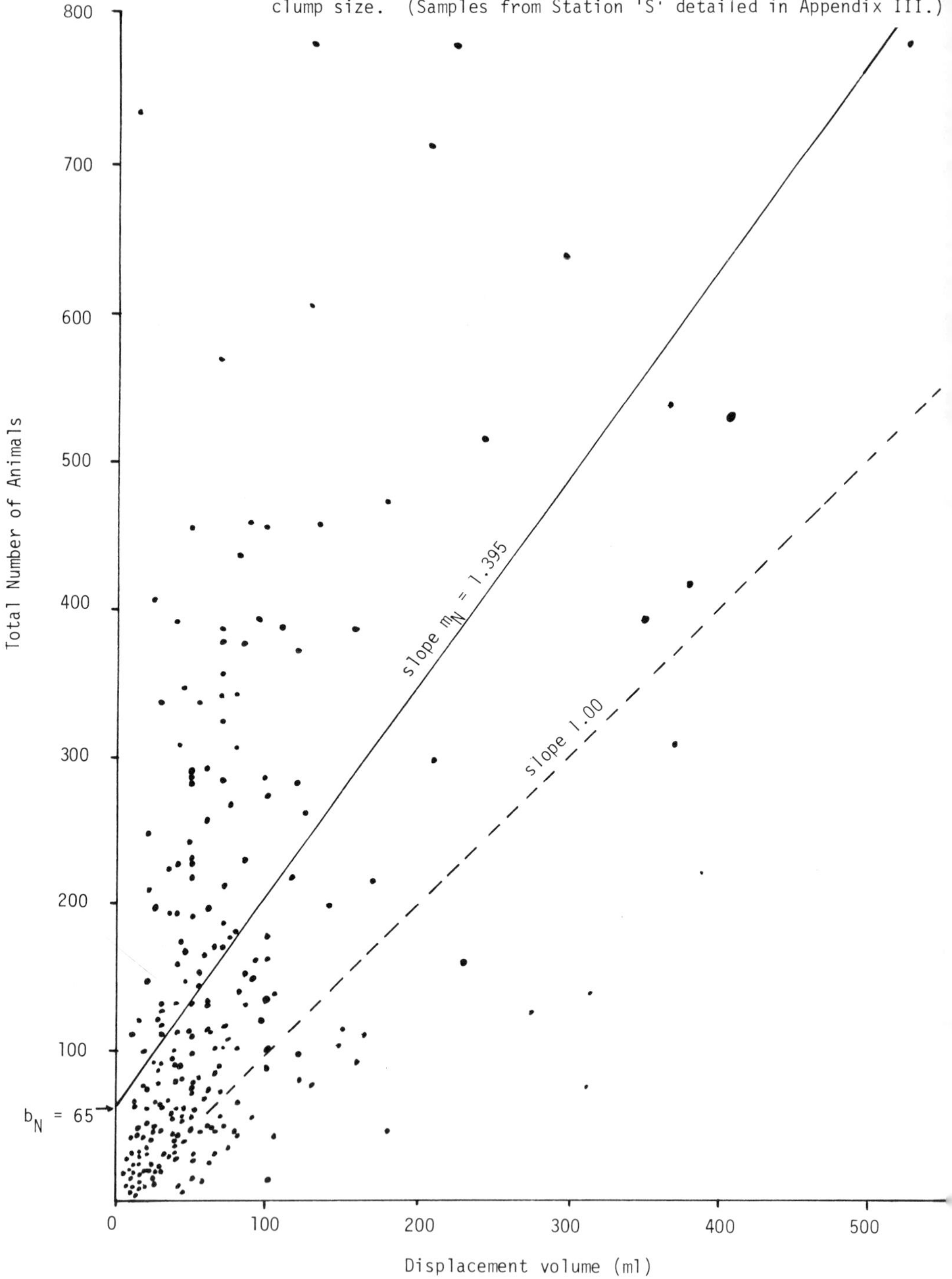

Fig. 2.6. Correlation of total number of motile macrofauna with *Sargassum* clump size for cruise data (See Appendix IV).

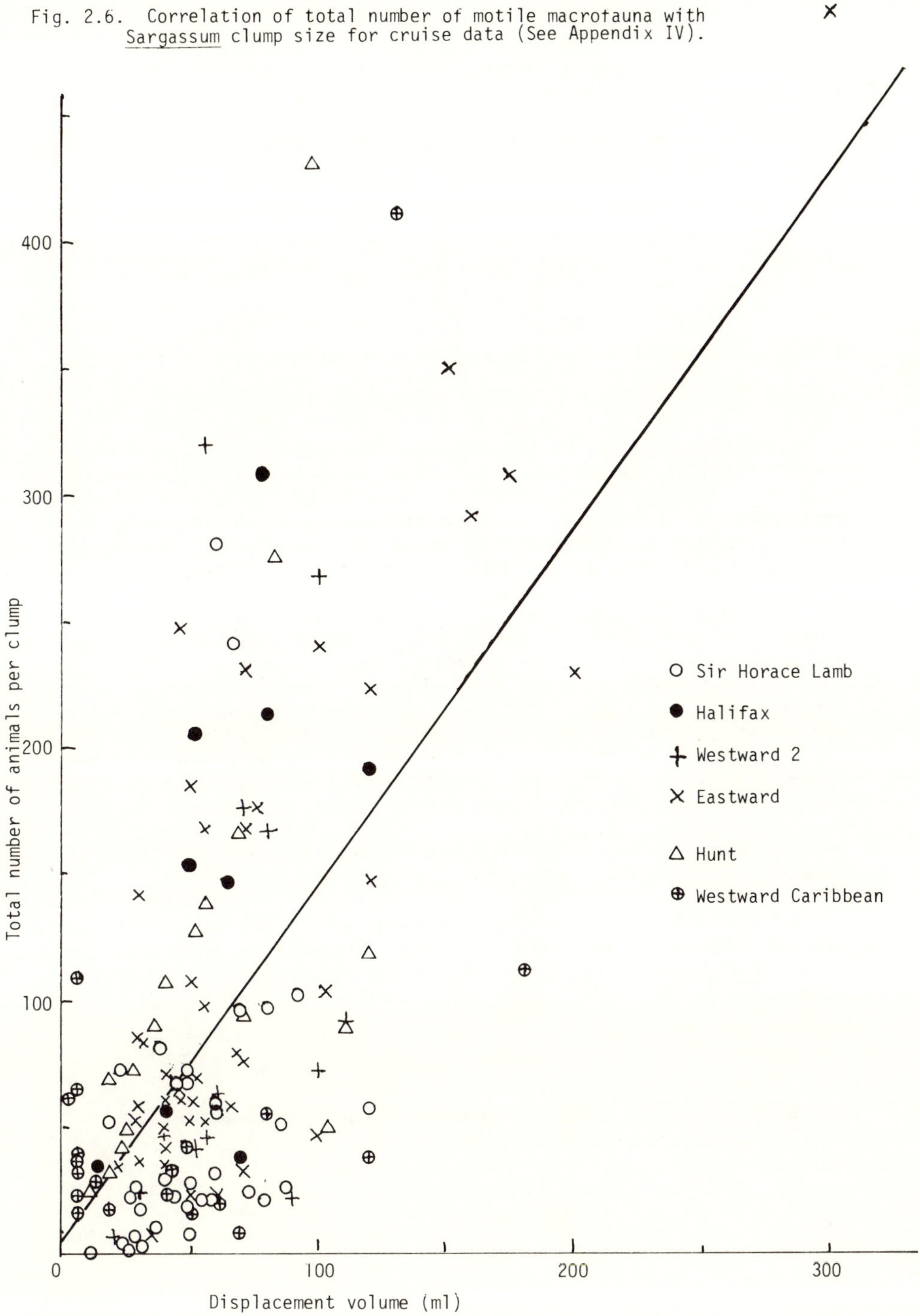

2.7 THE FOOD WEB

The animals inhabiting *Sargassum* can be categorized by their feeding habits as follows:

a) filter feeders - species that import food to the community from the plankton; e.g. hydroids, bryozoans, barnacles, tube worms;

b) omnivores - mainly scavengers feeding on the alga, detritus, and other organisms, e.g., most species of flatworms, polychaetes, amphipods, isopods, crabs and shrimp;

c) carnivores - feeding on other animals in the community; e.g., pycnogonid, gastropod, nudibranchs, crabs, fish;

d) detritus grazers - e.g., copepods.

Observations on the *Sargassum* community food web were conducted under the direction of Byron Morris by Joy Geiselman in 1972-73. Her report is included as Appendix VI. Although the work is preliminary, it is the most comprehensive to date, and has provided some new information on the food relationship of this community. These results are summarized in Figure 2.7.

No strict herbivores have been discovered within the community. There is considerable trophic overlap among the animals, even at the bottom trophic level, and above this level, the distinctions are even more blurred. Competition for food must occur. For the higher level omnivores and carnivores, there are major alternative food sources, and it is apparent that the limits of food availability within the *Sargassum* community have imposed a necessity for the larger predatory organisms (crabs, shrimp, and fish) to have the greatest number of alternative food sources available to them. For example, the most common crab (*Planes minutus*) seems to be a generalized feeder whose food is determined primarily by availability and success.

The similarity of food preferences and feeding types of the various species was deduced, from the information in Fig. 2.7 by cluster-analysis, using the Dices coefficient of similarity (Sokal and Rohlf, 1973). The Dices coefficient is defined as:

$$C_D = \frac{2\ AB}{A + B}$$

where AB = the number of possible food sources shared by species A and species B

A = the total number of food sources used by species A

B = the total number of food sources used by species B.

Fig. 2.7. Summary of food web data

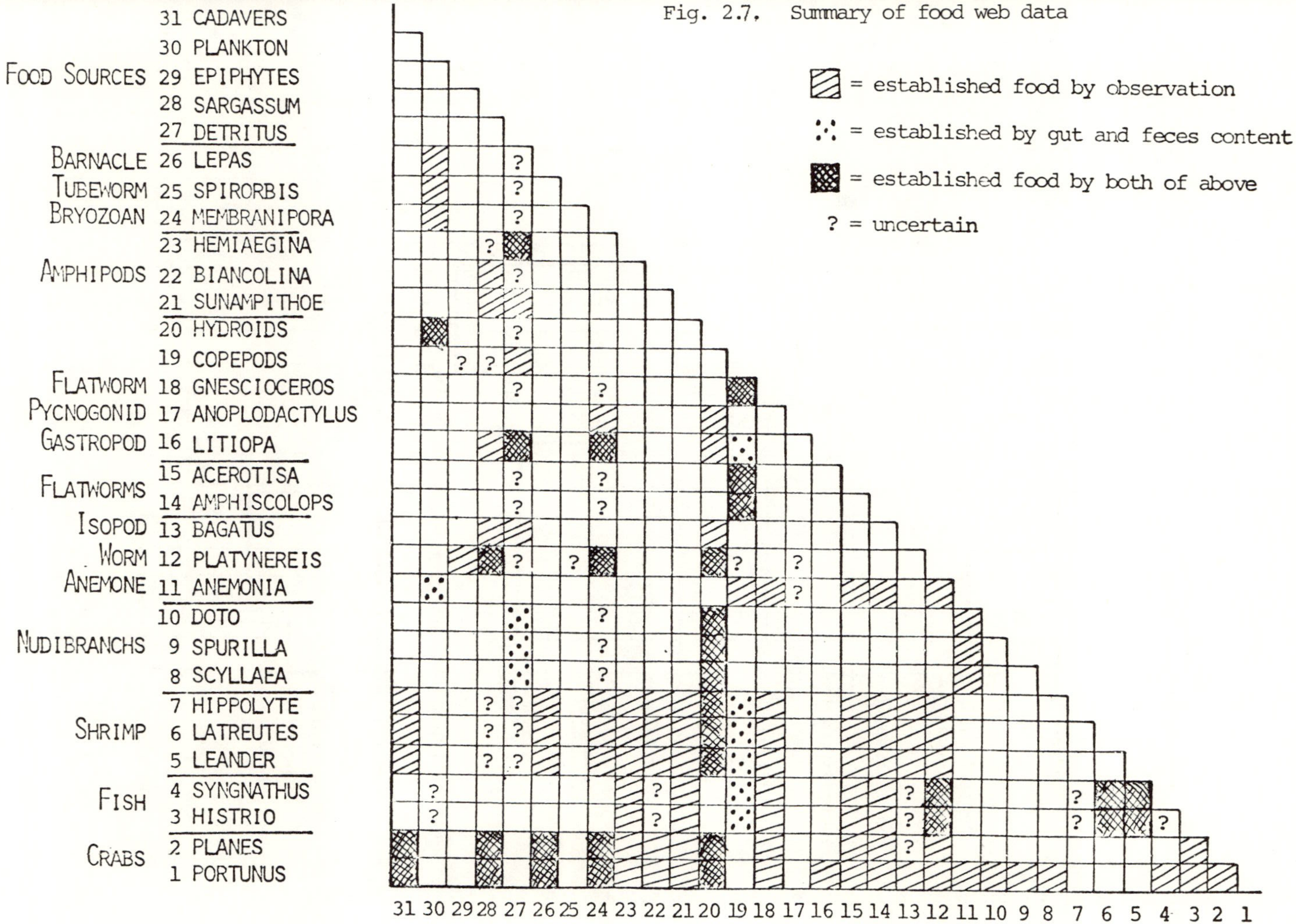

Fig. 2.8 Dendrogram of the similarities of food choice and feeding types within the <u>Sargassum</u> community

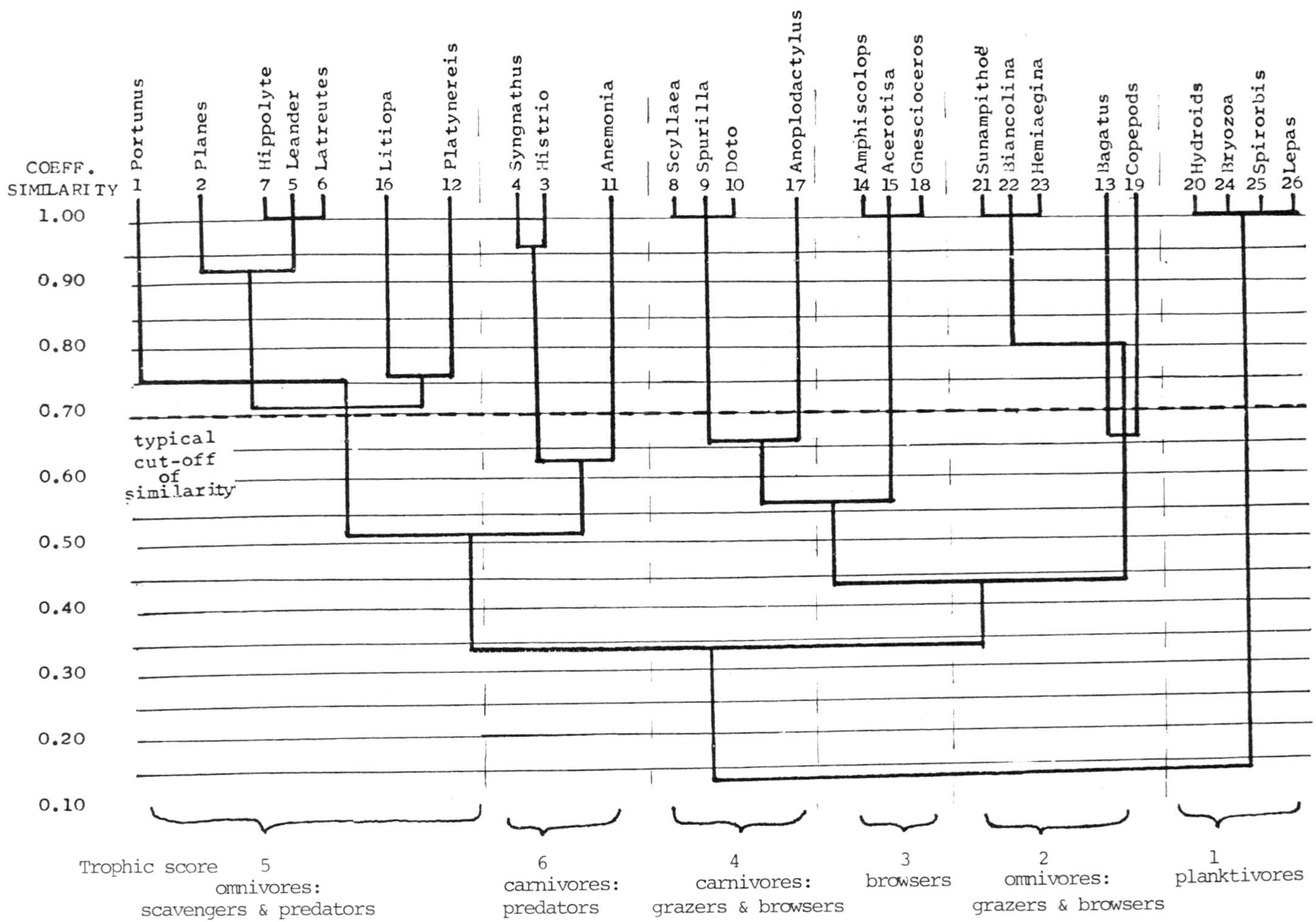

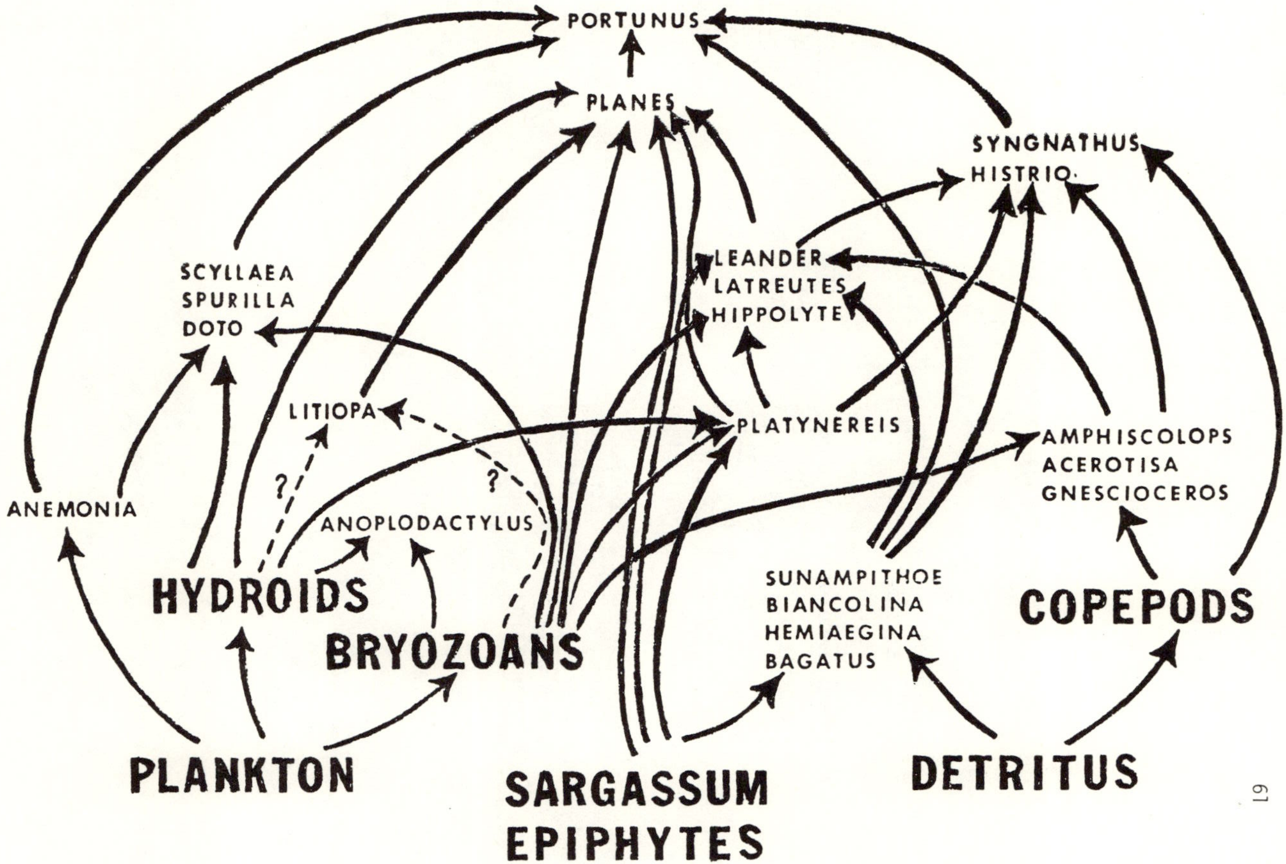

Fig. 2.9. Food web of the Sargassum community

Table 2.7. Seasonal Changes in the Occurrence and Abundance of 23 Common Species of Motile Macrofauna of the Pelagic *Sargassum* Community.

Season	Winter	Spring	Summer	Autumn
Months	J-F-M	A-M-J	J-A-S	O-N-D
# Species Present	21	21	21	23
Species Absent	*Histrio* *Amphiscolops*	*Syngnathus* *Amphiscolops*	*Hippolyte* *Spurilla*	None
Species at Maximum Abundance	*Spurilla* *Gnescioceros*	*Planes* *Scyllaea* *Doto* *Anoplodactylus* *Sunampithoë* *Biancolina* *Bagatus* *Hoploplana*	*Litiopa* *Histrio* *Amphiscolops* *Acerotisa*	*Portunus* *Latreutes* *Leander* *Platynereis* *Syngnathus* *Anemonia* *Hemiaegina*
Species at Minimum Abundance	*Portunus* *Planes* *Acerotisa* *Hoploplana* *Biancolina*	*Leander* *Latreutes* *Platynereis* *Anemonia* *Gnescioceros* *Hemiaegina*	*Doto* *Anoplodactylus* *Bagatus*	*Litiopa* *Scyllaea* *Sunampithoë*
Average # Common Species	14	15	19	19
Average # Individuals Per 100 cc dv of Sargassum	2705	2954	1370	3511

C_D can range from 0 (no similarity) to 1.0 (total similarity). The dendrogram (Fig. 2.8) shows the general division of feeding types and food resource similarities that exist among the Sargassum community members. Competition for food resources may occur among similar groups of animals.

The above information was used to develop a diagram of the food web of the community (Fig. 2.9). These relationships are qualitative only, and do not indicate the quantitative importance of a food source to the consumer. Additional work will be needed to quantify these relationships. To determine how the different species interact in the Sargassum community, further knowledge of the trophic relations of the important species is required. More direct observation of feeding and gut analysis will be needed. Quantitative biomass measurements (or estimates based on numbers and predetermined weights - Section 2.3) can be used to distinguish the important species in the food web. Knowledge of the standing stocks of the various species are prerequisite for energy-flow studies within the community.

Knowledge of the food web can help clarify the species association within the community. Indices of association (Southwood, 1972) between the various species can be determined. For example, Sorensen's (1948) Quotient of Similarity or Kendall's (1957; 1962) Rank Correlation Coefficient could show positive or negative correlations between species that would help identify predator-prey or competitive relationships. For example, the nudibranchs seem to be obligate feeders on the attached hydroids (Clark, 1975) and their abundance in the community would therefore logically correlate to hydroid abundance. Many similar links within the community can be statistically discerned and then tested by direct observations or the acquired knowledge of the community composition.

2.8 SEASONAL CHANGES

The Sargassum community undergoes annual successional cycles. In the Bermuda region of the Sargasso Sea, the late winter stocks of Sargassum are generally overgrown with epiphytes and epifauna. In the spring, new growth appears, with characteristically light-colored stipes and clumps relatively free of either attached or motile organisms. This begins the new annual cycle, and through summer and fall, subsequent stages exhibit increased colonization by attached organisms (initially hydroids, later followed by barnacles, forams, bryozoa, tubeworms, etc.) and increased variety of the motile fauna.

(The detailed growth studies on Sargassum muticum (Kane and Chamberlain, 1979; Chamberlain et al., 1979; Gorham, 1979) might be a useful starting point for further studies on the growth cycle of pelagic Sargassum species.)

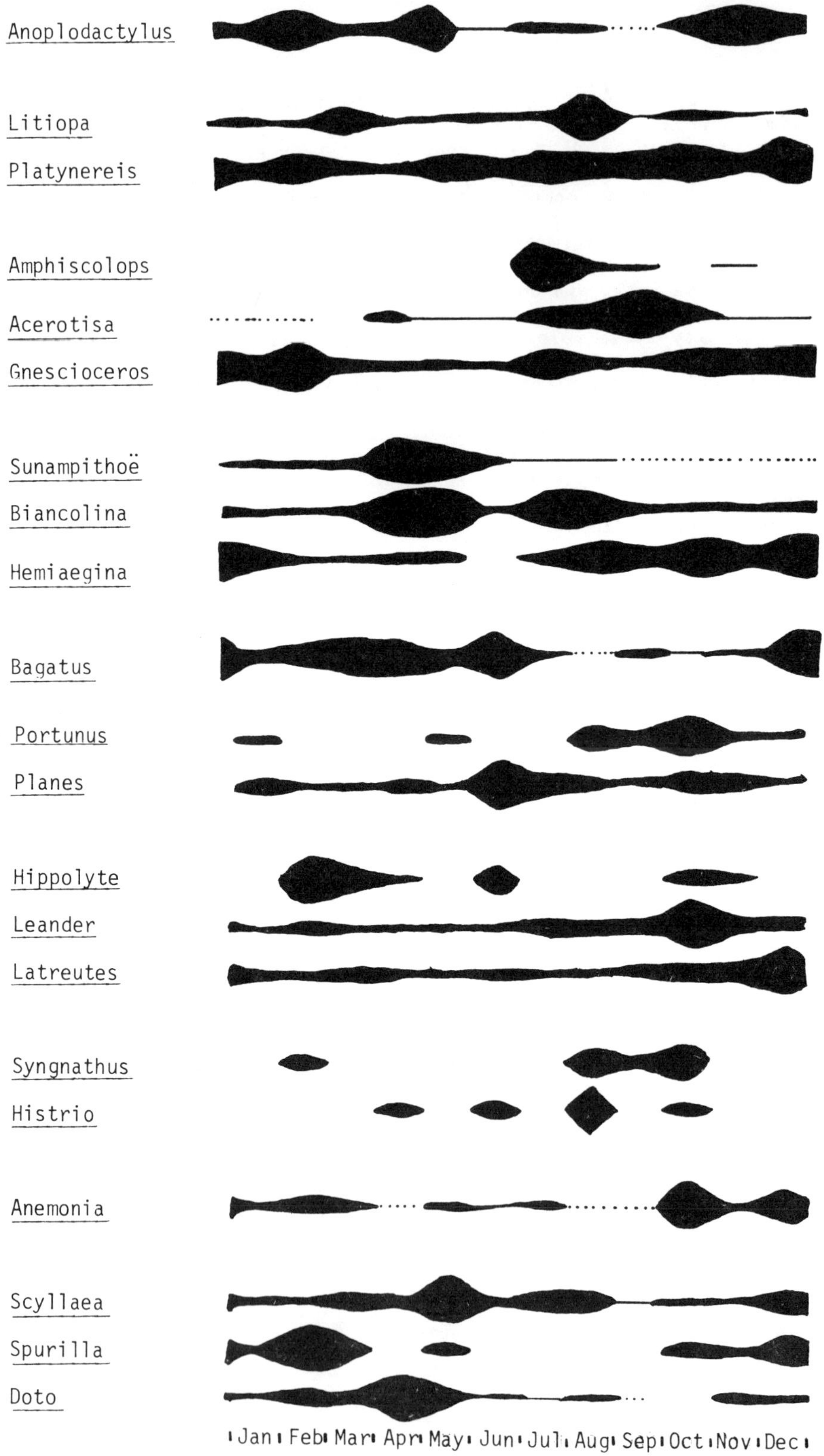
Anoplodactylus
Litiopa
Platynereis
Amphiscolops
Acerotisa
Gnescioceros
Sunampithoë
Biancolina
Hemiaegina
Bagatus
Portunus
Planes
Hippolyte
Leander
Latreutes
Syngnathus
Histrio
Anemonia
Scyllaea
Spurilla
Doto
Jan
Feb
Mar
Apr
May
Jun
Jul
Aug
Sep
Oct
Nov
Dec

Succession in the motile macrofauna is summarized in Table 2.7 and Fig. 2.10. This succession may proceed to a temporary "climax" stage in the fall characterized by the dominance of predators and detritus feeders. The climax stage must eventually atrophy as the alga dies, after which the annual cycle is repeated. The decline of the climax community is worth study as well, since it may provide useful insights into an aspect of community succession that has been little studied.

One aspect of succession that is evident on Sargassum is the variability of the sessile fauna on the different growth regions of maturing plants. Ryland (1974) recognized up to six different zones on the alga between the actively growing apical meristem and the eldest basal stem, and found that certain sessile species were associated with only certain growth zones of the alga. Thus, the bryozoan Membranipora tuberculata, the tubeworm Spirorbis corrugatus (= Janua formosus), and the blue-green alga Calothrix crustacea are abundant only on the older parts of the plant, while the hydroid Clytia noliformis fluorishes mainly on the new growth. Other sessile species possibly behave similarly, and influences of competition and predation must be involved. Predators on the sessile fauna, for example, must wait until the establishment of these colonies before they can become abundant.

Collection of Sargassum clumps of various ages on R/V Westward cruise W-58 by A.W. Stoner and H.S. Greening (Tables 2.8 and 2.9, and Fig. 2.11) confirmed trends expected in the associated motile macroinvertebrates with macrophyte age, increase in sessile fauna, and epiphytization.

All of the samples were taken in June and July, yet they show significant shifts in the relative abundance of macroinvertebrate species with apparent macrophyte age. The relative abundance of Latrentes fucorum increased with age in the Gulf Stream and decreased in the Sargasso Sea. Bagatus minutus showed a major decline with Sargassum age in the Gulf Stream and Gnescioceros sargassicola decreased in both geographic regions. Litiopa melanostoma showed major increases in the Sargasso Sea; other species were less affected by plant age and populations remained stable.

Examinations of Tables 2.2, 2.8 and 2.9 indicates that there is significant geographic variation in the motile associates of Sargassum. This geographic variation is currently under investigation by A.W. Stoner and H.S. Greening (1983).

← Fig. 2.10. Seasonal variation in the Sargassum fauna

Table 2.8. Motile invertebrates associated with Sargassum clumps from the Sargasso Sea. Values are numbers of individuals collected. Percentages are given in parenthesis. (Stoner and Greening, 1983)

Species	Age of Algae			
	Young	Moderate	Old	Overall
Litiopa melanostoma	10 (10.0)	130 (24.2)	83 (34.7)	223 (25.5)
Latreutes fucorum	42 (42.0)	115 (21.4)	43 (18.0)	200 (22.8)
Planes minutus	10 (10.0)	67 (12.4)	28 (11.7)	105 (12.0)
Sunampithoe pelagica	15 (15.0)	52 (9.7)	27 (11.3)	94 (10.7)
Bagatus minutus	5 (5.0)	59 (11.0)	13 (5.4)	77 (8.8)
Gnescioceros sargassicola	7 (7.0)	53 (9.8)	5 (2.1)	65 (7.4)
Platynereis dumerilli	4 (4.0)	16 (3.0)	13 (5.5)	33 (3.8)
Hoploplana grubei	2 (2.0)	9 (1.7)	12 (5.0)	23 (2.7)
Leander tenuicornis	4 (4.0)	7 (1.3)	3 (1.3)	14 (1.6)
Anoplodactylus petiolatus	1 (1.0)	11 (2.0)	1 (0.4)	13 (1.4)
Hemiaegina minuta		3 (0.5)	6 (2.5)	9 (1.0)
Luconacia incerta		5 (0.9)		5 (0.6)
Corambella depressa		4 (0.7)		4 (0.5)
Biancolina sp.		2 (0.4)	1 (0.4)	3 (0.3)
Doto pygmaea		2 (0.4)	1 (0.4)	3 (0.3)
Scyllaea pelagica		2 (0.4)	1 (0.4)	3 (0.3)
Hippolyte coerulescens			2 (0.9)	2 (0.2)
Portunus sayi		1 (0.2)		1 (0.1)
	100 (100)	538 (100)	239 (100)	877 (100)

NOTE: 40 clumps were collected at four stations in June and July 1981:

36°29'N 69°28'W
35°42'N 67°48'W
35°26'N 67°08'W
34°26'N 64°58'W

See Fig. 2.11.

Table 2.9. Motile invertebrates associated with *Sargassum* clumps from the Gulf Stream. Values are numbers of individuals collected and percentages.

Species	Age of Algae Young	Moderate	Old	Overall
Latreutes fucorum	19 (55.9)	571 (70.0)	20 (77.0)	610 (69.7)
Bagatus minutus	6 (17.6)	83 (10.2)	1 (3.8)	90 (10.3)
Leander tenuicornis	2 (5.9)	49 (6.0)	1 (3.8)	52 (5.9)
Portunus sayi	1 (2.9)	39 (4.8)	1 (3.8)	41 (4.7)
Platynereis dumerillii	4 (11.8)	26 (3.1)	3 (11.6)	33 (3.8)
Gnesciocceros sargassicola	2 (5.9)	18 (2.2)		20 (2.3)
Hemiaegina minuta		12 (1.5)		12 (1.4)
Litiopa melanostoma		8 (1.0)		8 (0.9)
Anoplodactylus petiolatus		5 (0.6)		5 (0.5)
Corambella depressa		3 (0.4)		3 (0.3)
Doto pygmaea		2 (0.2)		2 (0.2)
	34 (100)	816 (100)	26 (100)	876 (100)

Note: 38 Clumps were collected near the north wall of the Gulf Stream at two stations in June and July 1981: 37°25'N, 71°30'W and 38°05'N, 64°02'W. See Fig. 2.11. Data of Stoner and Greening (1983).

In the spring when much of the *Sargassum* is new and actively growing, the community structure is simple and restricted. Diversity builds up as the clumps age and the full range of growth zones develop. Likewise, those members restricted to the actively growing portions of the plant must die out with the senescence of the *Sargassum* clump.

We have observed these effects in our samples. The colonizing, stoloniferous, erect and unbranched hydroids *Clytia* and *Zanclea* characterize the apical growth regions, but on the older stipes and blades, the branched and erect larger hydroids such as *Aglaeophenia*, *Plumularia* and *Obelia* plus the other sessile organisms overgrow the simpler hydroids and either smother or outcompete them (Appendix V). This factor in the community succession of the *Sargassum* biota is an interesting direction for further study.

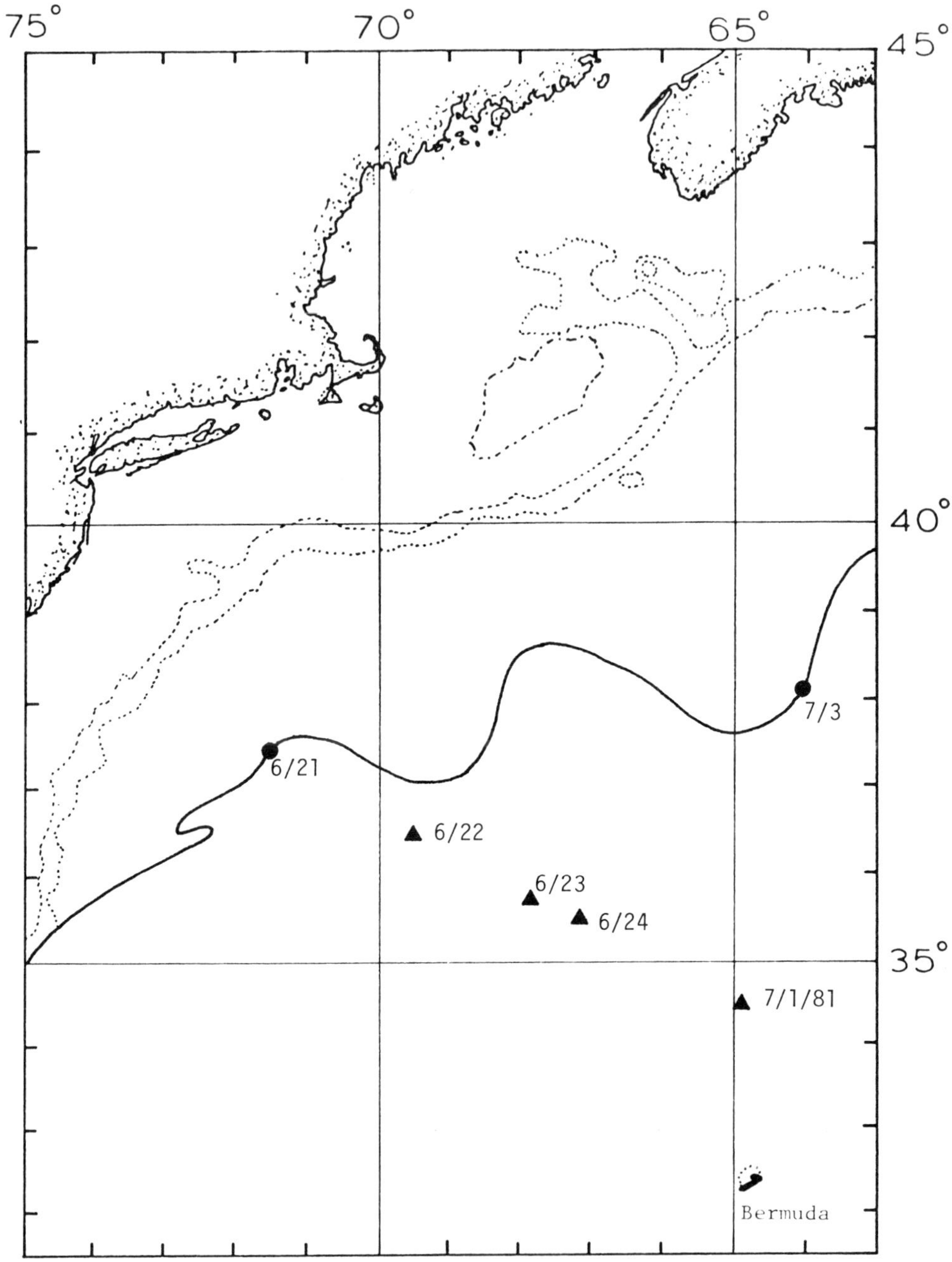

Fig. 2.11. *Sargassum* collection sites on R/V *Westward* cruise W-58. Triangles = Sargasso Sea sites. Circles = Gulf Stream sites. The position of the north wall of the Gulf Stream is shown for 24 June 1981, as reported by the National Weather Service.(Stoner and Greening, 1983)

2.9 EFFECTS OF POLLUTION

As mentioned in section 1.6, the occurrence of petroleum residues in *Sargassum* and in the motile macrofauna of the *Sargassum* community is well-documented elsewhere (Burns and Teal, 1973; Morris, Cadwallader, Geiselman and Butler, 1976 - see also Appendices VI and VIII), but this evidence is obtained by solvent extraction and gas chromatographic analysis, not from observations of community structure and function.

The original idea on which this project was based was to compare the population of a number of common species on communities associated with high tar lump density, or in which the *Sargassum* contained relatively high petroleum hydrocarbon concentrations, with populations on those communities relatively free of pollution.

This did not work out for several reasons. First, as collections proceeded, it became apparent that tar lumps did not stay with a given clump of alga, but collected temporarily in adjacent water, where both were brought together by a downwelling convergence. Therefore the happenstance of catching a certain weight of tar in the same net with a certain amount of *Sargassum* did not indicate whether or not the *Sargassum* and tar were intimately associated over an extended period, or whether the act of collection had brought them together.

This problem might have been overcome by chemical analysis of the actual alga which supported a given community. Although preservation of a sample for hydrocarbon analysis (by freezing, homogenization and solvent extraction) is not compatible with preservation (with formalin) for identification of motile macrofauna, currently available capillary gas chromatography could yield an adequate hydrocarbon profile from a small portion of the *Sargassum* clump. However, this has not yet been attempted.

Even if the amount of petroleum residue in the algal tissues could be accurately known, however, there would still be complications from direct uptake of tar particles by organisms. In the course of food-web studies (Appendix VII; see also Morris *et al.*, 1976), it was observed that *Portunus sayi* and *Planes minutus* ingested small tar lumps along with other food; black particles were found in the fecal pellets of *Litiopa melanostoma* and the gut contents of *Hemiaegina minuta*; and high concentrations of petroleum residues were found in the gills of the crabs *Planes* and *Portunus* (See Appendix VIII). Such direct uptake of hydrocarbons from the water in the form of tar lumps or particles could occur during a time period short compared to that required for absorption by algal tissues, and so might not be reflected in an analysis of *Sargassum* alone.

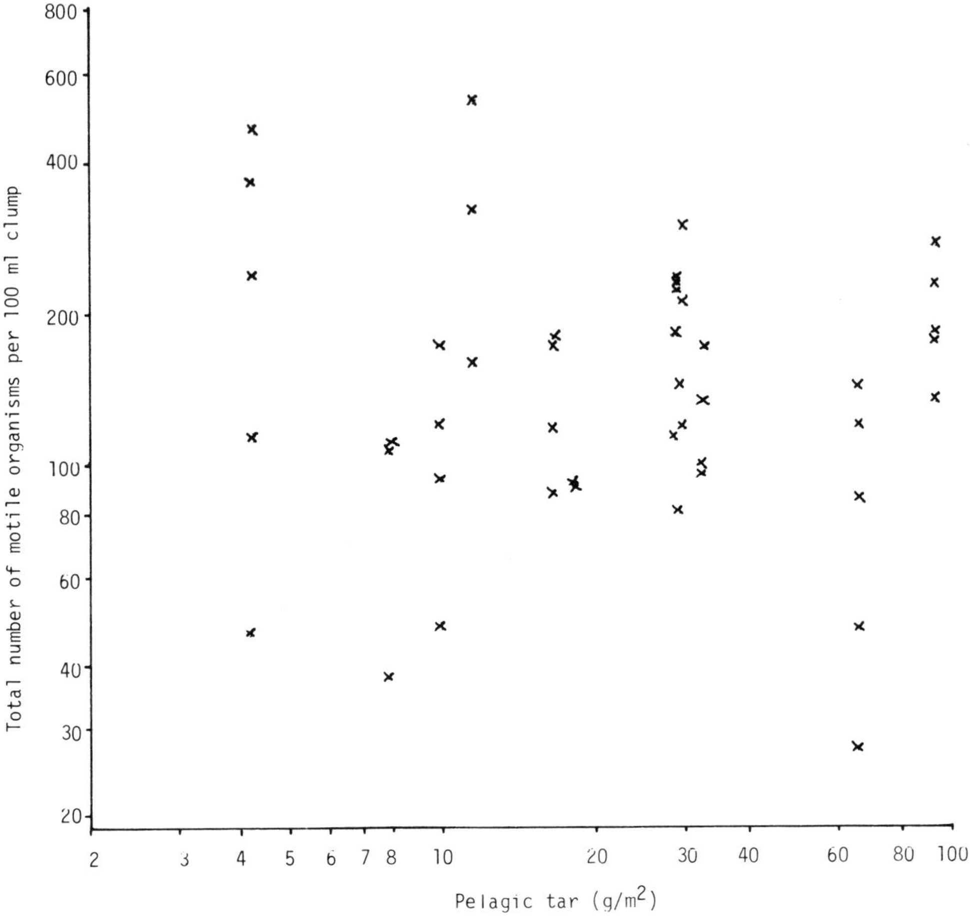

Fig. 2.12. Correlation of total number of motile organisms per 100 ml clump of Sargassum with occurrence of pelagic tar in the same vicinity. Data from RV Eastward cruise June-July 1973. Full data are given in Appendix IV.

Finally, the number and nature of motile macrofauna species found on a clump of Sargassum varied tremendously from sample to sample (Appendices III and IV), even when these were taken at the same location and within a period of an hour. This extreme variability made the confidence of comparisons correspondingly low, and thus it became apparent that any effects of petroleum would not be detected by correlation and regression analysis, but by chemical analysis alone. Some of these analytical results were published (Morris *et al.*, 1976 - reproduced as Appendix VI) and the remainder are presented in Appendix VIII).

The set of data which seemed most likely to yield evidence for a correlation between the amount of tar and the structure of the Sargassum community were the data taken on board the R/V Eastward in June-July 1973 (Appendix IV). Tar quantities were measured by neuston tows at approximately the same location where Sargassum was dipnetted for faunal identification and counting. In Figure 2.12, the total number of motile macrofauna (normalized to 100 ml displacement volume) are shown as a function of the tar concentration. There is no significant trend. Data for various individual species, and data from other cruises, likewise show a wide variability in each sample, and no detectable trend with the associated tar concentration.

NOTE:

The following bibliography contains references cited in the various appendices, as well as in the text of Sections 1 and 2. In addition, selected references (not cited in the text) are given to the natural history of the Sargasso Sea, the physical oceanography of the region, and related studies of other *Sargassum* communities.

All authors and subjects in the bibliography (including the small bibliography at the end of Appendix VI) are included in the general index at the back of this book.

3. SELECTED BIBLIOGRAPHY

Adams, J.A. 1960. A contribution to the biology and post larval development of the Sargassum fish, *Histrio histrio* (Linnaeus), with a discussion of the *Sargassum* complex. Bulletin of Marine Science of the Gulf and Carribbean 10(1):55-82.

Agardh, J.G. 1889. Species sargassorum Australie. Kgl. Svenska Vet.-Akad. Handl. Stockholm 23(3).

Backus, R.H., G.R. Flierl, D.R. Kester, D.B. Olson, P.L. Richardson, A.C. Vastano, P.H. Wiebe and J.H. Wormuth. 1981. Gulf Stream cold core rings: Their Physics, Chemistry and Biology. Science 212:1091-1100.

Baylor, E.R. and W.H. Sutcliffe, Jr. 1963. Dissolved organic matter in seawater as a source of particulate food. Limnol. Oceanogr. 8:369-371.

Bergh, R. 1871. Mollusken des Sargassomeeres. Zoologisch-Botanische gesellschaft Vienna, Verhandlungen 21:1273-1308.

Bidleman, T.F. and C.E. Olney. 1974. Chlorinated hydrocarbons in the Sargasso Sea atmosphere and surface water. Science 183:516-518.

Bidleman, T.F., C.P. Rice and C.E. Olney. 1976. High molecular weight hydrocarbons in the air and sea: Rates and mechanisms of air/sea transfer. In *Marine Pollutant Transfer*, H. Windom and R.A. Duce, eds. D.C. Heath, Lexington, MA. pp. 323-351.

Blake, N.J. and D.L. Johnson. 1976. Oxygen production-consumption of the pelagic *Sargassum* community in a flow-through system with arsenic additions. Deep-Sea Research 23(8):773-778.

Blumer, M., G. Souza and J. Sass. 1970. Hydrocarbon pollution of edible shellfish by an oil spill. Marine Biology 5:195-202.

Boalch, G.T. and G.W. Potts. 1977. The first occurrence of *Sargassum muticum* (Yendo) Fensholt in the Plymouth area. J. Mar. Biol. Assoc. U.K. 57:29-31.

Borgesen, F. 1914. The species of *Sargassum* around the coast of the Danish West Indies. Mindeskrift 8, Japetus Steenstrup 32, Copenhagen; Dansk Botanisk Arkiv. 2(2).

Bortone, S.A., P.M. Hastings and S.B. Collard. 1977. The pelagic Sargassum ichthyofauna of the eastern Gulf of Mexico. Northeast Gulf Science 1(2):60-67.

Bray, J.R. and J.T. Curtis. 1957. An ordination of the upland forest communities of southern Wisconsin. Cited by Field and McFarlane 1968.

Brillouin. L. 1956. Science and Information Theory. Cited by Lloyd et al., 1968.

Brylinsky, M. 1977. Release of dissolved organic matter by some marine macrophytes. Marine Biology 39(3):213-220.

Burkenroad, M.D. 1939. Hydroids on pelagic Sargassum vegetation of the Western North Atlantic. Bulletin of the Bingham Oceanographic Collection, Peabody Museum of Natural History, Yale University 6:23-25. (See Parr 1939).

Burns, K.A. and J.M. Teal. 1973. Hydrocarbons in the pelagic Sargassum community. Deep-Sea Research 20:207-211.

Butler, J.N. and B.F. Morris. 1974. Quantitative monitoring and variability of pelagic tar in the North Atlantic. Proc. Marine Pollution Monitoring (Petroleum) Symposium. National Bureau of Standards Special Publication No. 409, pp. 75-78.

Butler, J.N., B.F. Morris and J. Sass. 1973. Pelagic tar from Bermuda and the Sargasso Sea. Special Publication No. 10, Bermuda Biological Station, 346 pp.

Caron, D.A., P.G. Davis, L.P. Madin and J.M. Sieburth. 1982. Heterotrophic bacteria and bactivorous protozoa in oceanic macroaggregates. Science 218:795-797.

Carpenter, E.J. 1970. Diatoms attached to floating Sargassum in the western Sargasso Sea. Phycologia 9(3/4):271-274.

Carpenter, E.J. 1972. Nitrogen fixation by a blue-green epiphyte on pelagic Sargassum. Science 178:1207-1209.

Carpenter, E.J. and J.L. Cox, 1974. Production of pelagic Sargassum and a blue-green epiphyte in the western Sargasso Sea. Limnology and Oceanography 19(3):429-436.

Carr, A. and A.B. Meylan. 1980. Evidence of passive migration of green turtle hatchlings in Sargassum. Copeia 2:366-368.

Chamberlain, A.H.L., J. Gorham, D.F. Kane et al. 1979. Laboratory studies on Sargassum muticum II. Apical dominance. Botanica marina 22(1):11-19.

Clark, K.B. 1975. Nudibranch life cycles in the Northwest Atlantic and their relationship to the ecology of fouling communities. Helgol. Wiss. Meeresunters 27(1):28-69.

Collins, F.S. 1917. The Sargasso Sea. Rhodora 19:77-84.

Colton, J.B., D.E. Smith and J.W. Jossi. 1975. Further observations on a thermal front in the Sargasso Sea. Deep-Sea Research 22:433-439.

Conover, J.T. and J. McN. Sieburth. 1964. Effect of Sargassum distribution on its epibiota and antibacterial activity. Botanica Marina 6(1/2):147-157.

Cordero, P.A. 1980. Phycological observations VII. On the occurrence and phyco-geographical distribution of Sargassum duplicatum and Halicoryne wrightii in the Phillipines. Publications of the Seto Marine Biological Laboratory 25(1-4):27-38.

Cousteau, J.-Y. 1970. Statement on global marine degradation. Biological Conservation 4:61-65.

Craik, A.D.D. and S. Leibovich. 1976. A rational model for Langmuir circulations. J. Fluid. Mech. 73:401-426.

Crozier, W.J. 1918. Note on the coloration of Planes minutus. American Naturalist 52:262-263.

Culliney, J.L. 1970. Measurements of reactive phosphorus associated with pelagic Sargassum in the northwest Sargasso Sea. Limnology and Oceanography 15:304-306.

David, P.M. 1965. The neuston net: A device for sampling the surface fauna of the ocean. J. Mar. Biol. Assoc. U.K. 45:313-320.

Deacon, G.E.R. 1942. The Sargasso Sea. Geographical Journal 99(1):16-28.

Dietrich, G. 1965. General Oceanography, New York, Interscience (Wiley).

Dooley, J.K. 1972. Fishes associated with the pelagic Sargassum complex, with a discussion of the Sargassum community. Contributions in Marine Science, University of Texas Marine Science Institute 16:1-32.

Faller, A.J. and A.H. Woodcock. 1964. The spacing of windrows of Sargassum in the ocean. Journal of Marine Research 22(1):22-29.

Faller, A.J. 1978. Experiments with controlled Langmuir circulations. Science 201:618-620.

Farlow, W.G. 1914. The vegetation of the Sargasso Sea. Proc. Am. Phil. Soc. 53:257-262.

Field, J.G. and G.McFarlane. 1968. Numerical methods in marine ecology I. A quantitative "similarity" analysis of rocky shore samples in False Bay, South Africa. Zoologica Africana 3(2):119-137.

Fine, M.L. 1970. Faunal variation on pelagic Sargassum. Marine Biology 7(2):112-122.

Fletcher, R.L. and S.L. Fletcher. 1975a. Studies on the recently introduced brown alga Sargassum muticum (Yendo) Fensholt. I. Ecology and reproduction. Botanica Marina 18(3):149-156.

Fletcher, R.L. and S.L. Fletcher. 1975b. Studies on the recently introduced brown alga Sargassum muticum (Yendo) Fensholt. II. Regenerative ability. Botanica Marina 18(3):157-162.

Fletcher, R.L. 1980. Studies on the recently introduced brown alga Sargassum muticum (Yendo) Fensholt. III. Periodicity in gamete release and "incubation" of early Germling stages. Botanica Marina 23(7):425-432.

Fletemeyer, J.R. 1978. Underwater tracking evidence of neonate loggerhead sea turtles seeking shelter in drifting Sargassum. Copeia 1:148-149.

Gargett, A.E., T.B. Sanford and T.R. Osborn. 1979. Surface mixing layers in the Sargasso Sea. J. Physical Oceanography 9(6):1090-1111.

Germaine, L. 1935. La mer des Sargasses. Bull. Inst. Oceanog. Monaco 671:1-23.

Gorham, J. 1979. Laboratory growth studies on Sargassum muticum III. Effects of auxins and anti-auxins on extension growth. Botanica marina 22(5):273-280.

Green, J. 1963. A Biology of Crustacea. Northumberland Press, London, 180 pp.

Gregg, M.C. and T.B. Sanford. 1980. Signatures of mixing from the Bermuda slope, the Sargasso Sea and the Gulf Stream. J. Physical Oceanography 10(1):105-127.

Greve-Jensen, L. 1974. Anatanais normani (Richardson) found near Bermuda and notes on other Anatanais species. Sarsia 55:115-120.

Hagan, D.E., D.B. Olson, J.E. Schmitz and A.C. Vastano. 1978. A comparison of cyclonic ring structures in the northern Sargasso Sea. Journal of Physical Oceanography 8(6):997-1008.

Hammond, A.L. 1974. Undersea storms: Experiments in the Atlantic (MODE project). Science 185:244-247.

Hanson, R.B. 1977. Pelagic *Sargassum* community metabolism - carbon and nitrogen. Journal of Experimental Marine Biology and Ecology. 29(2):107-118.

Hartline, B.K. 1979. Polymode: Exploring the undersea weather. Science 205:571-573.

Hedgpeth, J.W. 1957. Marine Biogeography. In *Treatise on Marine Ecology and Paleoecology*, Vol. I. Marine Ecology. J.W. Hedgpeth, ed., Memoir no. 67, Geological Soc. of America. 1296 pp.

Hentschel, E. 1922 (1920). Uber den Bewuchs auf den treibenden Tangen der Sargassosee. Mitteilungen aus dem Zoologischen Staats Institut und Zoologischen Museum in Hamburg. Beiheft Zum Jahrbuch der Hamb. Wissenschaftl. Anstalten 38:1-26.

Hess, W. 1981. Private communication to J.N. Butler regarding sampling strategy and measures of central tendency.

Hitchcock, H.B. 1941. The coloration and color changes of the gulf-weed crab, *Planes minutus*. Biological Bulletin 80(1):26-30.

Howard, K.L. and R.J. Menzies. 1969. Distribution and production of *Sargassum* in the waters off the Carolina coast. Botanica Marina 12:244-254.

Hurka, H. 1971. Factors influencing the gas composition in the vesicles of *Sargassum*. Marine Biology 11:82-89.

Jennings, J.B. 1957. Studies on feeding, digestion, and food storage in free-living flatworms. Biol. Bull. 112:63-80.

Johnson, D.L. and P.L. Richardson. 1977. On the wind-induced sinking of *Sargassum*. J. Experimental Marine Biology and Ecology 28(3):255-267.

Johnson, D.L. and R.S. Braman. 1975. The speciation of arsenic and the content of germanium and mercury in members of the pelagic *Sargassum* community. Deep-Sea Research 22:503-507.

Jonas, R.B. and F.K. Pfaender. 1976. Chlorinated hydrocarbon pesticides in the western North Atlantic Ocean. Environmental Science and Technology 10(8):770-773.

Jorgensen, C.B. 1966. Biology of Suspension Feeding. Bay Tree Press (Pergamon), Oxford, 357 pp.

Joshi, G.V. and C.A. Gowda. 1975. Seasonal variation in chemical composition of Sargassum illicifolium (Grun) and Seawater. Indian J. Mar. Sci. 4(2):165-168.

Kane, D.F. and A.H.L. Chamberlain. 1979. Laboratory growth studies on Sargassum muticum I. Seasonal growth of whole plants and lateral sections. Botanica Marina 22(1):1-9.

Kendall, M.G. 1957. A Course in Multivariate Analysis. Griffin, London, 185pp.

Kendall, M.G. 1962. Rank Correlation Methods. Griffin, London, 199pp.

Kerr, R.A. 1981. Small eddies proliferating in the Atlantic. Science 213:632-634.

Khailov, K.M. and Z.P. Burlakova. 1969. Release of dissolved organic matter by marine seaweeds and distribution of their total organic production to inshore communities. Limnol. Oceanogr. 14:521-527.

Khailov, K.M. and Z.Z. Finenko. 1970. Organic macromolecular compounds dissolved in seawater and their inclusion into food chains. In Marine Food Chains, J.H. Steele, ed., Oliver and Boyd, Publishers, pp. 5-18.

Knap, A.H., T.M. Iliffe and J.N. Butler. 1980. Has the amount of tar on the open ocean changed in the past decade? Marine Pollution Bulletin 11:161-164.

Knight-Jones, P., E.W. Knight-Jones, C.H. Thorp and P.W.G. Gray. 1975. Immigrant spirorbids (Polychaeta Sedentaria) on the Japanese Sargassum at Portsmouth England. Zoologica Scripta 4:145-149.

Kohlmeyer, J. 1971. Fungi from the Sargasso Sea. Marine Biology 8(4):344-350.

Kracht, R. and F.W. Tesch. 1981. Progress report on the Eel Expedition of R.V. Anton Dohrn and R.V. Friedrich Heincke to the Sargasso Sea 1979. Environ. Biol. Fish. 6(3/4):371-375.

Krummel, O. 1891. Die nordatlantische Sargassosee. Petermann's Geogr. Mitteilungen, Heft VI 37:129-141.

Kuntze, O. 1881. Revision von Sargassum und das sogenannte sargasso meer. A. Englers Botanische Jahrbucher 1:191-239.

Lance, G.N. and W.T. Williams. 1966. A generalized sorting strategy for computer classification. Nature 212:218.

Langmuir, I. 1938. Surface motion of water induced by wind. Science 87:119-123.

Leetmaa, A. 1977. Effects of the winter of 1976-77 on the northwestern Sargasso Sea. Science 198:188-189.

Leibovich, S. and S. Paolucci. 1980. The Langmuir circulation instability as a mixing mechanism in the upper ocean. J. Physical Oceanography 10(2):186-207.

Lloyd, M., J.H. Zar and J.R. Karr. 1968. On the calculation of information theoretical measures of diversity. The American Midland Naturalist 79(2):257-272.

Lowry, J.K. 1974. A new species of the Amphipod Biancolina from the Sargasso Sea. Trans. Amer. Micros. Soc. 93(1):71-78.

Lukashev, Yu. F. and A.M. Chernyakova. 1979. Variability of nitrate fields associated with the passage of eddy formations. Oceanology 19(4):409-413.

MacArthur, R.H. and MacArthur, J.W., 1961. On bird species diversity. Ecology 42:594-8.

Margalef, R. 1967. The food web in the pelagic environment. Helgolander Wiss. Merresunters. 15:548-559.

Markham, J.C. 1977. Distribution and systematic review of the bopyrid isopod Probopyrinella latreuticola (Gissler, 1882). Crustaceana 33(2):189-197.

Mather, R.S., C. Rizos and R. Coleman. 1979. Remote sensing of surface ocean circulation with satellite altimetry. Science 205:11-17.

McCartney, M.S., L.V. Worthington and W.J. Schmitz, Jr. 1978. Large cyclonic rings from the northeast Sargasso Sea. J. Geeophys. Research 83(C2):901-914.

McDowell, S.E. and H.T. Rossby. 1978. Mediterranean Water: An intense Mesoscale Eddy off the Bahamas. Science 202:1085-1087.

Menzel, D.W. and J.H. Ryther. 1961. Nutrients limiting the production of phytoplankton in the Sargasso Sea, with special reference to iron. Deep-Sea Research 7:276-281.

Merrifield, M.P. 1878. Gulf-Weed. Nature 18:708-711.

Meyen, J. 1838. Jahresberichte uber die resultate der Arbeiten in Felde der physiologische Botanik v.d. Jahre 1837. Weigmanns Archiv fur Naturgeschichte, Vierter Jahrgang Zweiter Band. Berlin.

Miller, M.C. 1961. Distribution and food of the nudibranch mollusca of the Isle of Man. J. Anim. Ecol. 30:95-116.

MODE Group. 1978. The mid-ocean dynamics experiment. Deep-Sea Research 25:859-910.

Molinari, R.L., S. Baig, D.W. Behringer, G.A. Maul and R. Legeckis. 1977. Winter intrusions of the loop current. Science 198:505-506.

Morris, B.F. 1971. Petroleum: Tar quantities floating in the northwestern Atlantic taken with a new quantitative neuston net. Science 173:430-432.

Morris, B.F. 1973. Two new species of Macrochiron Brady 1872 and a new record for Pseudomacrochiron Parvum (A. Scott, 1909) in the northwest Atlantic. Crustaceana 24(1):57-71.

Morris, B.F. and J.N. Butler. 1973. Petroleum residues in the Sargasso Sea and on Bermuda beaches. Proc. 1973 Conference on Prevention and Control of Oil Spills, American Petroleum Institute, Washington, DC. pp. 521-529.

Morris, B.F., J. Cadwallader, J. Gieselman and J.N. Butler. 1976. Transfer of petroleum and biogenic hydrocarbons in the Sargassum community. In Marine Pollutant Transfer, H.L. Windom and R.A. Duce, eds., D.C. Heath, Lexington, MA, pp. 235-259. Reproduced as Appendix VI of this book.

Morris, B.F. and D.D. Mogelberg, 1973. Identification manual to the pelagic Sargassum fauna. Bermuda Biological Station, Special Publication No. 11, 63 pp.

Morris, B.F. and E. Schroeder. 1973. Hydrographic observations in the Sargasso Sea off Bermuda. May 1967 - Feb. 1973. Bermuda Biological Station, Special Publication No. 12, 102 pp.

Mukai, H. 1969. Life histories of the shrimps in the Sargassum region. Hiroshimam-Daigaku Seibutsugakkaishi (Bull. Biol. Soc. Hiroshima Univ.) 35:7-13.

Mukai, H. 1971. The phytal animals on the thalli of Sargassum serratifolium in the Sargassum region, with reference to their seasonal fluctuations. Marine Biology 8(2):170-182.

Naylor, E. 1955. The diet and feeding mechanism of Idotea. J. Mar. Biol. Assoc. U.K. 34:347-355.

Norton, T.A. 1977a. The growth and development of Sargassum muticum (Yendo) Fensholt. J. Exp. Mar. Biol. and Ecol. 26(1):41-53.

Norton, T.A. 1977b. Ecological experiments with Sargassum muticum. J. Mar. Biol. Assoc. U.K. 57:33-43.

Nuzzi, R. 1973. Effects of water soluble extracts of oil on phytoplankton. Proc. Joint Conf. Prev. Control Oil Spills. American Petroleum Institute. pp. 809-813.

Olson, D.B. 1978. Gulf Stream ring coalescence with the Gulf Stream off Cape Hatteras. Science 202:971-972.

Ortner, P.B., E.M. Hurlburt and P.H. Wiebe. 1979. Phytohydrography, Gulf Stream rings, and herbivore habitat contrasts. J. Exp. Marine Biology and Ecology 39(2):101-124.

Parker, C.E. 1971. Gulf Stream rings in the Sargasso Sea. Deep-Sea Research 18:981-993.

Parr, A.E. 1939. Quantitative observations on the pelagic Sargassum vegatation of the western North Atlantic. Bulletin of the Bingham Oceanographic Collection, Peabody Museum of Natural History, Yale University, 6(7):1-94.

Pettibone, M.H. 1963. Marine polychaete worms of the New England region. Smithsonian Inst., Bulletin No. 227, 356 pp.

Prat, H. 1935. Remarques sur la faune et la flore associees aux sargasses flottantes. Le Naturaliste Canadien. 62:120-129.

Prat, H. 1940. Observations bionomiques sur les rivages Atlantiques de L'Amerique du nord et les Iles Voisines. Mem. Soc. Biogeogr. 7:253-277.

Pratt, R., H. Mautner, G.M. Gardner, Y.H. Sha and J. Dufrenoy. 1951. Report on antibiotic activity of seaweed extracts. J. Amer. Pharmaceutical Assoc. 40(1):575-579.

Prince, J.S. 1980. Ecology of Sargassum pteropleuron Grunow (Phaeophyceae, Fucales) in the waters off south Florida. Phycologia 19(3):190-193.

Pocklington, R. 1972. Secular changes in the ocean off Bermuda. J. Geophys. Res. 77(33):6604-6607.

Renaud-Mornant, J. 1971. Tardigrades Marins de Bermudes. Bull. du Museum National D'Histoire Naturelle 42(6):1268-1276.

Risebrough, R.W., B.W. deLappe and W. Walker II. 1976. Transfer of higher molecular weight chlorinated hydrocarbons to the marine environment. In Marine Pollutant Transfer, H.L. Windom and R.A. Duce, eds. D.C. Heath, Lexington, MA. pp. 261-321.

Robinson, A.R. 1982. Dynamics of ocean currents and circulation: Results of POLYMODE and related investigations. Topics in Ocean Physics 80:3-29.

Ryland, J.S. 1974. Observations on some epibionts of gulf-weed, Sargassum natans (L. Meyen). J. Exper. Mar. Biol. and Ecol. 14(1):17-25.

Ryther, J.H. 1956. The Sargasso Sea. Scientific American 194, No. 1:98-104.

Sanders, H.L. 1968. Marine Benthic Diversity: A comparative study. The American Naturalist 102(925):243-282.

Saunders, C.G. 1966. Dietary analysis of Caprellids. Crustaceana 10:314-316.

Schoener, A. and G.T. Rowe. 1970. Pelagic Sargassum and its presence among the deep-sea benthos. Deep-Sea Research 17:923-925.

Schroeder, E., H. Stommel, D.W. Menzel and W.H. Sutcliffe, Jr. 1959. Climatic stability of 18-degree water at Bermuda. J. Geophys. Res. 64:363-366.

Schroeder, E. and H. Stommel. 1969. How representative is the series of Panulirus stations of monthly mean conditions off Bermuda? Prog. Oceanog. 5:31-40.

Sieburth, J.M. and A. Jensen. 1969. Studies on algal substances in the sea. II. The formation of Gelbstoff (humic material) by exudates of Phaeophyta. J. Exp. Mar. Biol. Ecol. 3:275-289.

Sieburth, J.M. and J.T. Conover. 1965. Sargassum tannin, an antibotic which retards fouling. Nature (London) 208:52-53.

Sisson, R.F. 1976. Adrift in a raft of Sargassum. National Geographic (Feb., 1976) 149(2):188-199.

Skutch, A.F. 1926. On the habits and ecology of the tube-building amphipod Ampithoe rubricata montagu. Ecology 7:481-502.

Smith, K.L., K.A. Burns and E.J. Carpenter. 1973. Respiration of the pelagic Sargassum community. Deep-Sea Research 20:213-217.

Smith, K.L. 1973. Energy transformations by the Sargassum fish Histrio histrio (L.). J. Exp. Mar. Biol. and Ecol. 12:219-227.

Sokal, R.R. and F.J. Rohlf. 1973. Introduction to Biostatistics. W.H. Freeman and Co., San Francisco, 368 pp.

Sorensen, T. 1948. A method of establishing groups of equal amplitude in plant sociology based on similarity of species content and its application to analyses of the vegetation on Danish commons. Biol. Skr. (K. danske vidensk. Selsk. N.S.) 5:1-34.

Southwood, T.R.E. 1972. Ecological Methods. Chapman and Hall, London, pp. 326-353.

Spindler, M. 1980. The pelagic gulf weed Sargassum natans as a habitat for the benthic forminifera Planorbulina acervalis and Rosalina globularis. N.Jb. Geol. Palaontol. Mh. 9:569-580.

Stommel, H. 1949. Trajectories of small bodies sinking slowly through convection cells. J. Marine Research 8:24-29.

Stoner, A.W. 1983. Pelagic Sargassum: Evidence for a major decrease in biomass. Deep-Sea Research, in press.

Stoner, A.W. and H.S. Greening. 1983. Faunal studies of Sargassum. In course of publication.

Sutcliffe, W.H., E.R. Baylor and D.W. Menzel. 1963. Sea surface chemistry and Langmuir circulation. Deep-Sea Research 10:233-243.

Sutcliffe, W.H., R.W. Sheldon, A. Prakash and D.C. Gordon. 1971. Relations between wind speed, Langmuir circulation and particle concentration in the ocean. Deep-Sea Research 18:639-643.

Taniguchi, K. and H. Ohkubo. 1975. Sublittoral algal communities on the southeast coast of Sado Island Japan. Nihon Kaiku Suisan Kenkyujo Niigata 26:57-66.

Tararam, A.S. and Y. Wakabara. 1981. Mobile fauna - especially Gammaridae - of Sargassum cymosum. Marine Ecol. Prog. Ser. 5(2):157-163.

Taylor, W.R. 1975. A pelagic Sargassum from the Western Atlantic. Contr. Univ. Mich. Herbarium 11(2):73-75.

Teal, J. and M. Teal. 1975. The Sargasso Sea. Atlantic-Little Brown, 216 pp.

Thulin, G. 1942. Ein neuer mariner Tardigrad. Goteborgs K. Vetensk. -o. VitterhSamh. Handl. 2(5):1-10.

Timmermann, G. 1932. Biogeographische untersuchungen uber die lebensgemeinschaft des treibenden golfkrautes. Z. Morphol. Okol. 25:288-335.

Vastano, A.C. and D.E. Hagan. 1977. Observational evidence for transformation of tropospheric waters within cyclonic rings. J. Physical Oceanography 7(6):938-943.

Volostnykh, B.V. 1979. Forms of phosphorus in the surface microlayer of the western Sargasso Sea. Oceanology 19(1):44-46.

Voorhis, A.D. 1969. The horizontal extent and persistence of thermal fronts in the Sargasso Sea. Deep-Sea Research 16(suppl.):331-337.

Voorhis, A.D. and J.B. Hersey. 1964. Oceanic thermal fronts in the Sargasso Sea. J. Geophysical Research 69:3809-3814.

Watt, K.E.F., 1964. Comments on fluctuations of animal populations and measures of community stability. Canadian Entomologist 96:1434-42.

Watts, D.R. and D.B. Olson. 1978. Gulf Stream ring coalescence with the Gulf Stream off Cape Hatteras. Science 202:971-972.

Weis, J.S. 1968. Fauna associated with pelagic Sargassum in the Gulf Stream. American Midland Naturalist 80(2):554-558.

Winge, O. 1923. The Sargasso Sea, its boundaries and vegetation. Report on the Danish Oceanogrphical Expeditions 1908-1910, to the Mediterranean and adjacent seas 3(2):3-34.

Withers, R.G., W.F. Farnham, S. Leavey, N.A. Jephson, J.M. Haythorn and P.W.G. Gray. 1975. The epibionts of Sargassum muticum in British waters. Marine Biology 31:79-86.

Woodcock, A.H. 1944. A theory of surface water motion deduced from the wind-induced motion of Physalia. J. Mar. Research 5:196-205.

Woodcock, A.H. 1950. Subsurface pelagic Sargassum. J. Marine Research 9(2):77-92.

Wunsch, C. 1977. Determining the general circulation of the oceans: A preliminary discussion. Science 196:871-875.

Yeatman, H.C. 1962. The problem of dispersal of marine littoral copepods in the Atlantic Ocean, including some redescriptions of species. Crustaceana 4:253-272.

APPENDIX I

Data on abundance of *Sargassum*
by Parr (1939) and Stoner (1982)

LIST OF HAULS MADE WITH PELAGIC WEED COLLECTOR

The following list is a definitive account with regard to the amounts of the dominant forms of pelagic weeds obtained in the individual hauls. That is to say, it presents the final figures and notes concerning the presence and abundance, or the absence, of *S. natans* (*I*) and *S. fluitans* (*III*), or of *S. fluitans* as a whole (*III + X*) where no distinction has been made between the two forms in which this species may occur. The list is also final with regard to the abundance of any other type of *pelagic Sargassum*, wherever such other forms were obtained in sufficient quantity to be recorded by weight. In addition to these definitive *quantitative* figures one will also find entries of all notes so far made, whether on shipboard or during the subsequent examination of the preserved material, concerning the mere presence of the rarer forms in amounts too small to be weighed. These non-quantitative records of the mere presence of a form cannot, however, be guaranteed to be entirely complete, in view of the vast amounts of weeds which had to be sorted on board, and thereafter discarded, and of the likewise very large amount preserved. It is quite likely that the reexamination of the preserved material alone, planned in connection with the final systematic analysis after sufficient benthonic collections have also been obtained, will increase the number of records of rarer weeds present in individual hauls.

There is also considerable variation in the details given in the field-records of each haul. When made under instructions in the absence of the writer[1] the standard entries are simply the weights of "fine-leafed" (*I*) and of "coarse-leafed" (*III*) weeds, which, together with a total preservation of all unusual-looking weeds, and a random sample of the dominant ones, provides ample information for the purposes of this investigation.

When a small catch was preserved in its entirety, the weight of the separate forms was subsequently determined, after sorting in the laboratory, as the wet weight in formalin, which will not differ significantly from the wet weight of the fresh material. The weight of such small samples determined in the laboratory could naturally be recorded in more accurate figures than those used in the weighing of large samples on board (usually nearest ¼ pound). In other instances, when difficulties in sorting arose on board in the absence of the writer, the total weight of the descriptive category "coarse-leafed" or "large-leafed," was taken on board and entered in the field-record, with a specially large random sample preserved, from which the relative amounts of the forms combined in this category were subsequently determined in the laboratory. In the case of the hauls made in the Gulf of Mexico during 1935 (Nos. 172–188), *S. natans VIII* and *S. fluitans* were commonly combined in the field-records under the heading of "broad-leaved" weeds, and the weight recorded under this head-

[1] Hauls Nos. 1–113 and 141–172 were all made and sorted under the supervision of the writer with the assistance of the ship's company.

ing has only subsequently been divided according to the relative abundance of the two forms in the large random samples. In the Gulf of Mexico the true *S. fluitans* generally occurred only in such fragmentary (although not quantitatively insignificant) material that it was not found practical to attempt to distinguish between its two forms. The discovery of the actual identity between *S. fluitans III* and *X* (on 1934 winter-cruise material) also lessened the emphasis upon a separate sorting of these two forms. In the 1935 records *X* has therefore only been entered as a notation of its presence in the quantitative record of *S. fluitans* given under *III* (the always predominant form) when some well-developed plants of the *X* form were conspicuous in the catch.

The following example may serve as an interpretation of the entries used in the records of the individual hauls.

Example:

#51. 1/27. 23°23.5′ N. 69°57.5′ W. 19.0 Mi. 44.00 I. 5.00 III. II. V. X.

Read:

Haul No. 51. January 27. Median latitude 23°23.5′ N. Median longitude 69°57.5′ W. Length of haul through the water (log-distance) 19.0 nautical miles.[1] 44.00 pounds of *S. natans* (*I*) (wet weight). 5.00 pounds of *S. fluitans* (*III*). *Sargassum II*, *V*, and *X* present in small amounts.

1933

SOUTH OF BERMUDA.

Haul	Date	Latitude	Longitude	Distance	I	III		
#1.	2/13.	31°25.5′ N.	66°45′. W.	3.0 Mi.	.50 I.	2.00 III.		
#2.	2/13.	31°24′ N.	66°52.5′ W.	8.0 Mi.	2.75 I.	9.75 III.		
#3.	2/13.	31°22′. N.	67° 3′. W.	10.0 Mi.	3.75 I.	13.50 III.		

SARGASSO SEA. NASSAU TO ANEGADA PASSAGE.

Haul	Date	Latitude	Longitude	Distance	I	III		
#4.	2/26.	25°38′ N.	75°16′ W.	14.1 Mi.	4.00 I.	6.50 III.	II. X.	
#5.	2/26.	25°36′ N.	75° 0′ W.	15.9 Mi.	12.00 I.	16.50 III.	2.00 X.	II.
#6.	2/26.	25°34′ N.	74°41.5′ W.	16.5 Mi.	12.25 I.	12.75 III.	1.25 X.	
#7.	2/26.	25°36.5′ N.	74°23′ W.	14.3 Mi.	8.50 I.	15.00 III.		
#8.	2/27.	25°46′ N.	72°35′ W.	14.4 Mi.	5.25 I.	15.50 III.	1.25 X.	II.
#9.	2/27.	25°46′ N.	72°23′ W.	8.0 Mi.	19.25 I.	56.00 III.	2.00 X.	
#10.	2/27.	25°45.5′ N.	72°15′ W.	6.7 Mi.	7.00 I.	4.00 III.		
#11.	2/27.	25°44.5′ N.	71°57′ W.	13.0 Mi.	.25 I.	8.00 III.		
#12.	2/28.	25°31.5′ N.	69°23′ W.	15.2 Mi.	21.00 I.	17.50 III.	2.50 X.	
#13.	2/28.	25°30′ N.	69° 5′ W.	17.0 Mi.	28.50 I.	22.00 III.	1.75 X.	.25 II
#14.	2/28.	25°29′ N.	68°49′ W.	13.0 Mi.	36.00 I.	28.50 III.	3.50 X.	
#15.	3/1	25°17.5′ N.	66°41.5′ W.	14.1 Mi.	19.75 I.	30.25 III.	.50 X.	
#16.	3/1.	25°17′ N.	66°23′ W.	17.2 Mi.	9.50 I.	10.50 III.	.50 X.	
#17.	3/1.	25°16′ N.	66° 7.5′ W.	7.1 Mi.	17.75 I.	35.00 III.	1.00 X.	
#18.	3/3.	23°19′ N.	62°30′ W.	16.7 Mi.	24.50 I.	20.25 III.	.25 X.	
#19.	3/3.	23° 5′ N.	62°40′ W.	9.4 Mi.	17.00 I.	20.50 III.	VIII.	
#20.	3/4.	22° 2.5′ N.	62°43′ W.	12.7 Mi.	4.50 I.	10.50 III.		
#21.	3/4.	21°51′ N.	62°46′ W.	9.2 Mi.	Less than 1 lb. total haul			
#22.	3/4.	21°42.5′ N.	62°44′ W.	6.7 Mi.	3.50 I.	30.00 III.		
#23.	3/5.	21° 8′ N.	62°13′ W.	18.0 Mi.	16.50 I.	16.25 III.		
#24.	3/5.	20°56′ N.	62° 8′ W.	10.7 Mi.	13.75 I.	16.75 III.		

[1] Not necessarily identical with the geographical length of the haul in the presence of significant currents.

Station	Date	Lat.		Long.		Distance	I.	III.		
1933										
CARIBBEAN.										
#25.	3/13.	16° 1′	N.	64°28.5′	W.	14.0 Mi.	No weed			
#26.	3/27.	17°25′	N.	71°40′	W.	21.4 Mi.	1.00 I.	1.00 III.		
#27.	3/27.	17°30′	N.	72° 6′	W.	31.5 Mi.	1.00 I.	1.00 III.		
CAYMAN SEA NEAR WINDWARD PASSAGE.										
#28.	4/18.	20° 0′	N.	74°19′	W.	11.0 Mi.	.50 I.	.50 III.		
#29.	4/18.	19°51′	N.	74°36′	W.	27.8 Mi.	2.00 I.	.50 III.		
1934										
CAPE COD TO BERMUDA.										
#30.	1/22.	36°21′	N.	66°39′	W.	5.9 Mi.	.70 I.	2.40 III.	V.	
#31.	1/22.	36°18′	N.	66°21′	W.	36.8 Mi.	.50 I.	11.00 III.		
SARGASSO SEA. BERMUDA TO TURKS ISLAND.										
#32.	1/23.	31°57′	N.	65°52′	W.	14.5 Mi.	21.50 I.	12.50 III.	X.	
#33.	1/23.	31°44′	N.	66° 2′	W.	14.0 Mi.	6.50 I.	5.25 III.		
#34.	1/24.	30° 9′	N.	66°51′	W.	8.0 Mi.	18.50 I.	1.00 III.		
#35.	1/24.	29°96′	N.	66°53′	W.	19.7 Mi.	30.00 I.	5.00 III.	II.	
#36.	1/24.	29°38′	N.	67° 0′	W.	14.8 Mi.	46.50 I.	10.00 III.	II.	
#37.	1/24.	29°26′	N.	67° 9′	W.	16.8 Mi.	33.25 I.	3.00 III.	1.00 II.	
#38.	1/24.	29°11′	N.	67°17′	W.	15.2 Mi.	51.00 I.	5.00 III.	II.	X.
#39.	1/25.	27°46′	N.	68° 6′	W.	19.4 Mi.	49.00 I.	4.25 III.	.50 II.	X.
#40.	1/25.	27°38′	N.	68°10′	W.	5.8 Mi.	108.00 I.	8.00 III.	2.50 II.	X.
#41.	1/25.	27°28.5′	N.	68°13′	W.	5.7 Mi.	71.00 I.	6.25 III.	1.00 II.	X.
#42.	1/25.	27°21.5′	N.	68°16′	W.	7.5 Mi.	123.00 I.	6.75 III.	2.25 II.	VIII.
#43.	1/25.	27°18′	N.	68°18′	W.	2.6 Mi.	118.00 I.	6.25 III.	2.50 II.	V.
#44.	1/25.	27°13′	N.	68°20′	W.	7.7 Mi.	16.00 I.	1.75 III.	II. V.	X.
#45.	1/25.	27° 5.5′	N.	68°23′	W.	5.8 Mi.	1.25 I.	.75 III.		
#46.	1/26.	25°50.5′	N.	68°56′	W.	18.0 Mi.	127.00 I.	9.50 III.	.60 II. V.	.80 X. VIII.
#47.	1/26.	25°33′	N.	68° 2′	W.	19.7 Mi.	100.00 I.	4.25 III.	1.50 II.	.90 X. VIII.
#48.	1/26.	25°12.5′	N.	69°12′	W.	20.0 Mi.	46.00 I.	5.00 III.	1.10 II.	V. X
#49.	1/26.	24°56.5′	N.	69°19′	W.	13.0 Mi.	79.00 I.	6.00 III.	.90 II.	.60 X.
#50.	1/27.	23°48′	N.	69°46′	W.	30.5 Mi.	33.00 I.	8.75 III.	II. V.	X.
#51.	1/27.	23°23.5′	N.	69°57.5′	W.	19.0 Mi.	44.00 I.	5.00 III.	II. V.	X.
#52.	1/27.	23° 5′	N.	70° 8′	W.	24.3 Mi.	89.00 I.	10.00 III.	II. V.	X.
#53.	1/27.	22°46′	N.	70°20′	W.	18.2 Mi.	128.00 I.	14.00 III.	II. V.	X.
#54.	1/27.	22°28′	N.	70°32′	W.	23.5 Mi.	50.00 I.	3.00 III.	II. V.	X.
#55.	1/27–28.	21°58′	N.	70°52′	W.	46.0 Mi.	94.00 I.	4.00 III.	II. V.	X.
TURKS ISLAND TO WINDWARD PASSAGE.										
#56.	1/28.	21°37′	N.	71° 9′	W.	7.0 Mi.	3.00 I.	.80 III.		
#57.	1/28.	21°27′	N.	71°15′	W.	14.6 Mi.	16.00 I.	1.10 III.		
#58.	1/28.	21°13′	N.	71°21′	W.	18.4 Mi.	72.00 I.	2.25 III.	II-VIII.	X.
#59.	1/28.	20°56′	N.	71°38′	W.	32.4 Mi.	120.00 I.	7.00 III.	II.	
#60.	1/28.	20°42′	N.	72° 4′	W.	19.1 Mi.	1.00 I.	1.50 III.	VIII.	
CARIBBEAN I. WINDWARD PASSAGE TO PANAMA.										
#61.	1/29.	19°53′	N.	73°27′	W.	43.3 Mi.	12.00 I.	4.00 III.	II. V.	
#62.	1/29.	19°24′	N.	74°10′	W.	69.2 Mi.	3.00 I.	.80 III.	VIII.	
#63.	1/29.	18°57.5′	N.	74°44′	W.	17.6 Mi.	1.00 I.	.20 III.	II.	
#64.	1/29–30.	18°26.5′	N.	75°23′	W.	76.0 Mi.	10.00 I.	2.00 III.		
#65.	1/30.	17°52′	N.	76°16′	W.	46.3 Mi.	3.50 I.	Few sprigs III.		
#66.	2/2.	17°35′	N.	76°39′	W.	17.4 Mi.	2.40 I.	.30 III.	.20 II.	X.
#67.	2/2(–3).	17°20′	N.	76°36′	W.	15.0 Mi.	2.50 I.	.40 III.	II. VIII.	X.

1934

CARIBBEAN I. WINDWARD PASSAGE TO PANAMA.—*Cont.*

#68.	2/3.	17° 3′	N.	76°30′	W.	16.7 Mi.	3.00 I. 2.00 III.
#69.	2/3.	16°33′	N.	76°27′	W.	43.6 Mi.	.24 I. .16 III. IX?
#70.	2/3(–4).	15°53′	N.	76°30′	W.	40.3 Mi.	1.20 I. 1.10 III.
#71.	2/4.	15° 2′	N.	76°39′	W.	35.7 Mi.	.10 I.
#72.	2/4–5.	14°12′	N.	76°50′	W.	67.1 Mi.	One sprig I.
#73.	2/5.	13°14′	N.	77°50′	W.	40.3 Mi.	One sprig I.
#74.	2/5.	12°50′	N.	77° 3′	W.	16.8 Mi.	.30 I. II.
#75.	2/5–6.	12° 2.5	N.	78°10′	W.	66.0 Mi.	I.
#76.	2/6.	11°17′	N.	78°45′	W.	41.5 Mi.	VIII.
#77.	2/6.	10°43′	N.	79° 6′	W.	37.1 Mi.	Fragment I.
#78.	2/7.	10°13′	N.	79°21′	W.	27.8 Mi.	No weed.

CARIBBEAN II. PANAMA TO EAST POINT OF JAMAICA.

#79.	2/14.	9°35′	N.	79°57′	W.	9.6 Mi.	One sprig III.
#80.	2/14.	9°44′	N.	80° 1′	W.	12.5 Mi.	No weed.
#81.	2/14.	10° 1′	N.	80°16′	W.	34.0 Mi.	VIII.
#82.	2/15.	10°34′	N.	80°48′	W.	56.1 Mi.	4.00 VIII.
#83.	2/15.	10°55′	N.	81°34′	W.	45.1 Mi.	.70 VIII.
#84.	2/16.	11°55′	N.	82°35′	W.	18.2 Mi.	Empty
#85.	2/16.	12° 9′	N.	82°50′	W.	11.0 Mi.	Empty
#86.	2/17.	13°23′	N.	82° 9′	W.	7.0 Mi.	Empty
#87.	2/17.	13°26′	N.	82° 0′	W.	16.4 Mi.	Empty
#88.	2/17.	13°21′	N.	81°38′	W.	26.2 Mi.	No *Sargassum*.
#89.	2/18.	13°22′	N.	81° 5′	W.	24.5 Mi.	No *Sargassum*.
#90.	2/18.	13°36′	N.	80° 8′	W.	31.0 Mi.	I. III.
#91.	2/18–19.	13°59′	N.	79°47′	W.	33.9 Mi.	Empty
#92.	2/19.	14°27′	N.	79°28′	W.	37.8 Mi.	3.75 I. .25 III. II. VIII.
#93.	2/19.	14°55′	N.	79°16′	W.	25.0 Mi.	2.10 I. .80 III. II.
#94.	2/20.	15°40′	N.	78°50′	W.	36.5 Mi.	5.50 I. .25 III. II. VIII.
#95.	2/20.	16° 8′	N.	78°39′	W.	17.8 Mi.	6.00 I. .50 III. VIII.
#96.	2/20.	16°24′	N.	78°27′	W.	20.0 Mi.	I. II.
#97.	2/21.	17° 1′	N.	78° 0′	W.	35.8 Mi.	1.50 I. 1.30 III. II. VIII.
#98.	2/21–22.	17°53′	N.	78° 0′	W.	60.5 Mi.	9.00 I. 1.60 III. II. V. VIII. IX.

CAYMAN SEA. EAST POINT OF JAMAICA TO YUCATAN CHANNEL.

#99.	2/22.	18°28′	N.	78°40′	W.	38.8 Mi.	2.00 I. 1.70 III. II. V. VIII.
#100.	2/22–23.	19°19′	N.	79°21′	W.	74.8 Mi.	2.70 I. 1.20 III. II. V. VIII.
#101.	2/23.	19°55′	N.	79°41′	W.	16.5 Mi.	1.10 I. .40 III. II. V. VIII.
#102.	2/23.	20°11′	N.	79°31′	W.	19.5 Mi.	.40 I. .40 III. II. VIII.
#103.	2/23.	20°28′	N.	79°23′	W.	19.4 Mi.	1.30 I. 2.20 III.
#104.	2/24–25.	21°15′	N.	80° 9′	W.	111.8 Mi.	I. II. III.
#105.	2/26.	19°22′	N.	81°31′	W.	19.0 Mi.	.80 I. II. III. VIII.
#106.	2/26–27.	19°11′	N.	82° 0′	W.	39.0 Mi.	7.50 I. 2.50 III. II. V. VIII.
#107.	2/27.	19° 5′	N.	82°42′	W.	28.9 Mi.	2.20 I. II. III. VIII.
#108.	2/27.	19° 2′	N.	83°20′	W.	40.6 Mi.	2.50 I. 1.50 III.
#109.	2/27–28.	19°20′	N.	83°52′	W.	29.5 Mi.	1.50 I. III.
#110.	2/28.	20° 2′	N.	83°56′	W.	39.5 Mi.	.50 I. II. III. VIII.
#111.	2/28.	21°21′	N.	84°18′	W.	42.2 Mi.	2.40 I. 1.40 III.

GULF STREAM NORTH OF MIAMI, FLORIDA.

#112.	3/8–9.	27°40′	N.	79°41′	W.	174.0*Mi.	60.00 I. 13.00 III. II. V. VIII. X.
#113.	3/9.	30° 5′	N.	79°12.5′	W.	52.4 Mi.	64.00 I. 8.20 III. 1.90 X.

* Geograph. length 240

1934

BETWEEN CAPE COD AND BERMUDA.

#114. 7/18–19. 39°47′ N. 68° 7′ W. 48.2 Mi. 1.20 I. .55 III.

#115. 7/19. 38°59′ N. 67°47′ W. 30.5 Mi. Only 2 small plants altogether.

#116. 7/19–20. 37°59′ N. 67°24′ W. 75.2 Mi. } 46.50 I. 3.00 III.

#117. 7/20. 37°33′ N. 67° 9′ W. 18.7 Mi. }

#118. 7/20. 37°19′ N. 67° 2′ W. 8.0 Mi. 4.25 I.

#119. 7/20. 36°58′ N. 66°42′ W. 38.8 Mi. } 23.75 I. III.

#120. 7/20–21. 36°21′ N. 65°58′ W. 62.1 Mi. }

#121. 7/22–23. 36° 3′ N. 65°52′ W. 53.5 Mi. 25.00 I.

#122. 7/23–24. 37° 0′ N. 67°30′ W. 102.9 Mi. 15.75 I. 4.00 III.

#123. 8/18. 40°21′ N. 69°47′ W. 59.4 Mi. 5.50 I. 2.25 III.

#124. 8/18. 39°38′ N. 69°20′ W. 43.8 Mi. 28.75 I. 6.25 III.

#125. 8/21. 38°29′ N. 68°54′ W. 71.6 Mi. 4.00 I. 1.50 III. VIII.

#126. 8/21–22. 37°33′ N. 67°56′ W. 44.7 Mi. 105.00 I. 20.25 III.

#127. 8/22. 37°17′ N. 67°31′ W. 7.2 Mi. 1.00 I. .25 III.

#128. 8/22. 37° 9′ N. 67°37′ W. 9.0 Mi. 43.00 I. 3.75 III.

#129. 8/22. 36°34′ N. 67°29′ W. 61.3 Mi. 24.50 I. 2.25 III.

#130. 8/22–23. 35°56′ N. 67°21′ W. 15.5 Mi. 5.00 I. .25 III.

#131. 8/26–27. 36° 4′ N. 68°12′ W. 92.3 Mi. 13.25 I. .75 III.

#132. 8/27. 36°53′ N. 68°58′ W. 29.1 Mi. 21.50 I. 4.25 III.

#133. 8/27–28. 37°50′ N. 69°25′ W. 94.5 Mi. 14.50 I. 4.50 III.

#134. 8/28. 38°46′ N. 69°49′ W. 27.6 Mi. 5.25 I. 2.25 III.

#135. 8/28. 39°18′ N. 70° 4′ W. 41.9 Mi. 12.50 I. 9.00 III.

1935

CAPE COD TO BERMUDA.

#136. 1/21. 38° 2.5′ N. 68°10′ W. 24.8 Mi. 1.75 I. 1.25 III. X.

#137. 1/21–22. 37°33′ N. 68° 0′ W. 35.9 Mi. .75 I. 1.25 III.

#138. 1/22. 35°53′ N. 67°17′ W. 28.8 Mi. 1.75 III.

#139. 1/23. 33°42′ N. 65°31′ W. 30.9 Mi. 1.75 I. 6.50 III.

#140. 1/23. 33° 7.5′ N. 65° 2.5′ W. 49.9 Mi. 4.75 I. 14.50 III. II.

SARGASSO SEA. BERMUDA TO TURKS ISLAND.

#141. 2/6. 29°56′ N. 65°55′ W. 17.65 Mi. 13.75 I. 7.25 III. 1.00 II. VIII. X.

#142. 2/6. 29°38′ N. 66° 4′ W. 21.5 Mi. 22.00 I. 11.00 III. 4.00 II-VIII. V. 1.00 X.

#143. 2/6. 29° 6′ N. 66°23′ W. 44.8 Mi. 12.00 I. 11.00 III. 1.50 II. 2.00 X. VIII.

#144. 2/7. 28°30′ N. 66°41′ W. 25.0 Mi. 16.50 I. 87.20 III. II. 1.50 X. VIII.

#145. 2/7. 28° 3′ N. 66°56′ W. 28.6 Mi. 18.00 I. 10.00 III. 1.50 X. 1.50II. VIII. V.

#146. 2/7. 27°35′ N. 67°17′ W. 33.1 Mi. 16.00 I. 12.00 III. 3.50 II. 5.50 X. VIII.

#147. 2/7. 26°57′ N. 67°44′ W. 55.5 Mi. 47.00 I. 21.00 III. 5.00 II. 6.50 X.

#148. 2/8. 26°11′ N. 68°10′ W. 46.0 Mi. 29.00 I. 12.00 III. 3.50 II. 3.50 X.

#149. 2/8. 25°37′ N. 68°25′ W. 29.4 Mi. 19.00 I. 2.50 X. II.

#150. 2/8. 24°55′ N. 68°42′ W. 17.3 Mi. 54.00 I. 7.5 III. 2.25 II. 6.00 X. IX.

#151.* 2/8–9. 24°35′ N. 68°54′ W. 24.3 Mi. 37.50 I. 9.00 III. 1.50 II. 4.00 X.

#152. 2/9. 23°58.5′ N. 69°27′ W. 39.5 Mi. 88.00 I. 35.00 III. 3.00 II. 17.50 X. VIII. IX?

#153. 2/9. 23°31.5′ N. 69°45′ W. 23.3 Mi. 23.00 I. 7.00 III. 2.00 II. 1.50 X. VIII.

#154. 2/9. 22°56′ N. 70°13′ W. 58.4 Mi. 17.00 I. 2.75 III. 1.25 II. 1.50 X.

#155. 2/10. 22° 4′ N. 70°40′ W. 57.9 Mi. 10.50 I. 15.00 III. 2.00 X. II. VIII.

TURKS ISLAND TO WINDWARD PASSAGE.

#156. 2/10. 21°17.5′ N. 71°15′ W. 49.4 Mi. 8.25 I. 2.50 III. 1.25 X.

#157. 2/10–11. 20°30′ N. 72°46.5′ W. 92.3 Mi. 14.00 I. 4.00 III. II. VIII. X.

* For account of subsurface nets used at this station see page 10.

1935

CAYMAN SEA. WINDWARD PASSAGE TO YUCATAN CHANNEL.

#158. 2/11–12. 19°57′ N. 74°14′ W. 69.3 Mi. 4.25 I. III. II. VIII. IX. X.
#159. 2/12–13. 19°24′ N. 76°30′ W. 172.6 Mi. 1.72 I. 4.20 III + X. .28 VIII. II.
#160. 2/13. 19°20′ N. 78°37′ W. 47.3 Mi. 1.10 I. 3.60 III + X. .62 VIII. .15 II. .30 IX.
#161. 2/13–14. 19°34′ N. 80°11′ W. 116.1 Mi. 1.70 I. 4.00 III + X. .40 VIII. II.
#162. 2/14. 19°53′ N. 81°54′ W. 67.9 Mi. .80 I. 2.20 III + X. 2.20 VIII. II. 1.32 IX.
#163. 2/14–15. 20°27′ N. 83° 2′ W. 78.5 Mi. 2.00 lb. Total haul. Details not available.
#164. 2/15. 21°12.5′ N. 84°20′ W. 89.0 Mi. .38 I. .81 III. 1.10 VIII. IX.

GULF OF MEXICO I. SOUTHEASTERN REGION BETWEEN YUCATAN BANK AND STRAITS OF FLORIDA.

#165. 2/16–17. 22°15′ N. 86°52′ W. 73.3 Mi. No pelagic weeds.
#166. 2/18. 24°15′ N. 85°51′ W. 24.4 Mi. .20 I. .75 III. .10 VIII. II.
#167. 2/18–19. 23°49′ N. 84°16′ W. 143.5 Mi. .40 I. 1.30 III. .20 VIII.

GULF OF MEXICO II. NORTHERN GULF.

#168. 3/5–6. 27°32′ N. 88°21′ W. 50.4 Mi. 4.50 I + III + VIII.
#169. 3/6. 26°55′ N. 88°21′ W. 40.9 Mi. 4.00 I + III + VIII.
#170. 3/6. 26°23′ N. 88°12′ W. 48.7 Mi. 33.50 I. Fragments III. VIII.
#171. 3/6. 25°54.5′ N. 88°18′ W. 1.4 Mi. I. VIII.
#172. 3/7. 25°43′ N. 88°23′ W. 10.9 Mi. 17.25 I. III. .50 VIII.
#173. 3/25. 28° 8′ N. 89°24′ W. 68.9 Mi. 45.25 I. 1.85 III. 1.90 VIII.
#174. 3/25–26. 27°30′ N. 89°47′ W. 35.5 Mi. 26.50 I. .95 III. 2.80 VIII.
#175. 3/26. 27°33′ N. 89°53′ W. 18.4 Mi. 17.00 I. 1.40 III. .35 VIII.
#176. 3/26. 27°57.5′ N. 89°53′ W. 21.9 Mi. 14.50 I. .50 III. .50 VIII.
#177. 3/26. 28°22.5′ N. 89°50′ W. 21.3 Mi. 12.25 I. .25 III. .50 VIII.
#178. 3/27. c.28°45′ N. c.89°50′ W. 42.3 Mi. 60.75 I. 4.50 III. 4.50 VIII.
#179. 3/27. 28°21′ N. 90°18′ W. 41.9 Mi. 33.75 I. 3.00 III. 3.00 VIII.
#180. 3/27–28. 27°40′ N. 91°24′ W. 92.3 Mi. 64.00 I. 11.75 III. + VIII.
#181. 3/28. 27°25′ N. 92°12′ W. 30.5 Mi. 11.00 I. .75 III. 1.00 VIII.
#182. 3/29. c.29° 7′ N. 92°41′ W. 27.6 Mi. 30.50 I. 2.75 III. 3.25 VIII.
#183. 3/29. 29° 7′ N. 93°19′ W. 40.2 Mi. 18.75 I. 4.90 III. 1.60 VIII. X.
#184. 3/31–4/1. 27°41′ N. 96°26′ W. 54.2 Mi. 27.50 I. 4.50 III. 3.00 VIII. X.
#185. 4/1. 27°15′ N. 96°30′ W. 14.4 Mi. 6.75 I. 2.00 III. .25 VIII.
#186. 4/2–3. 25°57′ N. 94°44′ W. 62.4 Mi. 13.00 I. 1.20 III. 4.80 VIII.
#187. 4/3–4. c.25°12′ N. c.94°6′ W. 64.1 Mi. 13.50 I. .80 III. 1.20 VIII.

GULF OF MEXICO III. YUCATAN BANK.

#188. 4/8–9. 24° 5′ N. 87°57′ W. 58.8 Mi. I. VIII.

GULF OF MEXICO IV. FLORIDA BANK.

#189. 4/11–12. 25°48′ N. 84° 5′ W. 86.4 Mi. 5.25 I. .60 III. .40 VIII. II.
#190. 4/12. 24°36′ N. 83° 2′ W. 46.0 Mi. 4.25 I. 1.30 III. .45 VIII.

STRAITS OF FLORIDA.

#191. 4/13. 23°41′ N. 81°52′ W. 77.7 Mi. 1.25 I. 1.75 III. II. VIII.
#192. 4/13–14. 24°42′ N. 80°33′ W. 59.4 Mi. 1.00 I. III. II. VIII.

GULF STREAM NORTH OF MIAMI, FLORIDA.

#193. 4/19–20. 28°10′ N. 79°42.5′ W. 135.3 Mi. 7.50 I. 6.50 III. + VIII.
#194. 4/20. 30°24′ N. 79°25′ W. 51.5 Mi. 3.75 I. 3.25 III. 1.00 VIII. X.

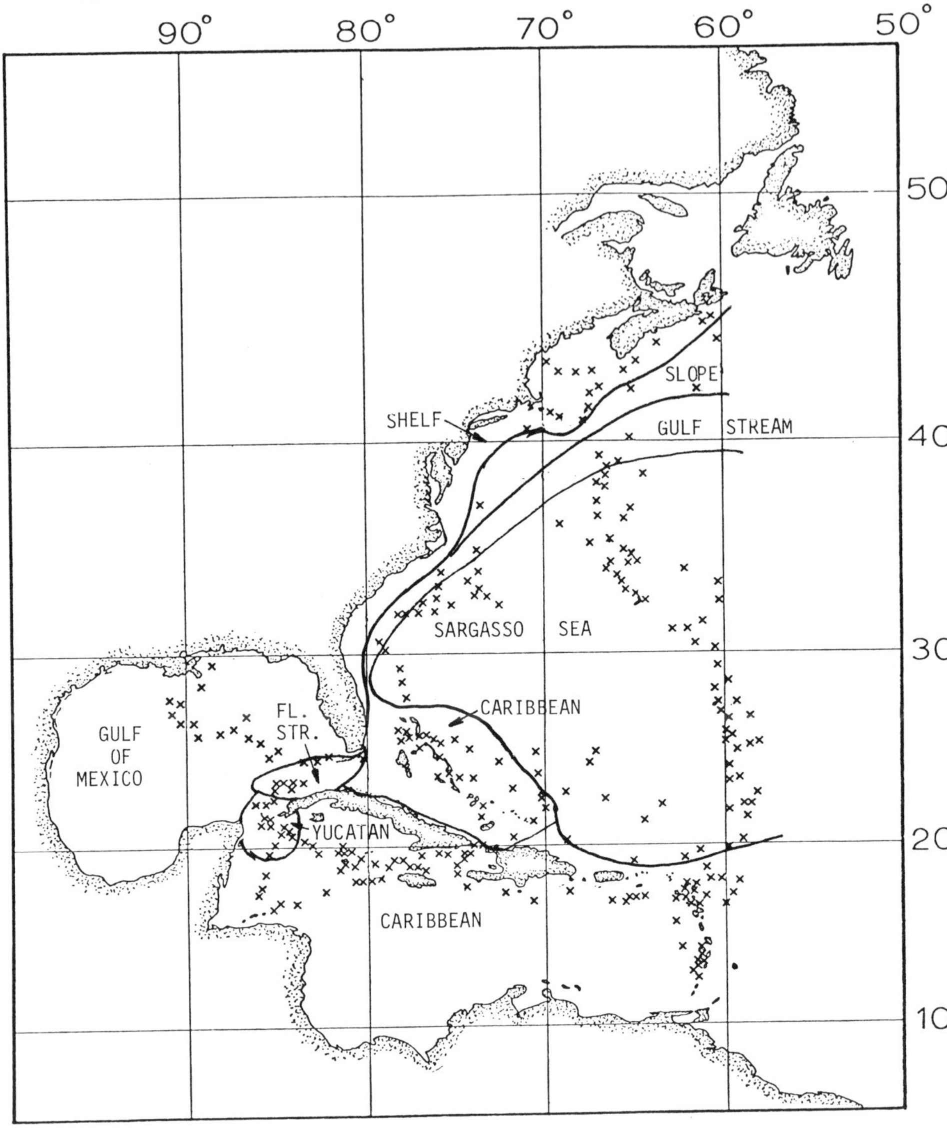

Station locations of the Westward surveys (Stoner, 1983) detailed in the following tables. (See also Fig. 1.8 and Table 1.4.) The boundaries are those used for the summary of Table 1.1.

Sargassum biomass values from R/V Westward cruises W-33 to W-61 (April, 1977 to January, 1982)

Station No.	Date	Latitude (N)	Longitude (W)	Sargassum (mg/m^2)
		Cruise W-33		
W33-N1	4/17/77	25°52'	77°40'	450
W33-N2	4/20/77	25°56'	76°59'	66
W33-N3	4/21/77	27°53'	77°49'	28
W33-N4	4/22/77	28°35'	77°58'	93
W33-N5	4/22/77	29°22'	78°15'	364
W33-N9	4/29/77	32°03'	77°44'	0
W33-N10	4/30/77	32°02'	77°01'	31
W33-N11	4/30/77	32°01'	76°02'	226
W33-N12	5/01/77	32°43'	75°48'	36
W33-N13	5/01/77	32°19'	75°16'	0
W33-N14	5/02/77	33°33'	74°19'	1120
W33-N15	5/02/77	33°57'	73°37'	90
W33-N16	5/03/77	33°02'	73°43'	60
W33-N17	5/04/77	32°50'	73°11'	145
W33-N18	5/04/77	32°27'	72°30'	11
W33-N19	5/05/77	32°51'	73°52'	217
W33-N20	5/06/77	35°04'	73°51'	245
		Cruise W-36		
W36-N1	10/19/77	34°01'	62°03'	28
W36-N2	10/20/77	33°21'	60°14'	220
W36-N3	10/21/77	32°29'	60°14'	21
W36-N4	10/22/77	30°07'	60°25'	12
W36-N5	10/23/77	28°12'	60°33'	190
W36-N6	10/23/77	27°30'	60°18'	31
W36-N7	10/24/77	27°07'	60°07'	27
W36-N8	10/25/77	25°37'	59°22'	0
W36-N9	10/25/77	25°00'	59°10'	180
W36-N10	10/26/77	23°32'	59°02'	29
W36-N11	10/27/77	22°09'	58°39'	0.3
W36-N12	10/29/77	18°41'	60°29'	0
		Cruise W-37		
W37-N1	11/30/77 to 1/11/78	22°31'	66°30'	0.6
W37-N2a		25°02'	67°03'	0.6
W37-N2b		24°33'	67°30'	0.1
W37-N3		20°23'	68°49'	0
W37-N4		18°53'	77°41'	0
W37-N5		18°54'	78°48'	51.7
W37-N6		18°51'	80°48'	0.1
W37-N7		19°05'	81°13'	0
W37-N8		19°04'	81°12'	0
W37-N9		20°07'	83°05'	23.4
W37-N10		20°16'	83°26'	4.0
W37-N11		21°14'	85°19'	39.2
W37-N12		24°20'	82°05'	6018.0

Station No.	Date	Latitude (N)	Longitude (W)	Sargassum (mg/m^2)
		Cruise W-42		
W42-N1	10/13/78	40°33'	70°48'	0
W42-N2	10/22/78	31°06'	62°31'	100
W42-N3	10/22/78	31°09'	61°52'	264
W42-N4	10/22/78	30°25'	61°28'	17.6
W42-N5	10/23/78	29°20'	60°27'	113
W42-N6	10/23/78	28°34'	59°51'	20.3
W42-N7	10/24/78	27°23'	59°09'	19.8
W42-N8	10/24/78	26°48'	58°30'	174
W42-N9	10/25/78	25°19'	58°30'	87.6
W42-N10	10/25/78	25°24'	58°12'	2.8
W42-N11	10/26/78	22°43'	58°18'	8.5
W42-N12	10/26/78	22°08'	58°25'	0
W42-N13	10/27/78	21°21'	58°41'	0
W42-N14	10/28/78	20°06'	59°00'	0
W42-N15	10/29/78	18°00'	59°10'	0
W42-N16	10/29/78	17°18'	59°30'	0
W42-N17	11/02/78	17°00'	62°06'	17.6
W42-N18	11/03/78	17°04'	62°05'	1.5
W42-N19	11/03/78	17°23'	62°02'	58.2
W42-N20	11/04/78	17°08'	62°02'	16.9
W42-N21	11/05/78	17°23'	61°54'	32.4
W42-N22	11/05/78	17°28'	61°58'	77.3
W42-N23	11/06/78	16°52'	61°45'	0
W42-N24	11/07/78	13°49'	61°12'	0
W42-N25	11/13/78	13°25'	61°25'	94.6
		Cruise W-44		
W44-N1	2/10/79	25°30'	75°12'	15.0
W44-N2	2/10/79	25°09'	74°10'	7.2
W44-N3	2/11/79	24°13'	72°28'	6.3
W44-N4	2/17/79	21°50'	70°00'	47.0
W44-N5	2/17/79	22°30'	70°06'	31.9
W44-N6	2/18/79	23°56'	70°20'	0
W44-N7	2/19/79	24°47'	70°28'	121.5
W44-N8	2/21/79	21°19'	70°35'	6.5
W44-N9	3/05/79	20°15'	71°44'	277.0
W44-N10	3/06/79	19°55'	73°23'	8.9
W44-N11	3/06/79	20°00'	70°00'	195.1
W44-N12	3/07/79	19°38'	74°28'	12.1
W44-N13	3/07/79	19°30'	75°30'	6.5
W44-N14	3/08/79	19°30'	75°57'	0
W44-N15	3/08/79	19°26'	77°04'	58.1
W44-N16	3/09/79	19°17'	78°33'	-
W44-N17	3/09/79	19°20'	80°17'	18.4
W44-N18	3/09/79	19°24'	81°22'	0
W44-N19	3/09/79	19°20'	81°24'	0
W44-N20	3/10/79	19°21'	81°25'	0
W44-N21	3/10/79	19°19'	81°27'	0
W44-N22	3/13/79	19°46'	82°89'	4.0
W44-N23	3/14/79	20°49'	84°21'	0
W44-N24	3/14/79	21°26'	85°23'	8.5
W44-N25	3/15/79	22°10'	86°00'	0

Station No.	Date	Latitude (N)	Longitude (W)	Sargassum (mg/m^2)
		Cruise W-45		
W45-N1	5/01/79	25°15'	77°10'	6.9
W45-N2	5/02/79	25°58'	77°49'	82.9
W45-N3	5/02/79	26°06'	78°08'	-
W45-N4	5/03/79	30°08'	79°03'	723
W45-N5	5/04/79	30°25'	79°20'	4520
W45-N6	5/05/79	31°58'	77°44'	217
W45-N7	5/05/79	32°27'	76°51'	7.8
W45-N8	5/06/79	33°16'	75°53'	-
W45-N9	5/06/79	33°56'	75°41'	1840
W45-N10	5/19/79	37°00'	73°30'	0
		Cruise W-48		
W48-N1	10/16/79	36°24'	65°17'	26.9
W48-N2	10/17/79	34°44'	65°04'	17.6
W48-N3	10/18/79	31°31'	61°04'	160.3
W48-N4	10/22/79	26°31'	59°58'	261.7
W48-N5	10/22/79	25°58'	59°46'	59.4
W48-N6	10/23/79	25°41'	58°58'	50.7
W48-N7	10/24/79	24°10'	59°24'	31.3
W48-N8	10/24/79	23°24'	59°25'	212.1
W48-N9	10/25/79	21°50'	59°30'	6.5
W48-N10	10/26/79	19°58'	59°34'	93.6
W48-N11	10/27/79	18°05'	60°20'	1.1
W48-N12	10/28/79	17°31'	61°42'	0
W48-N13	11/01/79	16°40'	61°45'	0
W48-N14	11/06/79	13°34'	61°13'	0
W48-N15	11/07/79	13°52'	61°07'	0
W48-N16	11/10/79	13°22'	61°15'	0
W48-N17	11/15/79	12°41'	61°24'	0.4
W48-N18	11/15/79	13°11'	61°31'	0
W48-N19	11/16/79	14°22'	62°17'	0
W48-N20	11/20/79	15°43'	62°51'	0.2
		Cruise W-50		
W50-N1	2/11/80	25°32'	76°18'	3670
W50-N2	2/14/80	23°35'	73°52'	183.5
W50-N3	2/15/80	23°05'	71°45'	18.3
W50-N4	2/17/80	22°53'	68°48'	29.7
W50-N5	3/01/80	18°12'	79°22'	47.5
W50-N6	3/01/80	18°08'	79°58'	2.7
W50-N7	3/04/80	16°55'	84°56'	57.2
W50-N8	3/09/80	17°40'	85°42'	34.0
W50-N9	3/09/80	18°22'	85°30'	91.7
W50-N10	3/10/80	19°45'	85°18'	63.1
W50-N11	3/10/80	20°08'	85°05'	190.0
W50-N12	3/11/80	20°49'	84°40'	75.6
W50-N13	3/12/80	20°58'	84°34'	216.4
W50-N14	3/12/80	21°19'	85°29'	51.8
W50-N15	3/13/80	22°52'	85°07'	8.1

Station No.	Date	Latitude (N)	Longitude (W)	Sargassum (mg/m^2)
		Cruise W-50 (continued)		
W50-N16	3/13/80	23°13'	85°04'	10.3
W50-N17	3/14/80	23°16'	84°38'	0
W50-N18	3/14/80	23°06'	84°11'	32.9
		Cruise W-52		
W52-N1	6/09/80	39°21'	66°39'	234
W52-N2	6/10/80	39°02'	65°44'	50.5
W52-N3	6/10/80	38°47'	66°18'	114
W52-N4	6/11/80	38°21'	66°23'	0
W52-N5	6/11/80	38°08'	66°45'	0
W52-N6	6/12/80	38°02'	66°39'	262
W52-N7	6/12/80	37°17'	66°54'	0
W52-N8	6/13/80	36°35'	66°48'	159
W52-N9	6/15/80	35°39'	66°09'	0
W52-N10	6/17/80	34°58'	65°08'	0
W52-N11	6/18/80	34°29'	64°38'	0
W52-N12	6/19/80	32°51'	64°33'	0
W52-N13	6/20/80	32°26'	64°14'	0
W52-N14	6/23/80	33°08'	65°17'	0
W52-N15	6/24/80	33°24'	65°24'	0
W52-N16	6/25/80	33°32'	65°44'	0
W52-N17	6/26/80	34°23'	66°11'	0
		Cruise W-53		
W53-N1	7/25/80	40°56'	68°57'	0
W53-N2	7/26/80	40°53'	67°38'	0
W53-N3	7/27/80	41°30'	67°29'	0
W53-N4	7/27/80	42°00'	67°23'	0
W53-N5	7/29/80	42°19'	65°38'	0
W53-N6	7/30/80	42°16'	64°49'	0
W53-N7	8/01/80	42°14'	61°10'	0
W53-N10	8/05/80	44°21'	60°02'	0
W53-N11	8/22/80	45°18'	60°34'	0
W53-N12	8/22/80	45°18'	60°50'	0
W53-N13	8/24/80	44°21'	63°29'	0
W53-N14	8/27/80	43°40'	64°39'	0
W53-N15	8/28/80	43°32'	65°02'	0
W53-N16	8/28/80	43°04'	65°17'	0
W53-N17	8/29/80	42°56'	67°04'	0
W53-N18	8/30/80	42°53'	67°56'	0
W53-N19	8/30/80	42°57'	69°00'	0
		Cruise W-55		
W55-N1	11/28/80	19°06'	64°51'	2.7
W55-N4	11/29/80	21°23'	64°24'	6.3
W55-N9	12/01/80	22°07'	63°23'	25.3
W55-N12	12/03/80	19°15'	62°04'	9.5

Station No.	Date	Latitude (N)	Longitude (W)	Sargassum (mg/m²)
		Cruise W-55 and W-55A	(continued)	
W55-N14	12/05/80	19°52'	61°20'	25.9
W55-N16	12/06/80	18°31'	60°59'	4.4
W55-N17	12/07/80	17°22'	60°50'	26.5
W55-N26	12/15/80	17°17'	64°41'	1.9
W55-N28a	12/16/80	17°07'	64°37'	0.6
W55-N28b	12/16/80	17°05'	65°53'	20.2
W55-N29	12/16/80	17°01'	65°59'	0
W55-N30	12/17/80	16°53'	66°14'	0.6
W55-N41	12/22/80	17°52'	74°29'	0.6
W55-N42	12/28/80	19°06'	79°10'	17.1
W55-N45	1/01/81	20°53'	84°12'	1030.0
W55-N47	1/02/81	22°39'	85°08'	0.6
W55A-4	1/11/81	24°41'	82°39'	0
W55A-7	1/14/81	24°52'	80°02'	1.3
		Cruise W-56		
W56-N1	2/07/81	25°48'	77°39'	1730.0
W56-N2	2/08/81	25°52'	76°45'	2.0
W56-N3	2/08/81	25°28'	75°53'	2210.0
W56-N4	2/09/81	24°49'	75°54'	240.0
W56-N5	2/12/81	24°03'	75°19'	3.1
W56-N6	2/12/81	23°35'	74°55'	3.1
W56-N7	2/13/81	22°10'	73°17'	6.8
W56-N8	2/14/81	21°38'	73°24'	112.0
W56-N9	2/17/81	19°38'	74°15'	3.1
W56-N10	2/17/81	19°29'	74°22'	6.2
W56-N11	2/17/81	18°58'	74°49'	100.0
W56-N12	2/18/81	18°30'	74°54'	1.0
W56-N13	2/19/81	-	-	-
W56-N14	2/23/81	18°40'	76°55'	7.8
W56-N15	2/23/81	18°42'	77°09'	17.0
W56-N16	2/26/81	18°27'	79°32'	7.6
W56-N17	2/26/81	18°05'	80°11'	5.2
W56-N18	2/27/81	18°05'	80°39'	0.5
W56-N19	2/28/81	17°31'	82°11'	6.8
W56-N20	3/01/81	16°54'	83°57'	0.5
W56-N21	3/02/81	16°39'	85°07'	0
W56-N22	3/08/81	17°27'	85°50'	19.0
W56-N23	3/08/81	17°51'	85°37'	4.7
W56-N24	3/11/81	21°22'	85°35'	65.0
W56-N25	3/12/81	22°15'	85°31'	138.8
W56-N26	3/14/81	23°28'	84°10'	315.0
W56-N27	3/14/81	23°21'	83°33'	0.5

Station No.	Date	Latitude (N)	Longitude (W)	Sargassum (mg/m²)
		Cruise W-58		
W58-N1	6/05/81	40°17'	69°26'	14
W58-N26	6/22/81	36°18'	69°00'	10
W58-N32	6/24/81	35°26'	67°07'	44
W58-N34	6/25/81	34°06'	66°12'	11
W58-N36	7/01/81	34°26'	64°58'	51
W58-N37	7/02/81	36°47'	65°07'	36
W58-N40	7/03/81	38°30'	64°14'	0
W58-N41	7/04/80	40°03'	64°56'	0
W58-N47	7/12/81	43°12'	69°43'	0
		Cruise W-60		
W60-N1	10/20/81	36°39'	63°05'	1381
W60-N2	10/22/81	32°18'	60°28'	71
W60-N4	10/24/81	31°03'	60°43'	113
W60-N5	10/25/81	30°28'	60°34'	845
W60-N6	10/27/81	26°48'	60°25'	70
W60-N7	10/28/81	24°59'	59°39'	0
W60-N8	10/29/81	23°52'	58°23'	65
W60-N9	10/30/81	22°29'	57°59'	183
W60-N10	11/02/81	20°24'	58°32'	0
W60-N11	11/03/81	18°45'	58°21'	0
W60-N12	11/05/81	16°57'	58°37'	0
W60-N13	11/06/81	15°27'	58°30'	1.1
W60-N14	11/07/81	14°06'	58°34'	0
W60-N15	11/08/81	13°10'	59°44'	0
W60-N17	11/12/81	12°42'	60°40'	0
W60-N19	11/14/81	13°05'	61°19'	0
W60-N20	11/23/81	18°14'	64°19'	0
		Cruise W-61		
W61-N1	12/04/81	17°37'	64°17'	8.4
W61-N3	12/05/81	16°35'	62°50'	29.1
W61-N5	12/06/81	15°40'	62°20'	0
W61-N8	12/08/81	14°37'	61°23'	0
W61-N9	12/08/81	14°21'	61°22'	0
W61-N12	12/10/81	11°31'	63°08'	0
W61-N14	12/13/81	10°41'	65°16'	0
W61-N15	12/14/81	11°18'	65°39'	0
W61-N19	12/19/81	12°46'	68°48'	1.2
W61-N20	12/19/81	12°19'	69°16'	0
W61-N25	12/22/81	15°32'	73°39'	0
W61-N29	12/24/81	16°36'	77°58'	0.9
W61-N31a	12/26/81	17°36'	82°28'	0.3
W61-N32	12/27/81	16°54'	85°03'	37.5
W61-N36	01/03/82	17°37'	85°35'	1.2
W61-N37	01/04/82	18°37'	85°20'	7.9
W61-N38	01/04/82	19°29'	84°57'	20.5
W61-N41	01/05/82	21°35'	85°31'	72.8
W61-N42	01/06/82	22°40'	85°36'	0.3

Station No.	Date	Latitude (N)	Longitude (W)	Sargassum (mg/m^2)
		Cruise W-61 (continued)		
W61-N43	1/07/82	23°19'	85°25'	42.1
W61-N44	1/07/82	24°00'	84°58'	17.3
W61-N45	1/08/82	24°27'	84°31'	56

APPENDIX II

MEIOFAUNA OF PELAGIC *SARGASSUM**

Ellen Deason

*Based on an unpublished 1972 paper. References are in Section 3 of the main text.

Fine (1970) summarized the work done on the Sargassum community to that date and pointed out how little the meiofauna had been studied. Thulin (1942) found a tardigrade, Styraconyx sargassi. Yeatman (1962) investigated the copepods, hypothesizing Sargassum as a means by which several of the species spread from North America to Europe. He collected copepods from Sargassum washings at six stations across the North Atlantic Ocean and found one species of Cyclopoida and six species of Harpacticoida. This study was undertaken to determine quantitatively the composition of the meiofauna of the Sargassum community.

METHODS AND ANALYSIS

The samples examined were collected by Morris on route to hydrostation 'S' 20 miles south of Bermuda. Single clumps were obtained with a dip net and preserved in 5% formalin. The motile macrofauna are described in Appendix III and sessile fauna are described in Appendix V. Samples were washed in 5% formalin and the motile fauna larger than 1 mm enumerated. Raft volume was measured by water displacement.

Ten samples were collected between 8 August 1971 and 1 November 1971. The samples were selected from available preserved samples in pairs of one S. fluitans clump and one S. natans clump of similar displacement volume from the same date's collections. The washings were divided with a Folsome Plankton Splitter to obtain a subsample of approximately 100 individuals. The individuals of the subsamples were identified to species, and numbers of females, males and immatures were recorded.

Diversity of the samples was calculated using a modification of Brillouin's (1956) measure of mean diversity:

$$H = \frac{C}{N}\{\log_{10} N! - \sum \log_{10} n_i!\} \qquad \text{(Lloyd et al, 1968)}$$

where N is the total number of individuals comprising n_i in each of s species where i = 1, 2, ..., s. The equation was used with Stirling's approximation for factorials:

$$\log_{10} n! \sim (n + 0.5)\log_{10} n - An + B \qquad \text{(Lloyd et al, 1968)}$$

where $A = \log_{10} e$ and $B = \log_{10} \sqrt{2\pi}$.

Similarity between occurrence of species was measured quantitatively using the coefficient of Czekanowski:

$$C_z = 2W/(A + B) \qquad \text{(Bray and Curtis, 1957)}$$

where A is the sum of all the individuals of a species in the samples, B is the sum of all the individuals of a second species from each sample. This results in a matrix of coefficients comparing every species to every other species quantitatively. This method is most useful with a few almost uniformly present species. Analysis on the basis of simple presence or absence of a species in a sample, would give no distinction.

The resulting matrix of Czekanowski coefficients was clustered and represented in a dengrogram. The clustering method used was the nearest neighbor method of Lance and Williams (1966). This joins two groups at the value of closest similarity between two individuals, one in each group. The use of the method with Czekanowski's coefficient was compared favorably to other clustering strategies by Field and McFarlane (1968).

RESULTS

The original observations on *Sargassum* samples are listed in Table 1. Table 2 contains data on the species composition of the meiofauna. Six species were found in all samples except the smallest. Two rarer species (*Paradactylopodia oculata* and *Amonardia* sp.) were restricted to larger clumps. *P. oculata* was found only on *S. fluitans*.

The total number of individuals on a *Sargassum* sample varied directly with the displacement volume of the sample (Figure 1). The regression coefficient of the line is 0.99, a much better correlation than found with the macrofauna in Table 2.5. Harpacticoid and cyclopoid populations correlated to raft volume with relationships of different slope (Figure 2).

Diversity index calculations (Figure 3) showed a general increase with raft volume with one conspicuous exception, the largest *S. fluitans* sample (348-1). The mean diversity was found to be 2.196 ± 0.279. This average is similar, but the increase is different from the motile macrofauna, which showed no increase in diversity with raft volume (Table 2.5).

The similarity coefficients are shown clustered in the form of dendrograms (Figures 4, 5, 6). There are distinct differences in affinity between populations found on *S. natans* and on *S. fluitans*.

DISCUSSION

All the meiofauna of these samples were copepods. In addition to the seven species described by Yeatman (1962), the autumn samples contained a species not before recorded on *Sargassum*, thought to be of the genus *Amonardia*.

While Fine (1970) and our group (See Chapter 2) found that the number of motile macrofauna in each sample fluctuated markedly throughout the year, the pattern in the meiofauna appeared to be more stable. The number of individuals increased linearly with raft displacement volume and there were no obvious differences in this correlation with date of collection or between the two *Sargassum* species.

The composition of the samples did vary depending on raft volume. As the raft size increased, the harpacticoid population initially increased faster than the cyclopoid did, but at large raft volumes, the cyclopoids increased in comparison to the harpacticoids. Perhaps the fact that "geographical" raft volume is not precisely the displacement volume influences the ability of the cyclopoids to maintain larger populations. The interleaf space as well as the algal surface increases as the displacement volume increases. This factor would presumably be more favorable for the larger cyclopoids.

The mean diversity (2.196) was lower than the 2.419 calculated by Fine (1968) for the larger fauna, but similar to the 2.203 for our macrofauna data (Table 2.5). It is still high considering the small number of species and the uniqueness of the habitat. While Fine (1968) found little variation in diversity with sample volume and we confirmed this for the macrofauna, the data for the meiofauna suggests an increase in diversity with raft volume. The very low diversity for the largest sample is due to the high percentage of the one cyclopoid species (*Macrochiron sargassi*); and it may not be representative.

The major differences in the species clusterings for the *S. natans* and *S. fluitans* samples are the absence of *Paradactylopodia oculata* on *S. natans* and the difference in the affinities of *Scutellidium longicauda* (Figures 4 and 5). The different grouping seems to reflect the high percentage (15.4%) of *Scutellidium longicauda* on *Sargassum natans* as compared to 6.6% on *S. fluitans*. There were greater similarities and larger reversals in general for *S. natans* than for *S. fluitans*.

The clustering of the combined data (Figure 6) is similar to that for *S. fluitans*. The tightest cluster reflects frequency of occurence and designates *Marcrochiron sargassi*, *Harpacticus gurneyi*, *Dactylopodia tisbiodes* and *Amonardia phyllopus* as the dominant species.

CONCLUSIONS

1. In the ten samples counted, the meiofauna were restricted to eight species of copepods.

2. There were no major differences between the diversity or number of individuals of the meiofauna of the two *Sargassum* species. One copepod species (*Paradactylopodia oculata*) was restricted to large *S. fluitans* samples and one (*Scutellidium longicauda*)was found in greater abundance on *S. natans*.

3. The only temporal change was the appearance of the unidentified species of *Amonardia* in the November samples.

4. Raft volume is important to size, composition, and diversity of the meiofauana population.

Table 1: COPEPODS FOUND ON *SARGASSUM*

Station No.	Sample	Date	Wet weight, g	55 *Macrochiron sargassi*	56 *Amonardia phyllopus*	57 *Harpacticus gurneyi*	58 *Paralaophonte congenera*	59 *Dactylopodia tisboides*	61 *Scutellidium longicauda*	unidentified spp.	Notes
346	1	6 Jul 71	95	52 ♀ ♂ with eggs	4	8	0	3	2		Small subsample of *Sargassum natans*. *Pseudomachrochiron parvum*: 1♀ found
348	1	5 Aug 71	323	2560: 960♀ 544♂ 1058j	224: 64♀ 64♂ 96j	930: 705♀ 160♂ 64j	64♀	224: 160♀ 64♂	96♀	128j	on *Sargassum fluitans*
348	5	5 Aug 71	230	576: 80♀ 16♂ 480j	448: 128♀ 32♂ 288j	576: 192♀ 64♂ 320j	224: 112♀ 48♂ 64j	432: 224♀ 144♂ 64	128: 48♀ no ♂ 80j	80j	on *Sargassum natans*
353	1	13 Oct 71	12	86: 30♀ 24♂ 32j	14: 6♀ 2♂ 6j	36: 10♀ 18♂ 8j	10: 6♀ 4♂ no j	78: 28♀ 26♂ 24j	10: 6♀ no ♂ 4j	16j	on *Sargassum fluitans*
	2		27	68: 22♀ 6♂ 40j	36: 6♀ 6♂ 24j	40: 18♀ 12♂ 10j	6: 2♀ 4♂ no j	56: 18♀ 16♂ 22j	46: 28♀ 8♂ 10j	8j	on *Sargassum natans*
	3		50	120: 48♀ 48♂ 24j	180: 16♀ 16♂ 148j	296: 36♀ 56♂ 104j	12♀	116: 36♀ 36♂ 44j	48: 20♀ 4♂ 24j	12j	on *Sargassum fluitans*
	4		15	68: 18♀ 20♂ 30j	30: 2♀ no ♂ 28j	32: 6♀ 4♂ 22j	2j	26: 10♀ 8♂ 8j	14: 8♀ 2♂ 4j	4j	on *Sargassum fluitans*
	5		79	288: 48♀ 48♂ 192j	136: 16♀ 40♂ 80j	160: 32♀ 80♂ 48j	16♀	80: 56♀ 24♂	112: 40♀ 40♂ 32j	32j	on *Sargassum natans*

Table 1 (cont.)

Station No.	Sample	Date	Wet weight, g	55 <u>Macrochiron sargassi</u>	56 <u>Amonardia phyllopus</u>	57 <u>Harpacticus gurneyi</u>	58 <u>Paralaophonte congenera</u>	59 <u>Dactylopodia tisboides</u>	61 <u>Scutellidium longicauda</u>	unidentified spp.	Notes
353	6	13 Oct 71	11	34: 12♀ 4♂ 18j	2: no ♀ 1♂ 1j	17 11♀ 3♂ 3j	none	5: 3♀ 2♂	13 5♀ 2♂ 6j		on <u>Sargassum natans</u>
354	4	1 Nov 71	76	520: 164♀ 104♂ 248j	76: 20♀ 8♂ 48j	64: 8♀ 4♂ 52j	16: 8♂ 8j	92: 28♀ 16♂ 48j	108: 28♀ 16♂ 64j	4j	on <u>Sargassum fluitans</u>. <u>Paradactylopodia</u> <u>Oculata</u>: 4♀, 4j (uncommon); <u>Amonardia</u> sp. 20♀
	5		121	552: 72♀ 88♂ 392j	120: 40♀ 80j	248: 88♀ 64♂ 96j	40: 24♀ 8♂ 8j	240: 112♀ 32♂ 96j	432: 120♀ 16♂ 296j		on <u>Sargassum natans</u>. <u>Amonardia</u> sp. 176:64♀24♂80j

Table 2. Species Composition of *Sargassum* Samples by Number of Individuals and Percentages

Sample	348-1	348-5	353-1	353-2	353-3	353-4	353-5	353-6	354-4	354-5
Sargassum species	F	N	F	N	F	F	N	N	F	N
Total Individuals*	4256	2464	250	258	684	176	816	72	912	1888
Macrochiron sargassi G.O. Sars, 1916	2560 63%	576 23%	86 34%	68 26%	120 18%	68 39%	288 35%	34 4%	260 57%	522 29%
Scutellidium longicauda Philippi, 1840	96 2%	128 5%	10 4%	46 18%	48 7%	14 8%	112 13%	13 18%	180 12%	432 23%
Harpacticus gurneyi Jakubisiak, 1933	928 23%	576 23%	36 14%	40 16%	196 30%	32 18%	160 19%	17 24%	64 7%	248 13%
Amonardia phyllopus G.O. Sars, 1906	224 6%	448 18%	14 6%	36 14%	180 27%	30 17%	136 16%	2 3%	76 8%	120 6%
Dactylopodia tisbiodes Claus, 1863	224 6%	432 18%	78 31%	56 22%	116 17%	26 15%	80 10%	3 4%	92 10%	240 13%
Paralaophonte congenera G.O. Sars, 1908	64 2%	224 9%	10 4%	6 2%	12 2%	2 1%	16 2%	--	16 2%	40 2%
Paradactylopodia oculata Gurney, 1927	32 1%	--	--	--	--	--	--	--	8 1%	--
Amonardia ??	--	--	--	--	--	--	--	--	20 2%	176 9%

*Includes unidentified immatures

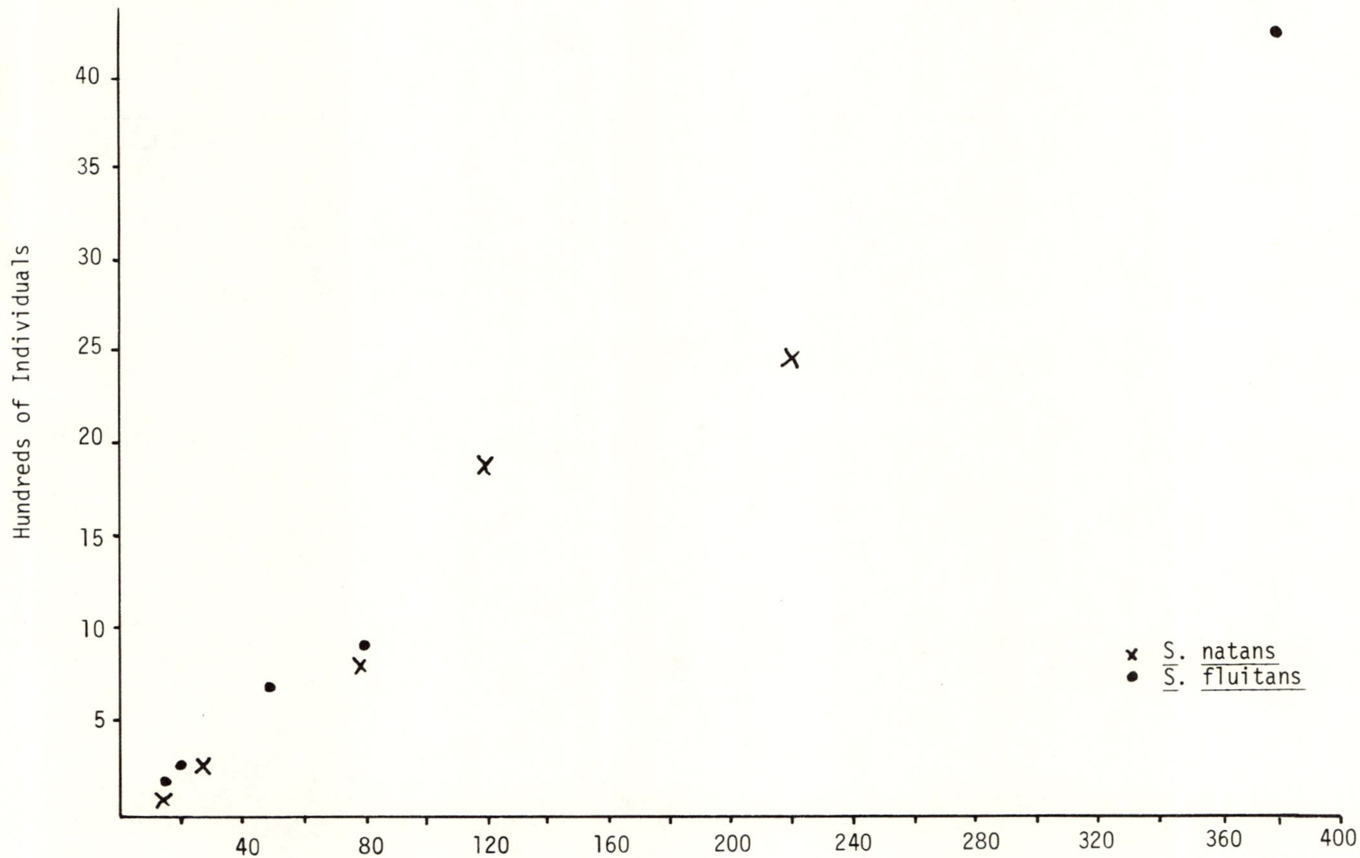

Fig. 1. Relationship of Number of Individuals to Raft Displacement Volume

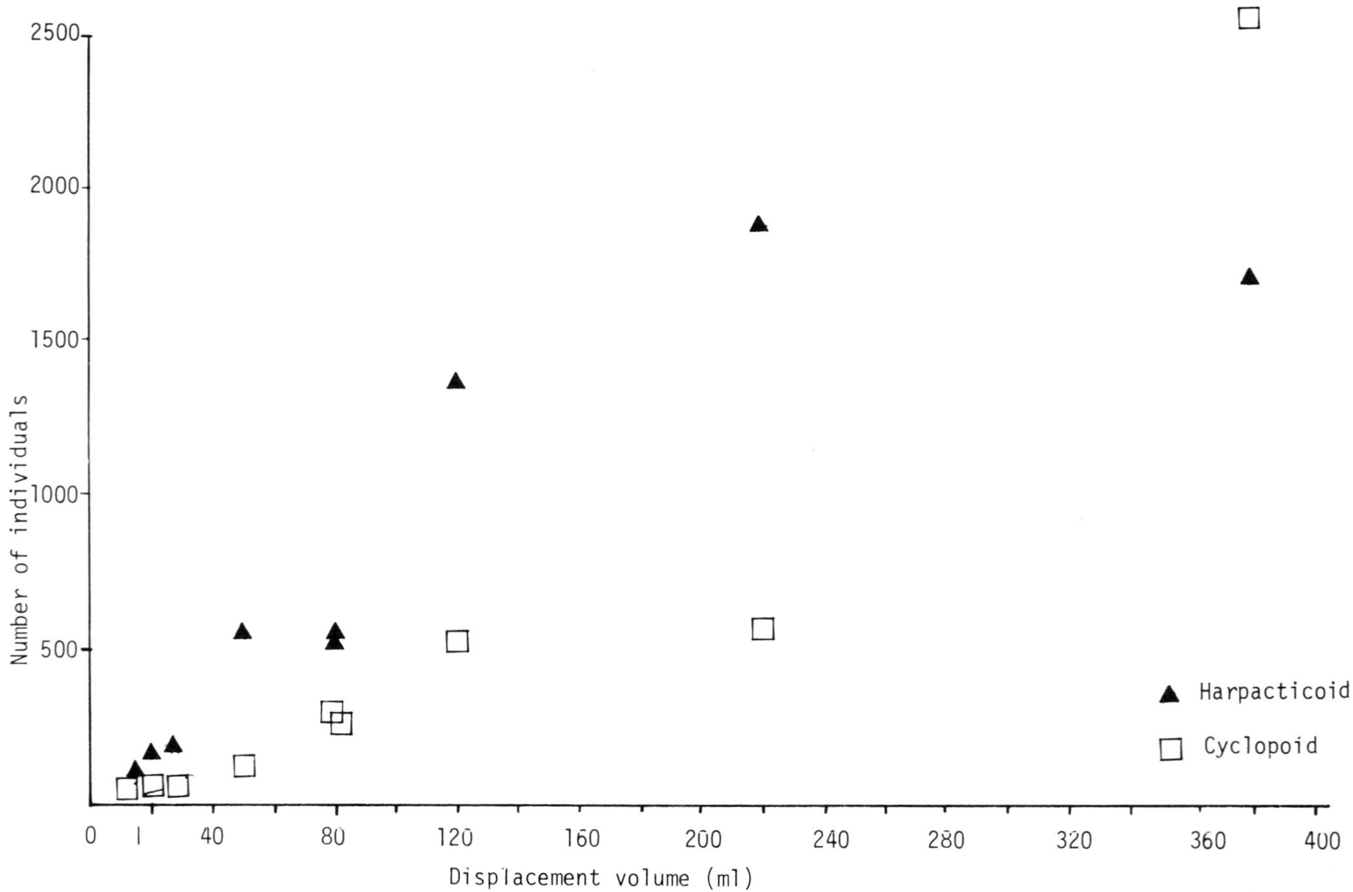

Figure 2. Relationship of Harpacticoid and Cyclopoid Populations to Raft Displacement Volume

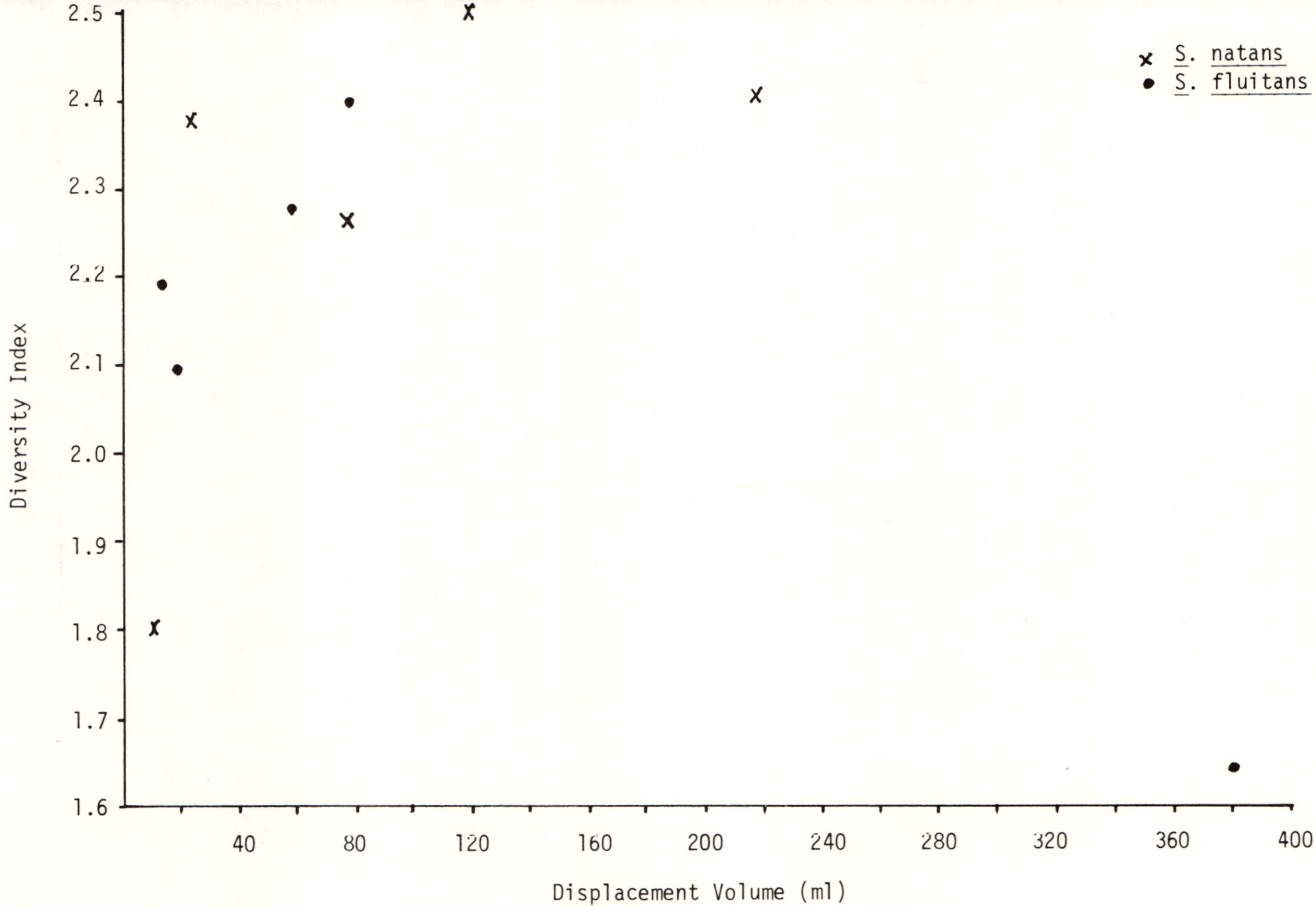

Fig. 3. Diversity of Meiofauna on Sargassum

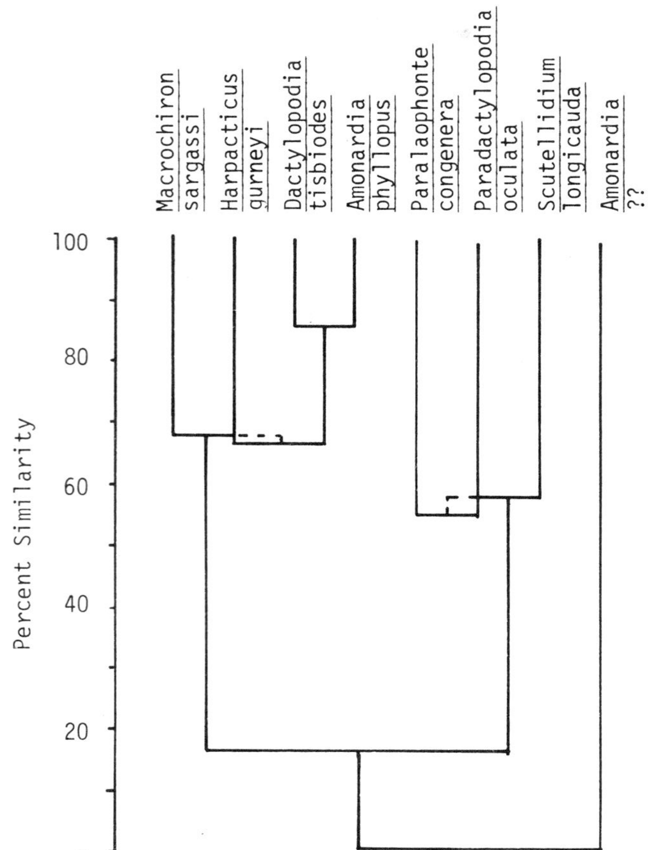

Fig. 4. Nearest Neighbor Clustering of Coefficients of Czekanowski for *Sargassum fluitans*

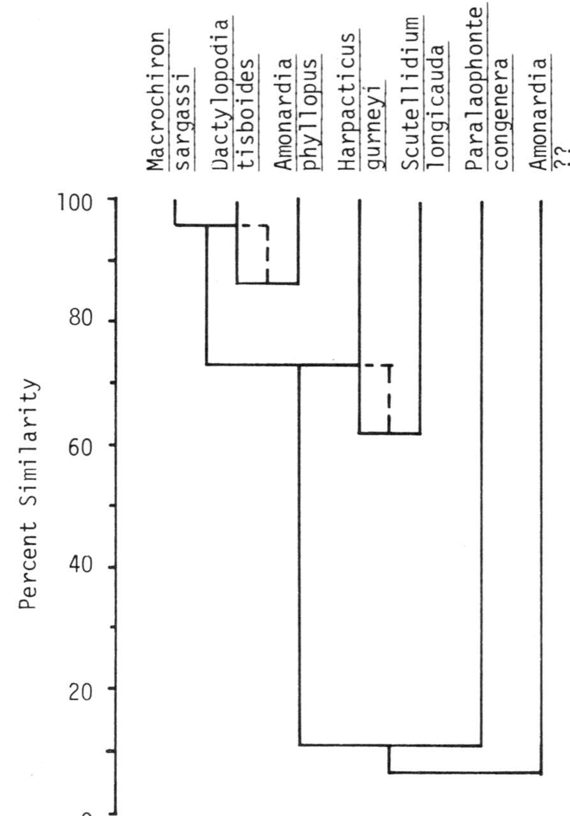

Fig. 5. Nearest Neighbor Clustering of Coefficients of Czekanowski for *Sargassum natans*

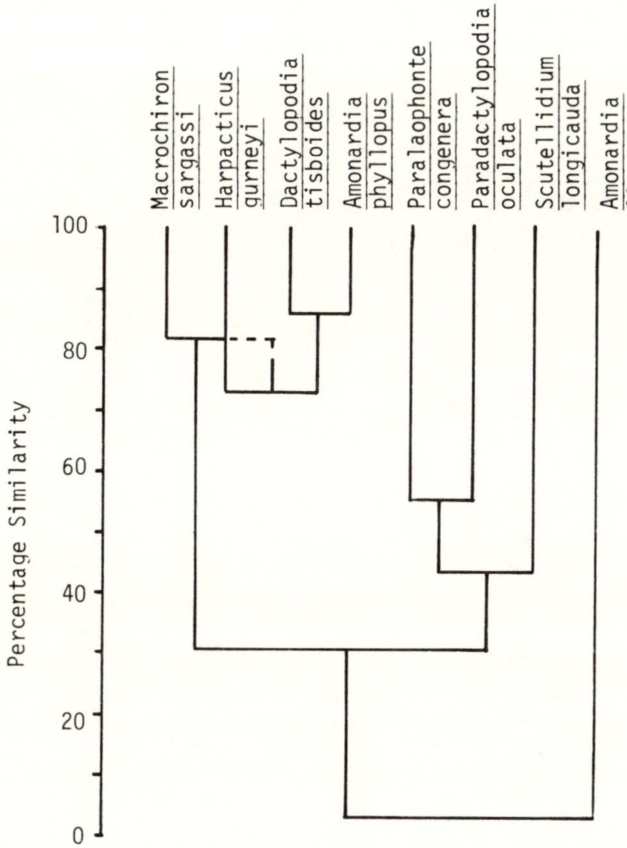

Fig. 6. Nearest Neighbor Clustering of Coefficients of Czekanowski for Combined *Sargassum* Samples

APPENDIX III

Motile Macrofauna on Samples from Station 'S'

Byron F. Morris
Deborah D. Mogelberg
Jill Cadwallader
and
numerous volunteers

MOTILE MACROFAUNA ON SARGASSUM

Station	346					347									
Date	6 July 71					17 July 71									
Sample No.	1	2	3	4	5	1	2	3	4	5	6	7	8	9	10
Wet weight, g	95	31	42	8	4	186	31	54	43	75	67	72	109	18	38
Displacement vol. ml	98	38	40	10	5	180	38	56	41	80	62	71	100	18	40
Sargassum species	---all *S. natans*---														
Fauna:															
44 Anemonia sargassensis	2		126			1				37	1		2	51	7
46 Acerotisa notulata	10	11	12			2			10	3	1	6	5		3
47 Hoploplana grubei			7						1						1
48 Gnesioceros sargassicola	63	24	34	11	2	7	6	8	17	24	9	14	8	11	16
51 Platynereis dumerillii	86	48	22	5	2	11			150	52	1	37	19	4	28
70 Bagatus minutus	44	17	138	12		13	1	2	63	196	22	31	17	16	76
73 Sunampithoë pelagica	2				1	1			1	14			3	2	
74 Biancolina sp.	1	1				1	1	2	2		1				
76 Hemiaegina minuta			12			1	14		9	14	1	18	1		77
80 Latreutes fucorum	8		11		2	3				13	6	8	7		
82 Leander tenuicornis	2								1	2	1	1			
83 Planes minutus	7	3	2	1		1		1	4	10	1	5	2	1	3
84 Portunus sayi															
87 Anoplodactylus petiolatus	24	2	21	8					1	71	1	3		12	5
89 Litiopa melanostoma	3	1	2			4		5	45	1	1	1	13		3
91 Corambella depressa		3													
95 Scyllaea pelagica		2	3	2		1			3		2	4	13	2	6
100 Histrio histrio															
101 Syngnathus pelagicus															
Other species (see notes)	1		3	3	1			1			2			1	3
Total number specimens	253	112	393	42	8	46	22	19	307	437	50	128	90	100	228

Notes: All samples were taken at Station "S", approx. 32°10'N, 64°30'W. Numbers are those of the Panulirus Hydrostation. This list of 19 species comprises more than 99% of the observed motile microfauna. Species numbers are those of Morris and Mogelberg (1973). Populations are the total of adult and juvenile specimens; details of these numbers and other observations are listed by station and sample in a separate table. Copepods were counted in only 11 samples; sessile fauna in only 20. These data are given in other tables.

MOTILE MACROFAUNA ON SARGASSUM

Station	348							349				353		
Date	5 Aug 71							18 Aug 71				13 Oct 71		
Sample No.	1	2	3	4	5	6	7	1	2	3	4	1	2	3
Wet weight, g	323	70	62	52	230	18	12	87	72	50	23	12	27	50
Displacement vol. ml	380	75	65	50	230	18	10	87	70	50	22	20	27	50
<u>Sargassum</u> species	F	N	N	N	N	N	N	N	N	N	N	F	N	F
Fauna:														
44 Anemonia sargassensis	3				4									
46 Acerotisa notulata	4		2	3	21	2	1							3
47 Hoploplana grubei														
48 Gnescioceros sargassicola	108	9	5	13	26	5	2	8	6	9	11	7	5	15
51 Platynereis dumerillii	52	2	22	60	62	13	1	20	31	3	18	8	5	53
70 Bagatus minutus	54	1			8	3		2	4		2			5
73 Sunampithoë pelagica	50	2			1					2				
74 Biancolina sp.	18	2			1	1	3		4					
76 Hemiaegina minuta	116	3		12	1	18		1	5		178			17
80 Latreutes fucorum	3	2		1	3			2	1			1	5	1
82 Leander tenuicornis					4			1			1		1	
83 Planes minutus	2							1				3		6
84 Portunus sayi	1							1		1			1	
87 Anoplodactylus petiolatus		7			1			4	1				1	1
89 Litiopa melanostoma	3	6	4	8	30			1				1	2	10
91 Corambella depressa														
95 Scyllaea pelagica				1			1	1						
100 Histrio histrio														
101 Syngnathus pelagicus														
Other species (see notes)		1		1										
Total number specimens	414	35	33	99	162	42	8	42	52	15	210	20	20	111

Note: F = S. fluitans, N = S. natans.

MOTILE MACROFAUNA ON SARGASSUM

	Station	353				354					355	
	Date	13 Oct 71				1 Nov 71					24 Nov 71	
	Sample No.	4	5	6	7	1	2	3	4	5	1	2
	Wet weight, g	15	79	11	163	15	277	113	76	121	210	38
	Displacement vol. ml.	13	79	12	165	20	275	105	80	122	210	36
	Sargassum species	F	N	N	N	F	N	N	F	N	N	N
	Fauna:											
44	Anemonia sargassensis										1	
46	Acerotisa notulata		2		3		2				13	3
47	Hoploplana grubei											
48	Gnescioceros sargassicola	3	10	2	13		8	3	2	3	64	8
51	Platynereis dumerillii	36	4		36	35	37	2	96	28	38	7
70	Bagatus minutus		2		6	16			25	2	20	4
73	Sunampithoë pelagica										1	
74	Biancolina sp.						2	1		1	4	1
76	Hemiaegina minuta					3		7	4		3	6
80	Latreutes fucorum	2	6		4	1	1				12	
82	Leander tenuicornis		2		1				2		1	1
83	Planes minutus	2	4			2	6	1	7	1	2	
84	Portunus sayi		1				1					
87	Anoplodactylus petiolatus		1		1		10	115	1	10	117	22
89	Litiopa melanostoma	6	14	1	48	2	58	8	5	36	9	2
95	Scyllaea pelagica					2			1		1	1
100	Histrio histrio											
101	Syngnathus pelagicus											
	Other Species:						2	1			11	
	Total Number Specimens:	49	46	3	112	61	127	138	143	81	297	55

Note: F = *S. fluitans*, N = *S. natans*.

MOTILE MACROFAUNA ON SARGASSUM

	Station	355				356							
	Date	24 Nov 71				2 Feb 72							
	Sample No.	3	4	5	6	1	2	5	6	7	8	9	10
	Wet weight, g	28	210	80	71								
	Displacement vol. ml.	25	210	80	74	80	40	60	25	15	20	80	30
	Sargassum species	N	N	N	N	N	N	N	N	N	N	N	N
	Fauna:												
44	Anemonia sargassensis		6										
46	Acerotisa notulata		5	2	11					2		2	
47	Hoploplana grubei		23	2									
48	Gnescioceros sargassicola	12	99	18	15	11	8	28	6	3	1	12	7
51	Platynereis dumerillii	12	28	2	23			2	2		1	2	
70	Bagatus minutus	5	2		2	1		7	23	1	1	10	3
73	Sunampithoë pelagica					10	13	7	2	2		10	
74	Biancolina sp.		2			2		1					
76	Hemiaegina minuta		2	2									
80	Latreutes fucorum	2	42	6			2					1	
82	Leander tenuicornis				5	1		2				1	
83	Planes minutus				1								
84	Portunus sayi												
87	Anoplodactylus petiolatus	13	278	28	45	14	12	52	28	9	4	56	9
89	Litiopa melanostoma	2	23		1		1		3		1		
95	Scyllaea pelagica		3									1	
100	Histrio histrio	1	1										
101	Syngnathus pelagicus												
	Other Species:	6	44	8	6	5	13	16	29		2	8	1
	Total Number Specimens:	54	558	68	109	44	49	115	93	17	10	103	20

Note: F = S. fluitans, N = S. natans.

MOTILE MACROFAUNA ON SARGASSUM

	Station	356			358							
	Date	2 Feb 72			5 Apr 72							
	Sample No.	11	12	13	1	2	3	4	5	6	7	8
	Wet Weight, g											
	Displacement vol. ml.	246	15	20	50	30	20	70	35	100	40	85
	Sargassum species	N	N	N	F	N	F	F	N	N	N	N
	Fauna:											
44	Anemonia sargassensis											
46	Acerotisa notulata				2	5	3	6		7	1	
47	Hoploplana grubei	5				3		2		4		
48	Gnescioceros sargassicola	75	5	9	12	6	13	11	5	20	9	17
51	Platynereis dumerillii	120		2	16	1	5	3	1	16		9
70	Bagatus minutus	40	13	15	26	2	11	106		60	5	6
73	Sunampithoë pelagica	20	2	10	133	76	80	83	17	113	49	28
74	Biancolina sp.	1		1			2	1		1		2
76	Hemiaegina minuta				57			2				
80	Latreutes fucorum	5	1						5		3	1
82	Leander tenuicornis	2							1			1
83	Planes minutus	3			1			1				1
84	Portunus sayi											
87	Anoplodactylus petiolatus	185	10	13			2	21	4	43	7	26
89	Litiopa melanostoma	7					1	5		3	4	
95	Scyllaea pelagica	3						1			1	1
100	Histrio histrio										1	
101	Syngnathus pelagicus											
	Other Species:	49		2	43	14	52	114	62	183	113	41
	Total Number Specimens:	515	31	52	290	107	169	356	95	450	193	133

Note: F = *S. fluitans*, N = *S. natans*.

MOTILE MACROFAUNA ON SARGASSUM

Station	360				361							
Date	29 Apr 72				12 May 72							
Sample No.	1	2	3	4	1	2	3	4	5	6	7	8
Wet Weight, g												
Displacement vol. ml.	40	21	12	45	50	75	180	30	75	60	35	98
Sargassum species	N	N	N	N	N	N	N	N	N	N	N	N
Fauna:												
44 Anemonia sargassensis												
46 Acerotisa notulata							1	1				1
47 Hoploplana grubei	3	14			1			1	1			3
48 Gnescioceros sargassicola	3	1	1	2	6	6	26		10	3	1	8
51 Platynereis dumerillii	6	1		7	4	6	43	8	11	5	2	4
70 Bagatus minutus	10	2	2	4	6	27	166	5	20	8	3	5
73 Sunampithoë pelagica	47	7	8	102	180	74	143	41	135	76	55	191
74 Biancolina sp.	2	3	1	6	14	10	12		2	1	1	7
76 Hemiaegina minuta							12					
80 Latreutes fucorum				1			1		4	1		
82 Leander tenuicornis							1					
83 Planes minutus	2		1	2	2	3	8	2	2	2	1	3
84 Portunus sayi						1						
87 Anoplodactylus petiolatus	2	1		5	12	41	55	1	45	21	1	35
89 Litiopa melanostoma				2				1	7	2	2	10
95 Scyllaea pelagica						2	3	2	2			
100 Histrio histrio						1			1			
101 Syngnathus pelagicus												
Other Species:	6	2	2	35	6	8	4	2	28	22	1	19
Total Number Specimens:	81	31	15	166	231	179	475	64	268	141	67	286

Note: F = *S. fluitans*; N = *S. natans*.

MOTILE MACROFAUNA ON SARGASSUM

	Station	364						364a*		
	Date	13 July 72						2 Aug 72		
	Sample No.	1	2	3	4	5	6	1	2	3
	Wet Weight, g									
	Displacement vol. ml.	39	20	50	15	15	50	60	5	50
	Sargassum species	N	N	N	N	N	N	N	N	F
	Fauna:									
44	Anemonia sargassensis									2
46	Acerotisa notulata							3	3	2
47	Hoploplana grubei									
48	Gnescioceros sargassicola	16	21	26	9	6	17		2	15
51	Platynereis dumerilii	10	9	12	2	1		14		30
70	Bagatus minutus			2						10
73	Sunampithoë pelagica	3	2	6	3	1	1			12
74	Biancolina sp.			2	4		1			1
76	Hemiaegina minuta									
80	Latreutes fucorum						1			
82	Leander tenuicornis						1			5
83	Planes minutes	6	1	3	1	3	5	2		4
84	Portunus sayi							1		
87	Anoplodactylus petiolatus	2			1	1	2			4
89	Litiopa melanostoma	2	2	6	1		2	30	38	200
95	Scyllaea pelagica	1				1	1			
100	Histrio histrio									
101	Syngnathus pelagicus									
	Other Species:							1	1	1
	Total Number Specimens:	40	33	57	21	13	31	51	44	286

Note: F = *S. fluitans*; N = *S. natans*.

*Not a regular hydrostation

MOTILE MACROFAUNA ON SARGASSUM

		365				366					367		
	Station	365				366					367		
	Date	6 Aug 72				8 Aug 72					13 Sept 72		
	Sample No.	1	2	3	4	1	2	3	4	5	1	2	3
	Wet weight, g.												
	Displacement vol. ml.	20	30	55	45	130	5	23	28	55	50	55	160
	Sargassum species	N	N	N	N	N	N	N	N	N	N	H	F
	Fauna:												
44	Anemonia sargassensis		3										2
46	Acerotisa notulata	1	8	8	4			1	3	3		3	1
47	Hoploplana grubei					7	1	7	4	7	1		1
48	Gnescioceros sargassicola	8	7	8	3	6		4	7	5	18	19	28
51	Platynereis dumerillii	22	5	21		29		11	48	39	12	114	21
70	Bagatus minutus					3		1	2	6		4	14
73	Sunampithoë pelagica		5		3				4	1			1
74	Biancolina sp.	2		1	1	5			3	12			1
76	Hemiaegina minuta			18				1	25		3	2	5
80	Latreutes fucorum		5	2		7	1	1		8	10	12	14
82	Leander tenuicornis												1
83	Planes minutus	2		2	1			1		1			3
84	Portunus sayi									1			
87	Anoplodactylus petiolatus	34	4	2	2	6			7	1			
89	Litiopa melanostoma	2		58	39	16	4	12	13	8	4		3
95	Scyllaea pelagica		2	17	7								
100	Histrio histrio			2									
101	Syngnathus pelagicus												
	Other Species:	4	7	6	1		2	1	6		1		
	Total Number Specimens:	75	46	145	61	79	8	40	122	92	49	154	95

Note: F = *S. fluitans*, N = *S. natans*, H = *S. hystrix*

MOTILE MACROFAUNA ON SARGASSUM

Station		367			368					369		
Date		13 Sept 72			26 Sept 72					10 Oct 72		
Sample No.		4	5	6	1	2	3	4	5	1	2	3
Wet weight, g												
Displacement vol. ml.		100	370	37	100	40	50	6	50	92	45	30
Sargassum species		F	H	N	N	N	N	N	F/N	F/N	N	N
	Fauna:											
44	Anemonia sargassensis						7	1		3		267
46	Acerotisa notulata	1					3				1	
47	Hoploplana grubei		2			1	1			3		
48	Gnesioceros sargassicola	25	92	19		3	21		6	90	7	18
51	Platynereis dumerillii	69	103	31	4		51	3	13	31	16	27
70	Bagatus minutus	31	72	2			19		2			4
73	Sunampithoë pelagica	2					1			1		1
74	Biancolina sp.		9		2						2	2
76	Hemiaegina minuta	2	11	2			84			1	6	3
80	Latreutes fucorum	2	5	3	2		2		2	12		2
82	Leander tenuicornis		2	1		1				3	2	3
83	Planes minutus	3	5	1			1			16	2	2
84	Portunus sayi				1?				1	1	1	1
87	Anoplodactylus petiolatus		3								1	2
89	Litiopa melanostoma	1	2		3	4		1	3			
95	Scyllaea pelagica											
100	Histrio histrio											1
101	Syngnathus pelagicus										2	1
	Other Species:		1				1			1		3
	Total number specimens:	136	307	59	12	9	191	5	25	162	40	337

Note: F = *S. fluitans*, N = *S. natans*, H = *S. hystrix*

MOTILE MACROFAUNA ON SARGASSUM

Station	369		370						371		
Date	10 Oct 72		23 Oct 72						3 Nov 72		
Sample No.	4	5	1	2	3	4	5	6	1	2	3
Wet weight, g											
Displacement vol. ml.	15	10	50	115	25	25	15	15	65	42	300
Sargassum species	N	N?	N+?	H	N	N	N	N	N	F	?
Fauna:											
44 Anemonia sargassensis	699	87	364	1	9	187	8	11	368		70
46 Acerotisa notulata	1		4					1		4	1
47 Hoploplana grubei				5		1	1				2
48 Gnescioceros sargassicola	14	6	27	80	8	11	9	5	43	31	210
51 Platynereis dumerillii		11	44	83	11	15	10	1	19	13	158
70 Bagatus minutus	3	3	2	15		12	1		4	15	60
73 Sunampithoë pelagica											2
74 Biancolina sp.							1		1		
76 Hemiaegina minuta			2	24	11	127	74	2	20	97	80
80 Latreutes fucorum	2	1	6	6	3	3	2	4	12	12	31
82 Leander tenuicornis			1	1					1	1	1
83 Planes minutus	1	2	1	2			1	1			1
84 Portunus sayi											
87 Anoplodactylus petiolatus	15	2	4		3	50	13	2	70	1	3
89 Litiopa melanostoma							1			1	5
95 Scyllaea pelagica											
100 Histrio histrio											
101 Syngnathus pelagicus											
Other Species:	1			1	2			1			15
Total Number of Specimens:	736	112	455	218	47	406	121	28	538	175	639

Note: F = S. fluitans; N = S. natans, H = S. hystrix

MOTILE MACROFAUNA ON SARGASSUM

	Station	371			372						373	
	Date	3 Nov 72			28 Nov 72						12 Dec 72	
	Sample No.	4	5	6	1	2	3	4	5	6	1	2
	Wet weight, g											
	Displacement vol. ml.	620	210	120	705	25	50	40	40	18	95	29
	Sargassum species	?	?	?	N	N	N	N	N	N	N	N
	Fauna:											
44	Anemonia sargassensis	141	26	11	608	120	22	33	8	50	145	2
46	Acerotisa notulata	2	1	1	1			2	2		1	
47	Hoploplana grubei	4	1	2	1					2	10	
48	Gnescioceros sargassicola	298	130	78	37	10	63	16	12	15	47	10
51	Platynereis dumerillii	355	296	69	27	10		20	8		20	18
70	Bagatus minutus	200	47	49	32	9	2	3		2	9	22
73	Sunampithoë pelagica	2										
74	Biancolina sp.				1						1	
76	Hemiaegina minuta	194	163	40	18	6	13	7			6	26
80	Latreutes fucorum	87	21	12	32	6	17	6	9	10	28	6
82	Leander tenuicornis	3	1	1	7		1	3	1	1	3	
83	Planes minutus	5	1		2						1	
84	Portunus sayi											
87	Anoplodactylus petiolatus	9	4	2	110	17	6	61	22	7	67	1
89	Litiopa melanostoma	11	1	1			1				1	2
95	Scyllaea pelagica										1	
100	Histrio histrio											
101	Syngnathus pelagicus											
	Other Species:	37	24	16		19	9	9	19	4	54	
	Total Number of Specimens:	1348	716	282	876	197	134	160	81	91	394	87

Note: F = *S. fluitans*; N = *S. natans*

MOTILE MACROFAUNA ON SARGASSUM

	Station	373								374	
	Date	12 Dec 72								27 Dec 72	
	Sample No.	4	5	6	7	8	9	10	11	1	2
	Wet weight, g										
	Displacement vol. ml.	70	68	70	80	58	12	12	12	45	26
	Sargassum species	N+?	N?	N+?	N+?	N	N	N	N	N	N
	Fauna:										
44	Anemonia sargassensis	46	250	172	29	20	3	14	11	105	17
46	Acerotisa notulata	1	1	1	1						
47	Hoploplana grubei				1		2	1			
48	Gnescioceros sargassicola	38	29	38	39	26	5	7	7	22	10
51	Platynereis dumerillii	70	59	38	41	26	1	7	9	40	9
70	Bagatus minutus	118	109	91	131	11	1	6	12	137	13
73	Sunampithoë pelagica	1	1	1	8						1
74	Biancolina sp.		7		1						
76	Hemiaegina minuta	18	100	3	24	13			19	20	
80	Latreutes fucorum	20	2	19	17	8	1	3	7	14	5
82	Leander tenuicornis							3			
83	Planes minutus	1	1	1	2	1				1	
84	Portunus sayi			1							
87	Anoplodactylus petiolatus	13	8	16	7	36	2	11	1	6	7
89	Litiopa melanostoma	6	1	1	2	3		1			1
95	Scyllaea pelagica						3	8			1
100	Histrio histrio										
101	Syngnathus pelagicus										
	Other Species:	10	2	4	4	25	2	3	1	2	3
	Total No. of Specimens:	342	570	386	307	169	20	64	67	347	67

Note: F = S. fluitans; N = S. natans

MOTILE MACROFAUNA ON SARGASSUM

	Station	375						376				
	Date	11 Jan 73						7 Feb 73				
	Sample No.	1	2	3	4	5	6	1	2	3	4	5
	Wet weight, g											
	Displacement vol. ml.	70	70	60	60	40	30	135	30	50	35	25
	Sargassum species	F	N	F	F	F+N	N	F?	N	N	N	?
	Fauna:											
44	Anemonia sargassensis	59	12	8	69	46	9	198		17	93	2
46	Acerotisa notulata											
47	Hoploplana grubei									1		1
48	Gnescioceros sargassicola	34	44	19	17	16	16	19	8	25	13	19
51	Platynereis dumerillii	46	9	27	21	19	1	34	17	18	24	47
70	Bagatus minutus	79	13	20	39	33	19	132	2	18	62	87
73	Sunampithoë pelagica	3	2	50	18	6	1	48		12	8	18
74	Biancolina sp.	1							2			
76	Hemiaegina minuta	54	1	4	26	9	20	14			10	6
80	Latreutes fucorum	7	3	1	2	1	1	5		10	8	4
82	Leander tenuicornis		1				1		1	1		
83	Planes minutus			4				1		2	1	
84	Portunus sayi											
87	Anoplodactylus petiolatus		75	1	2	1	19	2	4	48		
89	Litiopa melanostoma			1				1				
95	Scyllaea pelagica		1									
100	Histrio histrio											
101	Syngnathus pelagicus								2			
	Other Species:	1	53		1	2	6	4	83	67	4	13
	Total Number of Specimens:	284	214	625	195	133	93	458	119	219	223	197

Note: F = *S. fluitans*; N = *S. natans*

MOTILE MACROFAUNA ON SARGASSUM

	Station	377		378			380					
	Date	21 Feb 73		12 Mar 73			2 May 73					
	Sample No.	1	2	1	2	3	1	2	3	4	5	6
	Wet weight, g											
	Displacement vol. ml.	60	22	50	60	80	110	70	50	30	30	28
	Sargassum species	N	N	N	F	N	N	N	N	N	N	N
	Fauna:											
44	Anemonia sargassensis	7	111		61	197	107	6	2			
46	Acerotisa notulata											
47	Hoploplana grubei		1				1	7	19		6	8
48	Gnescioceros sargassicola	121	38	27	25	17	20	4	8	10	3	5
51	Platynereis dumerillii	38	20	3	24	23	119	11	2	6	10	1
70	Bagatus minutus	13	47	4	132	64		9	64	8	35	16
73	Sunampithoë pelagica	17	6	19	31	13	12	39	51	35	8	15
74	Biancolina sp.			1	1		2	2				1
76	Hemiaègina minuta				7		4					
80	Latreutes fucorum	7		1	4	13	6	2	2	3		3
82	Leander tenuicornis	1		1				1				
83	Planes minutus		1		1	2	4	1	2	1	2	
84	Portunus sayi											
87	Anoplodactylus petiolatus	32	10	11	3	7	67	21	31	38	8	3
89	Litiopa melanostoma	8		1		1	5	7	6	2	8	2
95	Scyllaea pelagica	1?		2		3	6	21	3	5	3	9
100	Histrio histrio											
101	Syngnathus pelagicus											
	Other Species:	19	15	4	4	2	6	40	92	19	28	16
	Total Number of Specimens:	264	249	74	293	342	359	171	282	127	111	79

Note: F = *S. fluitans*; N = *S. natans*

MOTILE MACROFAUNA ON SARGASSUM

	Station	382						383		
	Date	6 Jun 73						17 July 73		
	Sample No.	1	2	3	4	5	6	1	2	3
	Wet weight, g									
	Displacement vol. ml.	150	130	120	85	50	70	55	90	100
	<u>Sargassum</u> species	F	N	N	N	N	F	N	N	F
	Fauna:									
44	Anemonia sargassensis			30	8					
46	Acerotisa notulata	1					1		3	5
47	Hoploplana grubei	20	1	2	1	1		1	13	2
48	Gnescioceros sargassicola	6	7	17	2	1	7	4	6	7
51	Platynereis dumerillii	40	7	18	9	5	17	11	7	27
70	Bagatus minutus	18	495	220	143	142	44	4	2	
73	Sunampithoë pelagica	1	64	45	31	41		10	3	10
74	Biancolina sp.			1				3	7	10
76	Hemiaegina minuta									
80	Latreutes fucorum	3	6	8	4	3	12	2	3	3
82	Leander tenuicornis	1						1		1
83	Planes minutus	5	24	1	11	9	8	3	1	1
84	Portunus sayi									
87	Anoplodactylus petiolatus			5	4	2		2		
89	Litiopa melanostoma	9	2	6	12	23	9	6	11	14
95	Scyllaea pelagica				4		1			
100	Histrio histrio					1				
101	Syngnathus pelagicus									
	Other Species:	12		4	1		4			22
	Total Number of Specimens:	116	606	357	230	228	103	47	56	102

Note: F = <u>S</u>. <u>fluitans</u>; N = <u>S</u>. <u>natans</u>

MOTILE MACROFAUNA ON SARGASSUM

	Station	385								386			
	Date	8 Aug 73								3 Sept 73			
	Sample No.	1	2	3	4	5	6	7	8	1	2	3	4
	Wet weight, g												
	Displacement vol. ml.	11	32	38	69	39	38	21	9	56	62	44	65
	Sargassum species	------------------all S. natans------------------											
	Fauna:												
44	Anemonia sargassensis									1		2	
46	Acerotisa notulata			1	1	1	1	2	1	33	10	15	14
47	Hoploplana grubei	1		1	1					1	1	3	2
48	Gnescioceros sargassicola		3		1	4	1			16	5	5	5
51	Platynereis dumerillii	1	16	9	6	1?	1	12	10	13		38	11
70	Bagatus minutus		1					1				2	3
73	Sunampithoë pelagica		1	1	3		1	1		1			
74	Biancolina sp.			1		2						3	
76	Hemiaegina minuta												
80	Latreutes fucorum	2	2	1	2	2	1	1	2		8	8	10
82	Leander tenuicornis	1		1	1	1	1	1		1	1		
83	Planes minutus				1	1	2						1
84	Portunus sayi		1	1									
87	Anoplodactylus petiolatus												
89	Litiopa melanostoma	9	6	12	39	25	36	12	7	2		2	1
95	Scyllaea pelagica	1										1	
100	Histrio histrio	1				2	1						
101	Syngnathus pelagicus			3	1						1		
	Other Species:	2	1		1	1		30		1	1	3	1
	Total Number of Specimens:	18	31	31	57	40	45	60	20	69	26	82	48

MOTILE MACROFAUNA ON SARGASSUM

Station	386		387						388			
Date	3 Sept 73		10 Oct 73						14 Nov 73			
Sample No.	5	6	1	2	3	4	5	6	1	2	3	4
Wet weight, g												
Displacement vol. ml.	95	28	19	23	28	18	7	9	350	70	55	78
<u>Sargassum</u> species	N	N	N	N	N	N	N	N	N	N	N	N
Fauna:												
44 Anemonia sargassensis		1			6				6	24	7	9
46 Acerotisa notulata	54	11	2		10				1	1	4	2
47 Hoploplana grubei	1	1	1		12	1?	3	1	1	2	4	6
48 Gnescioceros sargassicola	27	4	1		1	2	2	1	127	35	18	43
51 Platynereis dumerillii	16		1	20	8	20	10		101	18	18	40
70 Bagatus minutus	3								64	15	21	9
73 Sunampithoë pelagica	1	2							34	1	1	1
74 Biancolina sp.			1				1		1	2	2	
76 Hemiaegina minuta						6	2		23	13	42	6
80 Latreutes fucorum	18	2	3		8	3	4		6	6	1	10
82 Leander tenuicornis	1	2	2			1			4			1
83 Planes minutus				3		1	3		5	12	4	5
84 Portunus sayi					1				1			1
87 Anoplodactylus petiolatus					1	10			6	45	22	23
89 Litiopa melanostoma		1	7		7	4	1		3	1	3	7
95 Scyllaea pelagica			1?		1?				1	1		1
100 Histrio histrio												
101 Syngnathus pelagicus												
Other Species:			1	1	11	3	2	3	55	12	12	18
Total Number of Specimens:	121	24	20	24	66	51	28	5	439	188	159	182

Note: F = <u>S</u>. <u>fluitans</u>; N = <u>S</u>. <u>natans</u>

MOTILE MACROFAUNA ON SARGASSUM

Station	388		389						390		
Date	14 Nov 73		26 Nov 73						7 Jan 74		
Sample No.	5	6	1	2	3	4	5	6	1	2	3
Wet Weight, g											
Displacement vol. ml.	60	100	35	70	48	170	40	48	130	120	140
Sargassum species	N	N	N	N	F	N	F	N	N	N	N
Fauna:											
44 Anemonia sargassensis	8	8	42	1?		3	1	52		6	21
46 Acerotisa notulata		1		1			1	1			
47 Hoploplana grubei	6	4		1						1	1
48 Gnescioceros sargassicola	14	69	16	37	35	100	30	29	38	27	60
51 Platynereis dumerillii	5	12	21	16	23	4	8	19	7	14	19
70 Bagatus minutus	4	3	18	4	18	4	13	46	4	9	53
73 Sunampithoë pelagica		2	1	1	2	1	13	3	2	2	14
74 Biancolina sp.	1?	1		2		2		1			
76 Hemiaegina minuta		44			18					7	1
80 Latreutes fucorum	1	1	24	9	3	10	15	8	13	6	5
82 Leander tenuicornis	1	1	1	3		4					5
83 Planes minutus	5	6	14	12	4	34	7	16	4	14	6
84 Portunus sayi						1					
87 Anoplodactylus petiolatus	8	11	52	275	6	31		53		3	5
89 Litiopa melanostoma	5	2			1	3		2	7	4	4
95 Scyllaea pelagica						2	1			2	2
100 Histrio histrio											
101 Syngnathus pelagicus											
Other Species:	22	13	7	24	5	34	3	30	3	3	11
Total Number of Specimens:	80	178	196	386	115	230	92	260	78	98	207

Note: F = *S. fluitans*; N = *S. natans*

MOTILE MACROFAUNA ON SARGASSUM

	Station	390			391					
	Date	7 Jan 74			21 Jan 74					
	Sample No.	4	5	6	1	2	3	4	5	6
	Wet Weight, g									
	Displacement vol. ml.	90	50	75	90	105	40	45	40	30
	Sargassum species	F+N	N	N	N	N	N	N	F	N+F
	Fauna:									
44	Anemonia sargassensis	2	1				1	15	10	1
46	Acerotisa notulata	2								
47	Hoploplana grubei				2					2
48	Gnescioceros sargassicola	49	14	16	35	15	8	29	21	14
51	Platynereis dumerillii	21	1		4	10	2	26	17	62
70	Bagatus minutus	176	12	3		1	5	37	32	34
73	Sunampithoë pelagica	26		1	1		1	14	3	
74	Biancolina sp.									3
76	Hemiaegina minuta	160	30					1	19	6
80	Latreutes fucorum	16	7	5	5	4	1	1	1	
82	Leander tenuicornis									
83	Planes minutus	3	4	8	4			2	1	1
84	Portunus sayi									1
87	Anoplodactylus petiolatus	2	3	2	2		4	13	1	
89	Litiopa melanostoma		1		97	14	39	8	7	9
95	Scyllaea pelagica						1			
100	Histrio histrio									
101	Syngnathus pelagicus									
	Other Species:	11	2	1				1	1	
	Total Number of Specimens:	468	75	36	150	44	62	147	113	133

Note: F = _S. fluitans_; N = _S. natans_

MOTILE MACROFAUNA ON SARGASSUM

Station		392					
Date		28 Jan 74					
Sample No.		1	2	3	4	5	6
Wet weight, g							
Displacement vol. ml.		52	68	50	35	25	37
Sargassum species		N	N	N	N	N	N
	Fauna:						
44	Anemonia sargassensis		2	8		1	
46	Acerotisa notulata						
47	Hoploplana grubei	1	1				
48	Gnescioceros sargassicola	12	12	9	5	3	13
51	Platynereis dumerillii	3		23			2
70	Bagatus minutus	25	13	54	14	1	60
73	Sunampithoë pelagica	2	3			1	2
74	Biancolina sp.						
76	Hemiaegina minuta			36	2		2
80	Latreutes fucorum	9	8		2		1
82	Leander tenuicornis						1
83	Planes minutus	4	2	1		2	1
84	Portunus sayi						
87	Anoplodactylus petiolatus		26	1	1	2	11
89	Litiopa melanostoma	3	1	3	6		
95	Scyllaea pelagica		1				
100	Histrio histrio						
101	Syngnathus pelagicus						
	Other Species:	2	6	6		?	3
Total Number of Specimens:		61	75	141	30	10	96

Note: F = *S. fluitans*; N = *S. natans*

MOTILE MACROFAUNA ON SARGASSUM

Station		393										
Date		2 Mar 74										
Sample No.		1	2	3	4	5	6	7	8	9	10	11	12
Wet weight, g													
Displacement vol. ml.		160	125	100	100	85	65	55	40	50	10	10	5
Sargassum species		F+N	N	F+N	N	N	N	N	N	N	N	N	N
	Fauna:												
44	Anemonia sargassensis												
46	Acerotisa notulata												
47	Hoploplana grubei	4	1	2	2	4	1	2	2	3		1	1
48	Gnescioceros sargassicola	11	5	5	10	5	4		1	10	1		
51	Platynereis dumerillii	15	1	4	23	3			1	26			
70	Bagatus minutus	263	52	58	108	132	55	3	19	68	9	1	
73	Sunampithoë pelagica	44	36	27	48	64	12	14	41	5	2	2	4
74	Biancolina sp.		1		1								
76	Hemiaegina minuta	15	2	4	2	15	3			1			
80	Latreutes fucorum	7	7	5	2	2	4			2			
82	Leander tenuicornis	1											
83	Planes minutus	1		1					1				
84	Portunus sayi												
87	Anoplodactylus petiolatus	7	23	8	6	19	5	13	6	5		2	
89	Litiopa melanostoma	9	112	39	62	136	74	106	20	10	5	23	12
95	Scyllaea pelagica	1	8	2	4	2		1		2	1		
100	Histrio histrio												
101	Syngnathus pelagicus												
	Other Species:	9	13	8	6	6	14	198	3	1	3	1	
Total Number of Specimens:		387	261	163	274	388	172	337	94	133	21	30	19

Note: F = *S. fluitans*; N = *S. natans*

MOTILE MACROFAUNA ON SARGASSUM

Station	394				398?						
Date	23 Apr 74				11 Feb 75						
Sample No.	1	2	3	4	1	2	3	4	5	6	7
Wet weight, g											
Displacement vol. ml.	230	130	70	40	25	68	65	15	65	50	85
Sargassum species	N	N	N	N	N	N	F+N	N	N	F?	N
Fauna:											
44 Anemonia sargassensis		2					1		3		1
46 Acerotisa notulata	14										
47 Hoploplana grubei	4	3		1							
48 Gnescioceros sargassicola	1	40	10	5	2	5	7		2	4	4
51 Platynereis dumerillii	14	28	10			15	4		19	14	8
70 Bagatus minutus	407	535	110	33	1	22	30	1	39	21	22
73 Sunampithoë pelagica	382	159	150	46	1	10	9	1	14	13	23
74 Biancolina sp.											
76 Hemiaegina minuta	1	23				27	14	2	5?	7	
80 Latreutes fucorum	28	18	7			5	4		2	9	10
82 Leander tenuicornis	1	1									
83 Planes minutus	6	3	3	2			1		1	1	
84 Portunus sayi											
87 Anoplodactylus petiolatus		1	7		4				5		58
89 Litiopa melanostoma	46	37	18	2	2	5	1	1	8		8
95 Scyllaea pelagica	8	9	2	2				1	1		
100 Histrio histrio	1										
101 Syngnathus pelagicus											
Other Species:	27	71	8	1	5	4	15		6	5	19
Total Number of Specimens:	940	930	325	92	15	93	86	6	104	74	153

Note: F = *S. fluitans*; N = *S. natans*

MOTILE MACROFAUNA ON SARGASSUM

	Station	399					
	Date	26 Feb 75					
	Sample No.	1	2	3	4	5	6
	Wet weight, g						
	Displacement vol. ml.	60	252	98	73	170	175
	Sargassum species	N	N	N	F	F	N
	Fauna:						
44	Anemonia sargassensis	2					131
46	Acerotisa notulata						
47	Hoploplana grubei						
48	Gnescioceros sargassicola	2	1	53	1		10
51	Platynereis dumerillii						1
70	Bagatus minutus	2		20	2	5	8
73	Sunampithoë pelagica						
74	Biancolina sp.			2		1	2
76	Hemiaegina minuta						
80	Latreutes fucorum	1	2	6	1		8
82	Leander tenuicornis		3	3			2
83	Planes minutus		1				
84	Portunus sayi						
87	Anoplodactylus petiolatus	2	1	15	5	13	31
89	Litiopa melanostoma	92	12	40	16	175	12
95	Scyllaea pelagica	1		1		2	
100	Histrio histrio						
101	Syngnathus pelagicus						
	Other Species:	13	3	8		80	5
	Total Number of Specimens:	115	23	148	25	276	210

Note: F = *S. fluitans*; N = *S. natans*

MOTILE MACROFAUNA:
Notes to tables of species counts

Station	Sample	Observations
346	1	Platynereis dumerilli: 85 juveniles 0.4-15 mm, 1 adult (25 mm with tube) ; Bopyrus sp.: 1 ♂ Bagatus minutus: 44 all sizes, with eggs Latreutes fucorum: 1 adult, 7 juv
	2	Platynereis dumerilli: 27 adults, 21 juv. One adult found in washed sample in mucous sac on Sargassum stem. Bagatus minutus: 15 juv, 2 adults with eggs Anoplodactylus petiolatus: 2 adults, 1 with eggs Corambella depressa: 3 ("uncommon")
	3	Platynereis dumerilli: 1 adult, 21 juv Bagatus minutus: 8 of 27 adults with eggs, 111 juv Hemiaegina minuta: 6 adult, 6 juv Latreutes fucorum: 1 of 11 with eggs Planes minutus: 1 adult, 1 juv Endeis spinosa: 1 ("common" according to literature, but not in these samples) Anoplodactylus petiolatus: 2 of 21 with eggs Larval brittle star: 2
	4	Platynereis dumerilli: 5 juv Bagatus minutus: 3 adults, 9 juv Anoplodactylus: 3 of 8 adults with eggs, 2 juv unidentified nudibranch "C": 3
	5	Platynereis dumerilli: 2 juv parasitic isopod (Probopyrinella latreuticola?): 1 Latreutes fucorum: 2 juv
347	1	Acerotisa notulata: 2 juv Gnesioceros sargassicola: 6 adults, 1 juv Hemiaegina minuta: 1 with eggs Latreutes fucorum: 2 adult, 1 juv
	2	Gnesioceros sargassicola: 3 adults, 3 juv Hemiaegina minuta: 1 of 5 adults with eggs, 9 juv
	3	Gnesioceros sargassicola: 5 adults, 3 juv parasitic isopod: 1
	4	Bagatus minutus: 16 adults, 47 juv Leander tenuicornis: 1 adult with eggs Scyllaea pelagica: 3 juv

Station	Sample	Observations
347	5	Platynereis dumerilli: 2 adults, 50 juv Bagatus minutus: 3 of 196 with eggs Latreutes fucorum: 2 adults with eggs, 11 juv Leander tenuicornis: 1 of 2 adults with eggs Planes minutus: 4 of 7 adults with eggs, 3 juv Anoplodactylus petiolatus: 3 of 71 with eggs
	6	Platynereis dumerilli: 1 juv Bagatus minutus: 6 adults, 16 juv, no eggs Parasitic isopod(Probopyrinella latreuticola on Latreutes or Grapsicepon edwardsi on Planes?): 2 Biancolina sp.: 1 juv Latreutes fucorum: 6 juv Leander tenuicornis: 1 juv Planes minutus: 1 juv Scyllaea pelagica: 2 juv
	7	Platynereis dumerilli: 2 adults, 35 juv Bagatus minutus: 6 adults, 15 juv Hemiaegina minuta: 15 adults, 3 juv Planes minutus: 1 adult, 4 juv Scyllaea pelagica: 1 adult, 3 juv
	8	Platynereis dumerilli: 9 adults, 10 juv Bagatus minutus: 1 of 3 adults with eggs, 14 juv Sunampithoë pelagica: 1 adult, 2 juv Latreutes fucorum: 7 juv Scyllaea pelagica: 13 juv
	9	Gnesioceros sargassicola: 7 adults, 4 juv Platynereis dumerilli: 4 juv Bagatus minutus: 1 of 7 adults with eggs, 9 juv Anoplodactylus petiolatus: 2 of 10 adults with eggs, 2 juv Scyllaea pelagica: 2 juv unidentified nudibranch "C": 1 (juv. Spurilla? see 372-5)
	10	Bagatus minutus: 4 of 9 adults with eggs, 67 juv parasitic isopod (Grapsicepon edwardsi?): 3 Hemiaegina minuta: 23♀ with eggs, 10 ♂, 42 juv Planes minutus: 1 of 3 adults with eggs Anoplodactylus petiolatus: 5 juv Scyllaea pelagica: 5 adults, 1 juv

Station	Sample	Observations
348	1	Acerotisa notulata: size 1.0-1.4 mm Gnecioseros sargassicola: size 1.0-4.0 mm Platynereis dumerilli: 2 adults, 52 juv, 0.4-11 mm Bagatus minutus: 3 ♀ with eggs, 0.7-1.9 mm Sunampithoë pelagica: 50 juv 2.2-4.5 mm, some ♀ w/eggs? Biancolina sp.: 9 ♀ w/eggs 1.2-2.0 mm, 9 juv Hemiaegina minuta: size 1.0-3.4 mm (♀ ♂ j) Latreutes fucorum: 2 ♀ w/eggs, 1 juv Portunus sayi: 1 ♀ w/eggs, carapace 24x47 mm Litiopa melanostoma: 3 (2.5, 3.8, 4.0 mm)
	2	Gnescioceros sargassicola: 9 juv Bagatus minutus: 1 juv Hemiaegina minuta: 1 adult, 2 juv Latreutes fucorum: 2 juv unidentified nudibranch "C": 1 (juv. Spurilla? see 372-5)
	3	no notes
	4	Platynereis dumerilli: 60 juv parasitic isopod: 2 ♂ Hemiaegina minuta: 3 adults (1 w/eggs), 9 juv Latreutes fucorum: 1 juv
	5	Anemonia sargassensis: 4 juv Platynereis dumerilli: 62 juv Bagatus minutus: 4 adults, 4 juv Hemiaegina minuta: 1 adult Latreutes fucorum: 2 adults(1 w/eggs), 1 juv Leander tenuicornis: 3 adults (1 w/eggs), 1 juv
	6	Platynereis dumerilli: 13 juv Bagatus minutus: 2 adults, 1 juv Hemiaegina minuta: 8 of 18 adults w/eggs
	7	Platynereis dumerilli: 1 juv Biancolina sp.: 2 adults, 1 juv
349	1	Gnescioceros sargassicola: 3 adults, 5 juv Platynereis dumerilli: 17 juv, 3 adults Latreutes fucorum: 2 juv Leander tenuicornis: 1 adult Planes minutus: 1 juv Portunus sayi: 1 juv Anoplodactylus petiolatus: 4 (1 w/eggs)
	2	Gnescioceros sargassicola: 2 adults, 4 juv Platynereis dumerilli: 7 adults, 24 juv Bagatus minutus: 4 juv Biancolina sp.: 2 adults, 2 juv Hemiaegina minuta: 1 adult, 4 juv Latreutes fucorum: 1 juv

Station	Sample	Observations
349	3	Platynereis dumerilli: 3 juv Portunus sayi: 1 juv
	4	Platynereis dumerilli: 18 juv Bagatus minutus: 1 adult w/eggs, 1 juv Hemiaegina minuta: 9 of 178 with eggs Leander tenuicornis: 1 juv
353	1	Latreutes fucorum: 1 juv Planes minutus: 3 juv
	2	Platynereis dumerilli: 5 juv Leander tenuicornis: 1 juv Portunus sayi: 1 juv
	3	Bagatus minutus: 1 of 5 w/eggs Hemiaegina minuta: 4 of 17 w/eggs (♀ ♂ j)
	4	Platynereis dumerilli: 17 adults, 19 juv Latreutes fucorum: 1 adult ♀ w/eggs, 1 juv
	5	Acerotisa notulata: 2 juv Platynereis dumerilli: 4 juv Bagatus minutus: 2 juv Latreutes fucorum: 4 adults w/eggs, 2 juv Leander tenuicornis: 1 adult, 1 juv Planes minutus: 1 adult, 3 juv Portunus sayi: 1 juv
	6	no notes
	7	Bagatus minutus: 1 adult, 5 juv Latreutes fucorum: 1 adult w/eggs, 3 juv Leander tenuicornis: 1 adult w/eggs Spirorbis common
354	1	Some of the samples from station 354 may consist of more than one clump of Sargassum. Sample 1 was 17 ml of S. fluitans and 3 ml of S. natans. Platynereis dumerilli: 2 adults, 33 juv Bagatus minutus: 11 adults (1 w/eggs), 4 juv Hemiaegina minuta: 3 (1 w/eggs) Scyllaea pelagica: 2 juv
	2	Probopyrinella latreuticola: 1 on L. fucorum host (juv) Planes minutus: 6 juv Portunus sayi: 1 juv nudibranch "A": 1 (Doto pygmaea? See 375-2)

Station	Sample	Observations
354	3	Platynereis dumerilli: 2 juv Biancolina sp.: 1 juv Hemiaegina minuta: 1 adult, 6 juv Planes minutus: 1 juv Anoplodactylus petiolatus: 115 (2 w/eggs) Abudefduf saxatilis: 1 juv ("transient": see Dooley 1972)
	4	Bagatus minutus: 12 adult (3 w/eggs), 13 juv Hemiaegina minuta: 4 (2 w/eggs) Leander tenuicornis: 2 w/eggs Scyllaea pelagica: 1 juv
	5	Platynereis dumerilli: 1 adult, 27 juv
355	1	Platynereis dumerilli: 21 adults, 17 juv Bagatus minutus: 8 adults (3 w/eggs), 12 juv Probopyrinella latreuticola (on L. fucorum host): 1 ♀ w/eggs, 1 ♂ Biancolina sp.: 4 (1 w/eggs) Hemiaegina minuta: 3 (1 w/eggs) Latreutes fucorum: 7 adults (all w/eggs), 5 juv (1 with parasitic isopod P. latreuticola) Leander tenuicornis: 1 juv Planes minutus: 2 juv Anoplodactylus petiolatus: 117 (16 w/eggs) Scyllaea pelagica: 1 juv unidentified nudibranch "A": 3 (Doto pygmaea? See 375-2) unidentified nudibranch "B": 6
	2	Platynereis dumerilli: 1 adult, 6 juv Bagatus minutus: 2 adults (1 w/eggs), 2 juv Hemiaegina minutus: 2 adults, 4 juv Leander tenuicornis: 1 adult ♀ w/eggs Anoplodactylus petiolatus: 22 (2 w/eggs) Scyllaea pelagica: 1 juv
	3	Bagatus minutus: 1 adult, 4 juv Lepas anatifera: 1 adult, 2 juv ("uncommon") Latreutes fucorum: 2 juv Doto sp.?: 2 ("common" but not in these samples) Scyllaea pelagica: 1 juv unidentified mollusc "slug": 1
	4	Platynereis dumerilli: 2 adults, 26 juv Lepas anatifera: 4 adults, 24 juv ("uncommon") Bagatus minutus: 2 juv Anatanais nordmani: 7 ("rare", not reported in literature) Biancolina sp: 2 juv Hemiaegina minuta: 2 juv Latreutes fucorum: 4 adults w/eggs, 38 juv Anoplodactylus petiolatus: 278 (15 w/eggs)

Station	Sample	Observations
355	4 (cont.)	Litiopa melanostoma: 4 adults, 19 juv Scyllaea pelagica: 3 juv unidentified nudibranch "A": 4 (Doto? See 375-2) unidentified nudibranch "B": 2 unidentified nudibranch "C": 3 Histrio histrio: 1 juv
	5	Probopyrinella latreuticola: 2 ♀, 2♂ on L. fucorum (eggs in sac) unidentified isopod "crescent moon shape": 2 Latreutes fucorum: 2 adults w/eggs, 4 juv Anoplodactylus petiolatus: 28 (1 w/eggs) unidentified nudibranch "A": 2
	6	Bagatus minutus: 2 juv Leander tenuicornis: 5 adults (2 w/eggs) Planes minutus: 1 juv unidentified nudibranch "A": 3 unidentified nudibranch "C": 3
356	1	Anoplodactylus petiolatus: 11 adults (1 w/eggs), 3 juv unidentified nudibranch "A": 5
	2	unidentified polyclad flatworm: 1 (possibly G. sargassicola? 8 of these observed) unidentified isopod: 1 ♀ Sunampithoë pelagica: 8 adults, 5 juv Latreutes fucorum: 2 adults w/eggs Anoplodactylus petiolatus: 12 (1 w/eggs) unidentified nudibranch "A": 8 adults, 1 juv unidentified ostracods: 2
	3	mixed 50% S. natans and 50% S. fluitans. Fauna not counted.
	4	mixed Sargassum species. Fauna not counted.
	5	Platynereis dumerilli: 2 juv Bagatus minutus: 2 adults, 5 juv Sunampithoë pelagica: 1 adult, 6 juv Leander tenuicornis: 2 adults Anoplodactylus petiolatus: 52, 1 w/eggs unidentified nudibranch "A": 2 adults, 8 juv unidentified nudibranch "B": 1 unidentified nudibranch "C": 5

Station	Sample	Observations
356	6	Platynereis dumerilli: 2 juv Bagatus minutus: 23 (1 w/eggs) Sunampithoë pelagica: 1 adult, 1 juv Anoplodactylus petiolatus: 28 (2 w/eggs) Litiopa melanostoma: 2 adults, 1 juv unidentified isopod "crescent moon shape": 1 unidentified nudibranch "A": 1 adult, 20 juv unidentified nudibranch "B": 1 unidentified nudibranch "C": 6
	7	Bagatus minutus: 1 juv Sunampithoë pelagica: 2 juv
	8	Platynereis dumerilli: 1 juv Bagatus minutus: 1 juv Anoplodactylus petiolatus: 4 (1 adult w/eggs) 2 unidentified animals
	9	Bagatus minutus: 10 (2 w/eggs) Latreutes fucorum: 1 adult w/eggs Anoplodactylus petiolatus: 56 (8 w/eggs) unidentified nudibranch "A": 1 unidentified nudibranch "C": 7
	10	Bagatus minutus: 3 juv Anoplodactylus petiolatus: 1 adult, 8 juv unidentified nudibranch "A": 1
	11	Platynereis dumerilli: 12 adult, 108 juv Bagatus minutus: 7 adult (3 w/eggs), 33 juv Sunampithoë pelagica: 9 adult, 11 juv Latreutes fucorum: 5 (3 w/eggs) Anoplodactylus petiolatus: 45 adults, 140 juv Scyllaea pelagica: 2 adults 1 juv unidentified nudibranch "A": 2 adults, 19 juv unidentified nudibranch "B": 1 adult unidentified nudibranch "C": 2 adults, 25 juv
	12	Gnescioceros sargassicola: 2 adults, 3 juv Bagatus minutus: 7 adults (3 w/eggs), 6 juv Sunampithoë pelagica: 1 adult, 1 juv Latreutes fucorum: 1 w/eggs Anoplodactylus petiolatus: 10 (2 w/eggs)
	13	Platynereis dumerilli: 1 adult, 1 juv Bagatus minutus: 6 adults (1 w/eggs), 9 juv Sunampithoë pelagica: 4 adults, 6 juv Anoplodactylus petiolatus: 13 juv unidentified nudibranch "A": 2 adults

Station	Sample	Observations
358	1	Platynereis dumerillii: 9 adults, 7 juv Bagatus minutus: 5 A w/eggs, 11 juv Sunampithoë pelagica: 70% juv Hemiaegina minuta: 44 adults, 13 juv Unidentified nudibranch A: 32 adults, 11 juv
	2	Idotea metallica (?): 1 Bagatus minutus: 1 adult, 1 juv Sunampithoë pelagica: 23 adults, 53 juv Unidentified isopod: 1 Unidentified nudibranch A: 9 adult, 3 juv
	3	Platynereis dumerillii: 2 adults, 3 juv Bagatus minutus: 2 adults, 9 juv Sunampithoë pelagica: 25 adults, 55 juv Biancolina sp.: 1 adult, 1 juv Anoplodactylus petiolatus: 2 juv Litiopa melanostoma: 1 juv Unidentified isopod: 12 Unidentified nudibranch A: 21 adult, 18 juv Unidentified nudibranch B: 1
	4	Platynereis dumerillii: 2 adult, 1 juv Bagatus minutus: 29 adults, 7 w/eggs, 77 juv (some very small, may be in meiofauna) Sunampithoë pelagica: 22 adults, 7 w/eggs, 61 juv (some very small, may be in meiofauna) Biancolina sp.: 1 juv Planes minutus: 1 adult w/eggs Anoplodactylus petiolatus: 21, 2 w/eggs Litiopa melanostoma: 3 adults, 2 juv Scyllaea pelagica: 1 juv Unidentified isopod: 5 Unidentified nudibranch A: 108 Unidentified nudibranch C: 1
	5	Gnescioceros sargassicola: 2 adult, 3 juv Sunampithoë pelagica: 8 adult, 9 juv Latreutes fucorum: 1 adult w/eggs, 4 juv Anoplodactylus petiolatus: 1 adult, 3 juv Unidentified isopod: 6 Unidentified nudibranch A: 156, 50% juv
	6	Platynereis dumerillii: 4 adults, 12 juv Bagatus minutus: 21 adults, 39 juv Sunampithoë pelagica: 19 adults, 94 juv Anoplodactylus petiolatus: 43, 8 w/eggs Litiopa melanostoma: 1 adult, 2 juv Corambella depressa: 6 (3 adults, 3 juv) Unidentified isopod: 4 Unidentified nudibranch A: 173 (60% juv)

Station	Sample	Observations
358	7	Bagatus minutus: 3 adults, 2 juv Sunampithoë pelagica: 18 adults, 31 juv Litiopa melanostoma: 4 juv Histrio histrio: 1 juv Unidentified isopod: 24 Unidentified nudibranch A: 89
	8	Platynereis dumerillii: 9 juv Bagatus minutus: 1 adult, 5 juv Sunampithoë pelagica: 7 adults, 21 juv Biancolina sp.: 1 adult, 1 juv Latreutes fucorum: 1 juv Planes minutus: 1 juv Anoplodactylus petiolatus: 26 (2 w/eggs) Corambella depressa: 1 Scyllaea pelagica: 1 juv Unidentified polychaete: 1 adult Unidentified isopod: 1 Unidentified nudibranch A: 12 adult, 25 juv Unidentified nudibranch C: 1
360	1	Platynereis dumerilli: 3 adults, 3 juv Bagatus minutus: 1 adult w/eggs, 9 juv Sunampithoë pelagica: 21 adults, 26 juv Biancolina sp.: 1 adult, 1 juv Planes minutus: 1 adult w/eggs, 1 juv Anoplodactylus petiolatus: 1 adult w/eggs, 1 juv Corambella depressa: 2 juv Unidentified isopod: 4
	2	Platynereis dumerillii: 1 juv Bagatus minutus: 1 adult w/eggs, 1 juv Sunampithoë pelagica: 7 juv Biancolina sp.: 3 juv Corambella depressa: 2
	3	Sunampithoë pelagica: 2 adults, 6 juv Biancolina sp.: 1 juv Planes minutus: 1 juv Corambella depressa: 2
	4	Platynereis dumerillii: 1 adult, 6 juv Idotea metallica: 1 Bagatus minutus: 2 adults (1 w/eggs), 2 juv Sunampithoë pelagica: 31 adults, 74 juv Biancolina sp.: 4 adults, 2 juv Latreutes fucorum: 1 juv Planes minutus: 2 adults Anoplodactylus petiolatus: 5 (2 w/eggs) Corambella depressa: 12 Unidentified isopod: 21 Unidentified nudibranch C: 1

Station	Sample	Observations
361	1	Platynereis dumerillii: 4 juv Bagatus minutus: 3 adults (1 w/eggs) 3 juv Sunampithoë pelagica: 46 adults, 134 juv Biancolina sp.: 6 adults, 8 juv Planes minutus: 2 juv Unidentified isopod: 5 unidentified nudibranch A: 1 juv
	2	Platynereis dumerillii: 6 juv Bagatus minutus: 18 adults (5 w/eggs) 9 juv Sunampithoë pelagica: 20 adults, 54 juv Biancolina sp.: 5 adults, 5 juv Planes minutus: 3 juv Portunus sayi: 1 juv Janthina janthina (?): 1 juv Scyllaea pelagica: 2 juv Histrio histrio: 1 juv Unidentified isopod: 7
	3	Platynereis dumerillii: 9 adults, 34 juv Bagatus minutus: 64 adults (12 w/eggs) 102 juv Sunampithoë pelagica: 47 adults, 96 juv Biancolina sp. 4 adults, 8 juv Latreutes fucorum: 1 juv Anoplodactylus petiolatus: 55 (2 w/eggs) Corambella depressa: 1 Planes minutus: 6 adults (3 w/eggs) 2 juv Scyllaea pelagica: 2 adults, 1 juv Unidentified isopod: 1 Unidentified nudibranch C: 2
	4	Platynereis dumerillii: 4 adults, 4 juv Bagatus minutus: 3 adults (2 w/eggs) 2 juv Sunampithoë pelagica: 19 adults, 21 juv Planes minutus: 1 adult, 1 juv Anoplodactylus petiolatus: 1 juv Scyllaea pelagica: 2 juv Unidentified isopod: 2
	5	Platynereis dumerillii: 11 juv Bagatus minutus: 7 adults, 13 juv Sunampithoë pelagica: 33 adults, 102 juv Biancolina sp.: 1 adult, 1 juv Latreutes fucorum: 4 juv Planes minutus: 2 juv Anoplodactylus petiolatus: 45 juv Scyllaea pelagica: 2 juv Histrio histrio: 1 juv Probopyrinella latreuticola: 2, 1 ♂, 1 ♀ w/eggs on host Latreutes fucorum Unidentified isopod: 25 Unidentified nudibranch B: 1

Station	Sample	Observations
361	6	Platynereis dumerillii: 5 juv Bagatus minutus: 6 adults (3 w/eggs) 2 juv Sunampithoë pelagica: 20 adults, 56 juv Latreutes fucorum: 1 juv Planes minutus: 2 juv Unidentified isopod: 11 Unidentified nudibranch C: 1
	7	Platynereis dumerillii: 2 juv Bagatus minutus: 2 adults (1 w/eggs) 1 juv Sunampithoë pelagica: 10 adults, 45 juv Planes minutus: 1 juv Anoplodactylus petiolatus: 1 juv Corambella depressa: 1 juv
	8	Platynereis dumerillii: 4 juv Bagatus minutus: 3 adults, 2 juv Sunampithoë pelagica: 30 adults, 161 juv Biancolina sp.: 3 adults, 4 juv Planes minutus: 3 juv Anoplodactylus petiolatus: 23 adults, 12 juv Corambella depressa: 2 juv Unidentified isopod: 13 Unidentified nudibranch C: 4
364	1	Gnescioceros sargassicola: 16 juv Platynereis dumerillii: 10 juv Planes minutus: 6 juv Anoplodactylus petiolatus: 1 w/eggs, 1 other Litiopa melanostoma: 1 adult, 1 juv Scyllaea pelagica: 1 juv
	2	Platynereis dumerillii: 9 juv Sunampithoë pelagica: 1 adult, 1 juv Planes minutus: 1 adult w/eggs
	3	Gnescioceros sargassicola: 26 juv Platynereis dumerillii: 12 juv Sunampithoë pelagica: 4 adults, 2 juv Biancolina sp.: 1 adult, 1 juv Planes minutus: 3 juv Litiopa melanostoma: 4 adults, 2 juv
	4	Gnescioceros sargassicola: 9 juv Platynereis dumerillii: 2 juv Biancolina sp.: 2 adults, 2 juv Planes minutus: 1 juv Anoplodactylus petiolatus: 1 juv
	5	Gnescioceros sargassicola: 6 juv Platynereis dumerillii: 1 juv Planes minutus: 3 juv Anoplodactylus petiolatus: 1 juv Scyllaea pelagica: 1 juv

Station	Sample	Observations
364	6	Gnescioceros sargassicola: 17 juv Biancolina sp.: 1 juv Latreutes fucorum: 1 juv Planes minutus: 1 adult w/eggs, 4 juv Anoplodactylus petiolatus: 2 juv Litiopa melanostoma: 2 juv Scyllaea pelagica: 1 juv
364A	1	Platynereis dumerillii: 1 adult, 13 juv Planes minutus: 2 juv Portunus sayi: 1 juv Unidentified nudibranch B: 1 juv Litiopa melanostoma (minute?): 30
	2	Amphiscolops sargassi (?): 1 juv Gnescioceros sargassicola: 2 juv Litiopa melanostoma (minute?): 38 (1 normal adult)
	3	Platynereis dumerillii: 1 adult, 29 juv Bagatus minutus: 3 adults, 7 juv Sunampithoë pelagica: 9 adults, 3 juv Biancolina sp.: 1 juv Latreutes fucorum: 2 adults w/eggs, 3 juv Planes minutus: 3 adults (2 w/eggs) 3 juv Anoplodactylus petiolatus: 4 juv Litiopa melanostoma (minute?) approx. 200, incl. 5 normal adults Unidentified nudibranch C: 1
365	1	Platynereis dumerillii: 22 juv Biancolina sp.: 1 adult, 1 juv Anoplodactylus petiolatus: 3 adults (1 w/eggs) 31 juv Planes minutus: 2 juv Litiopa melanostoma: 1 adult, 1 juv Unidentified nudibranch A: 3 (Doto? See 375-2) Unidentified nudibranch C: 1
	2	Platynereis dumerillii: 5 juv Sunampithoë pelagica: 3 adults, 2 juv Latreutes fucorum: 1 adult w/eggs, 4 juv Anoplodactylus petiolatus: 4 juv Scyllaea pelagica: 2 juv Unidentified anemone: 5 Unidentified nudibranch A: 2 juv
	3	Platynereis dumerillii: 21 juv Biancolina sp.: 1 juv Hemiaegina minuta: 18 juv Latreutes fucorum: 1 adult w/eggs, 1 juv Planes minutus: 2 juv Anoplodactylus petiolatus: 2 juv Litiopa melanostoma: 5 adults, 53 juv

Station	Sample	Observations
365	3	Scyllaea pelagica: 1 adult, 16 juv Histrio histrio: 2 juv Unidentified nudibranch A: 2 Unidentified nudibranch B: 1 Unidentified nudibranch C: 3
	4	Gnescioceros sargassicola: 3 juv Biancolina sp.: 1 juv Planes minutus: 1 juv Anoplodactylus petiolatus: 2 juv Litiopa melanostoma: 1 adult, 38 juv Scyllaea pelagica (?):7 juv Unidentified nudibranch C: 1
366	1	Platynereis dumerillii: 7 adults, 22 juv Biancolina sp.: 3 adults, 2 juv Anoplodactylus petiolatus: 3 adults, 3 juv (2 ♀ w/eggs) Litiopa melanostoma: 3 adults, 13 juv Latreutes fucorum: 5 adults, 2 juv (4 w/eggs)
	2	Latreutes fucorum: 1 adult w/eggs Litiopa melanostoma: 4 juv Doto pygmaea (?): 2
	3	Platynereis dumerillii: 5 adults, 6 juv Bagatus minutus: 1 juv Latreutes fucorum: 1 adult w/eggs Planes minuts: 1 juv Litiopa melanostoma: 6 adults, 6 juv Doto pygmaea (?): 1
	4	Platynereis dumerillii: 22 adults, 26 juv Sunampithoë pelagica: 2 adults, 2 juv Hemiaegina minuta: 25, incl 2 ♀ w/eggs Anoplodactylus petiolatus: 7 incl 1 w/eggs Litiopa melanostoma: 2 adults, 11 juv Doto pygmaea (?): 3
	5	Platynereis dumerillii: 8 adults, 31 juv Bagatus minutus: 3 adults, 3 juv Biancolina sp.: 4 adults, 8 juv Latreutes fucorum: 8 incl 5 ♀ w/eggs Planes minutus: 1 juv Portunus sayi: 1 juv Litiopa melanostoma: 1 adult, 7 juv
367	1	Platynereis dumerillii: 12 juv Hemiaegina miniuta: 2 adults, 1 juv Latreutes fucorum: 4 adults, 6 juv (2 ♀ w/eggs

Station	Sample	Observations
367	2	Platynereis dumerillii: 8 adults, 106 juv Bagatus minutus: 3 adults, 1 juv Hemiaegina minuta: 2 juv Latreutes fucorum: 6 adults (2 ♀ w/eggs), 6 juv
	3	Platynereis dumerillii: 4 adults, 17 juv Bagatus minutus: 14 incl 3 ♀ w/eggs Latreutes fucorum: 14 incl 5 ♀ w/eggs Planes minutus: 3 juv
	4	Platynereis dumerillii: 10 adules, 59 juv Bagatus minutus: 7 adults, 24 juv (1 ♀ w/eggs) Hemiaegina minuta: 2 juv Latreutes fucorum: 2 incl 1 ♀ w/eggs Planes minutus: 3 juv Litiopa melanostoma: 1 juv
	5	Platynereis dumerillii: 19 adults, 84 juv Bagatus minutus: 72 incl 8 ♀ w/eggs Latreutes fucorum: 3 ♀ w/eggs Litiopa melanostoma: 1 adult, 1 juv Unidentified nudibranch B: 1
	6	Platynereis dumerillii: 4 adults, 27 juv
368	1	Platynereis dumerillii: 4 juv Biancolina sp.: 1 adult, 1 juv Latreutes fucorum: 2 juv Portunus sayi: 1 juv Litiopa melanostoma: 1 adult, 2 juv
	2	Litiopa melanostoma: 1 adult, 3 juv
	3	Platynereis dumerillii: 10 adults, 41 juv Bagatus minutus: 19 incl 3 ♀ w/eggs Hemiaegina minuta: 8 adults, 76 juv Latreutes fucorum: 2 juv Planes minutus: 1 juv Unidentified nudibranch C: 1
	4	Platynereis dumerillii: 3 juv Litiopa melanostoma: 1 juv
	5	Platynereis dumerillii: 13 juv Bagatus minutus: 2 juv Latreutes fucorum: 2 juv Portunus sayi: 1 juv Litiopa melanostoma: 3 juv

Station	Sample	Observations
369	1	Platynereis dumerillii: 1 adult, 30 juv Hippolyte coerulescens: 1 ♀ w/eggs Latreutes fucorum: 12 incl 4 w/eggs Planes minutus: 16 juv Portunus sayi: 1 juv
	2	Platynereis dumerillii: 1 adult, 15 juv Hemiaegina minuta: 1 adult, 5 juv Planes minutus: 2 juv Portunus sayi: 1 juv
	3	Bagatus minutus: 4 adults incl 1 ♀ w/eggs Planes minutus: 2 juv Portunus sayi: 1 juv Latreutes fucorum: 2 juv Anoplodactylus petiolatus: 2 adults w/eggs Spurilla neapolitana: 3
	4	Latreutes fucorum: 2 incl 1 ♀ w/eggs Planes minutus: 1 juv Anoplodactylus petiolatus: 15 juv Spurilla neapolitana: 1
	5	Platynereis dumerillii: 2 adults, 9 juv Bagatus minutus: 1 ♀ w/eggs, 2 juv Latreutes fucorum: 1 juv Planes minutus: 2 juv Anoplodactylus petiolatus: 2 incl 1 ♀ w/eggs
370	1	Platynereis dumerillii: 4 adults, 40 juv Bagatus minutus: 1 adult, 1 juv Hemiaegina minuta: 2 juv Latreutes fucorum: 6 juv Leander tenuicornis: 1 juv Planes minutus: 1 juv Anoplodactylus petiolatus: 4 incl 2 ♀ w/eggs
	2	Bagatus minutus: 7 adults 8 juv Latreutes fucorum: 6 incl 1 ♀ w/eggs Planes minutus: 2 juv Unidentified nudibranch C: 1
	3	Hemiaegina minuta: 11 incl 1 ♀ w/eggs Hippolyte coerulescens: 1 ♀ w/eggs Anoplodactylus petiolatus: 3 incl 1 ♀ w/eggs Spurilla neopolitana: 1
	4	Bagatus minutus: 4 adults, 8 juv Latreutes fucorum: 3 juv
	5	Bagatus minutus: 1 juv Latreutes fucorum: 1 ♀ w/eggs, 1 juv Planes minutus: 1 juv

Station	Sample	Observations
370	6	Platynereis dumerillii: 1 juv Hippolyte coerulescens: 1 Planes minutus: 1 juv Anoplodactylus petiolatus: 3 juv
371	1	Platynereis dumerillii: 3 adults, 16 juv Bagatus minutus: 4 juv Latreutes fucorum: 12 incl 1 ♀ w/eggs Anoplodactylus petiolatus: 70 incl 17 w/eggs
	2	Platynereis dumerillii: 5 adults, 8 juv Hemiaegina minuta: 97 incl 12 ♀ w/eggs Latreutes fucorum: 3 ♀ w/eggs, 9 juv Leander tenuicornis: 1 juv Anoplodactylus petiolatus: 1 juv Litiopa melanostoma: 1 juv
	3	Bagatus minutus: 60 incl 5 ♀ w/eggs Latreutes fucorum: 7 adults incl 2♀ w/eggs, 26 juv. Leander tenuicornis: 1 juv Anoplodactylus petiolatus: 3 juv Litiopa melanostoma: 5 juv Unidentified nudibranch A: 2 Unidentified nudibranch C: 13
	4	Hippolyte coerulescens: 1 juv Latreutes fucorum: 87 incl 13 ♀ w/eggs Leander tenuicornis: 2 ♀ w/eggs, 1 juv Anoplodactylus petiolatus: 9 incl 1 w/eggs Unidentified nudibranch A: 1 adult, 2 juv Unidentified nudibranch C: 33
	5	Bagatus minutus: 47 incl 7 ♀ w/eggs Latreutes fucorum: 21 incl 4 ♀ w/eggs Leander tenuicornis: 1 juv Planes minutus: 1 juv Unidentified nematode or polychaete: 1 Unidentified nudibranch A: 1 Unidentified nudibranch C: 22
	6	Hemiaegina minuta: 40 incl 7 ♀ w/eggs Latreutes fucorum: 12 incl 2 ♀ w/eggs Leander tenuicornis: 1 ♀ w/eggs Anoplodactylus petiolatus: 2 juv Unidentified nudibranch C: 16
372	1	Bagatus minutus: 32 incl 4 ♀ w/eggs Biancolina sp.: 1 juv Hemiaegina minuta: 2 adults, 16 juv Latreutes fucorum: 32 incl 4 ♀ w/eggs Leander tenuicornis: 4 adults incl 1 ♀ w/eggs, 3 juv Planes minutus: 1 adult, 1 juv

Station	Sample	Observations
372	2	Latreutes fucorum: 2 adults, 4 juv Anoplodactylus petiolatus: 17 incl 5 ♀ w/eggs Unidentified nudibranch C: 19
	3	Latreutes fucorum: 17 juv Leander tenuicornis: 1 juv unidentified nudibranch C: 8 Endeis spinosa: 1
	4	Bagatus minutus: 2 adults w/eggs, 1 juv Latreutes fucorum: 1 adult, 5 juv Leander tenuicornis: 1 andult, 2 juv Anoplodactylus petiolatus: 61 incl 6 w/eggs Spurilla neopolitana: 3 Corambella depressa: 2 Unidentified nudibranch C: 4
	5	Platynereis dumerillii: 8 juv Latreutes fucorum: 1 adult w/eggs, 8 juv Spurilla neopolitana: 10 Unidentified nudibranch C: 9 = juv Spurilla?
	6	Bagatus minutus: 1 adult, 1 juv Latreutes fucorum: 1 adult, 9 juv Leander tenuicornis: 1 juv Endeis spinosa: 1 Unidentified nudibranch A: 1 Unidentified nudibranch C: 2
373	1	Platynereis dumerillii: 3 adults, 17 juv Bagatus minutus: 9 incl 2 w/eggs Hemiaegina minuta: 3 adults, 3 juv Latreutes fucorum: 8 adults incl 2 ♀ w/eggs, 20 juv Leander tenuicornis: 1 adult, 2 juv Anoplodactylus petiolatus: 67 incl 12 w/eggs Spurilla neopolitana: 7 Unidentified nudibranch C: 47
	2	Platynereis dumerillii: 3 adults, 15 juv Bagatus minutus: 6 adults incl 3 w/eggs, 16 juv Latreutes fucorum: 6 juv
	4	Platynereis dumerillii: 7 adults, 63 juv Bagatus minutus: 13 adults (1 w/eggs) 105 juv Hemiaegina minuta: 7 adults, 11 juv Latreutes fucorum: 3 adults (2 w/eggs) 17 juv Anoplodactylus petiolatus: 13 incl 2 ♀ w/eggs Litiopa melanostoma: 4 adults, 2 juv Corambella depressa: 1 Unidentified nudibranch C: 8 juv Endeis spinosa: 1

Station	Sample	Observations
373	5	Platynereis dumerillii: 11 adults, 48 juv Bagatus minutus: 45 adults (6 w/eggs), 64 juv Biancolina sp.: 4 adults, 3 juv Latreutes fucorum: 2 adults, incl 1 ♀ w/eggs Planes minutus: 1 juv Endeis spinosa: 1 Unidentified nudibranch C: 1
	6	Platynereis dumerillii: 6 adults, 32 juv Latreutes fucorum: 6 adults (5 w/eggs), 13 juv Planes minutus: 1 juv Litiopa melanostoma: 1 juv Unidentified nudibranch C: 4
	7	Sunampithoë pelagica: 6 adults, 2 juv Biancolina sp.: 1 juv Latreutes fucorum: 2 adults w/eggs, 15 juv Planes minutus: 2 juv Anoplodactylus petiolatus: 7 incl 2 ♀ w/eggs Unidentified nudibranch C: 4
	8	Platynereis dumerillii: 5 adults, 21 juv Bagatus minutus: 3 adults, 8 juv Latreutes fucorum: 4 adults (3 w/eggs), 4 juv Anoplodactylus petiolatus: 36 incl 4 w/eggs Unidentified nudibranch C: 20 Unidentified spionid larvae (Polydora sp.?): 2 Spurilla neapolitana: 3
	9	Platynereis dumerillii: 1 juv Bagatus minutus: 1 juv Latreutes fucorum: 1 juv Anoplodactylus petiolatus: 2 juv Scyllaea pelagica: 3 juv Unidentified nudibranch C: 2 juv
	10	Platynereis dumerillii: 7 juv Bagatus minutus: 6 juv Latreutes fucorum: 3 juv Leander tenuicornis: 1 adult, 2 juv Anoplodactylus petiolatus: 11 incl 2 w/eggs Litiopa melanostoma: 1 juv Scyllaea pelagica: 8 juv Unidentified nudibranch C: 3
	11	Platynereis dumerillii: 1 adult, 8 juv Bagatus minutus: 5 adults (1 w/eggs), 8 juv Latreutes fucorum: 3 adults (3 w/eggs), 4 juv Endeis spinosa: 1

Station	Sample	Observations
374	1	Platynereis dumerillii: 3 adults, 37 juv Latreutes fucorum: 9 adults (3 w/eggs) 5 juv Planes minutus: 1 juv Anoplodactylus petiolatus: 6 (1 w/eggs) Unidentified nudibranch A: 2 juv
	2	Platynereis dumerillii: 9 juv Bagatus minutus: 7 adults (3 w/eggs), 6 juv Anoplodactylus petiolatus: 7 incl 1 w/eggs Unidentified nudibranch C: 3
375	1	Platynereis dumerillii: 4 adults, 42 juv Bagatus minutus: 40 adults (7 w/eggs), 39 juv Sunampithoë pelagica: 1 adult,2 juv Biancolina sp.: 1 juv Latreutes fucorum: 3 adults (1 w/eggs), 4 juv Unidentified nudibranch C: 1 juv
	2	Platynereis dumerillii: 1 adult, 8 juv Bagatus minutus: 13 incl 2 w/eggs Sunampithoë pelagica: 1 adult, 1 juv Latreutes fucorum: 3 juv Scyllaea pelagica: 1 juv Spurilla neapolitana: 2 Unidentified nudibranch A (= Doto!): 51
	3	Platynereis dumerillii: 4 adults, 23 juv Sunampithoë pelagica: 3 adults, 47 juv Latreutes fucorum: 1 w/eggs Planes minutus: 4 adults incl 2 w/eggs Anoplodactylus petiolatus: 1 adult w/eggs Litiopa melanostoma: 1 juv
	4	Bagatus minutus: 15 adults, 24 juv Sunampithoë pelagica: 1 adult, 17 juv Latreutes fucorum: 2 juv Endeis spinosa: 1
	5	Bagatus minutus: 20 adults, 13 juv Sunampithoë pelagica: 2 adults, 4 juv Unidentified nudibranch C: 1 Endeis spinosa: 1
	6	Bagatus minutus: 2 adults (1 w/eggs), 17 juv Sunampithoë pelagica: 1 juv Anoplodactylus petiolatus: 19 incl 4 w/eggs Unidentified nudibranch C: 6
376	1	Bagatus minutus: 70 adults, 62 juv Sunampithoë pelagica: 14 adults, 34 juv Latreutes fucorum: 5 juv Planes minutus: 1 juv Endeis spinosa: 1 Unidentified nudibranch A: 1 Unidentified nudibranch C: 2

Station	Sample	Observations
376	2	Platynereis dumerillii: 17 juv Bagatus minutus: 2 juv Biancolina sp.: 1 adult, 1 juv Spurilla neapolitana: 9 Unidentified nudibranch A: 66 juv Unidentified nudibranch C: 8 Syngnathus pelagicus: 2 juv
	3	Platynereis dumerillii: 4 adults, 14 juv Sunampithoë pelagica: 3 adults, 9 juv Bagatus minutus: 10 adults, 8 juv Latreutes fucorum: 2 adults (2 w/eggs), 8 juv Anoplodactylus petiolatus: 48 (10 w/eggs) Spurilla neapolitana: 3 Unidentified nudibranch A: 57 Unidentified nudibranch C: 7
	4	Platynereis dumerillii: 6 adults, 18 juv Bagatus minutus: 14 adults, 48 juv Sunampithoë pelagica: 2 adults, 6 juv Latreutes fucorum: 2 adults (1 w/eggs), 6 juv Hippolyte coerulescens: 1 juv Planes minutus: 1 adult w/eggs Unidentified nudibranch A: 1 Unidentified nudibranch C: 2
	5	Platynereis dumerillii: 9 adults, 38 juv Sunampithoë pelagica: 7 adults, 11 juv Latreutes fucorum: 4 juv Unidentified nudibranch A: 1 Unidentified nudibranch C: 11 Unidentified nematode: 1
377	1	Platynereis dumerillii: 3 adults, 35 juv Bagatus minutus: 3 adults, 10 juv Sunampithoë pelagica: 5 adults, 12 juv Latreutes fucorum: 7 juv Endeis spinosa: 2 Spurilla neapolitana: 4 Unidentified nudibranch A: 4 Unidentified nudibranch C: 8 Unidentified spionid: 1
	2	Platynereis dumerillii: 2 adults, 18 juv Bagatus minutus: 14 adults, 33 juv Sunampithoë pelagica: 4 adults, 2 juv Anoplodactylus petiolatus: 10 (7 w/eggs) Unidentified nudibranch A: 9 Unidentified nudibranch C: 6
378	1	Platynereis dumerillii: 3 juv Bagatus minutus: 2 adults, 2 juv Sunampithoë pelagica: 7 adults, 12 juv Latreutes fucorum: 1 juv

Station	Sample	Observations
378	1 (cont.)	Anoplodactylus petiolatus: 11 incl 3 w/eggs Spurilla neapolitana: 2 Unidentified nudibranch A: 1 Unidentified nudibranch C: 1
	2	Platynereis dumerillii: 3 adults, 21 juv Sunampithoë pelagica: 8 adults, 23 juv Hippolyte coerulescens: 1 adult, 1 juv Latreutes fucorum: 1 adult, 3 juv Planes minutus: 1 adult w/eggs Endeis spinosa: 1 Spurilla neapolitana: 1
	3	Platynereis dumerillii: 4 adults, 19 juv Bagatus minutus: 20 adults, 44 juv Sunampithoë pelagica: 5 adults, 8 juv Latreutes fucorum: 1 adult w/eggs, 12 juv Planes minutus: 2 adults (1 w/eggs) Spurilla neapolitana: 1 Endeis spinosa: 1 Scyllaea pelagica: 1 adult, 2 juv
380	1	Sunampithoë pelagica: 5 adults, 7 juv Biancolina sp.: 1 adult, 1 juv Planes minutus: 4 juv Latreutes fucorum: 6 juv Spurilla neapolitana: 2 Unidentified nudibranch A: 3 Unidentified nudibranch C: 1
	2	Bagatus minutus: 2 adults (1 w/eggs), 7 juv Sunampithoë pelagica: 5 adults, 34 juv Latreutes fucorum: 2 juv Leander tenuicornis: 1 juv Planes minutus: 1 juv Anoplodactylus petiolatus: 21 (2 w/eggs) Scyllaea pelagica: 21 juv Unidentified nudibranch A(Doto): 35 Unidentified nudibranch C: 5
	3	Platynereis dumerillii: 2 juv Bagatus minutus: 7 adults (2 w/eggs), 57 juv Sunampithoë pelagica: 8 adult, 43 juv Latreutes fucorum: 2 juv Planes minutus: 2 juv Anoplodactylus petiolatus: 31 (6 w/eggs) Scyllaea pelagica: 3 juv Unidentified nudibranch A: 89

Station	Sample	Observations
380	4	Platynereis dumerillii: 6 juv Bagatus minutus: 3 adults (2 w/eggs), 5 juv Sunampithoë pelagica: 4 adults, 31 juv Latreutes fucorum: 2 adults w/eggs, 1 juv Planes minutus: 1 juv Anoplodactylus petiolatus: 38 incl 13 w/eggs Scyllaea pelagica: 5 juv Unidentified nudibranch A: 9 Unidentified nudibranch C: 10
	5	Hoploplana grubei: 6 juv Platynereis dumerillii: 10 juv Bagatus minutus: 7 adults, 28 juv Sunampithoë pelagica: 2 adults, 6 juv Planes minutus: 2 juv Litiopa melanostoma: 1 adult, 7 juv Spurilla neapolitana: 1 Unidentified nudibranch A: 6 juv Unidentified nudibranch C: 21 juv
	6	Hoploplana grubei: 3 adults, 5 juv Platynereis dumerillii: 1 juv Sunampithoë pelagica: 1 adult, 14 juv Bagatus minutus: 4 adults (2 w/eggs), 12 juv Latreutes fucorum: 1 adult w/eggs, 2 juv Anoplodactylus petiolatus: 3 juv Litiopa melanostoma: 2 juv Scyllaea pelagica: 9 juv Unidentified nudibranch A: 16 juv
382	1	Hoploplana grubei: 20 juv Platynereis dumerillii: 40 juv Bagatus minutus: 18 juv Latreutes fucorum: 3 juv Planes minutus: 5 juv Litiopa melanostoma: 1 adult, 8 juv Unidentified nudibranch A: 10 juv Unidentified nudibranch C: 2
	2	Sunampithoë pelagica: 19 adults, 45 juv Planes minutus: 1 adult w/eggs, 23 juv Litiopa melanostoma: 2 juv
	3	Platynereis dumerillii: 4 adults, 14 juv Sunampithoë pelagica: 11 adults, 34 juv Hippolyte coerulescens: 1 adult w/eggs, 1 juv Latreutes fucorum: 8 adults (4 w/eggs), 4 juv Planes minutus: 1 adult w/eggs, 12 juv Anoplodactylus petiolatus: 5 juv Litiopa melanostoma: 6 juv Unidentified nudibranch C: 2

Station	Sample	Observations
382	4	Platynereis dumerillii: 9 juv Sunampithoë pelagica: 11 adults, 20 juv Hippolyte coerulescens: 1 adult w/eggs Latreutes fucorum: 1 adult, 3 juv PLanes minutus:2 adults (1 w/eggs) 9 juv Anoplodactylus petiolatus: 4 (1 w/eggs) Litiopa melanostoma: 1 adult, 11 juv
	5	Platynereis dumerillii: 2 adults, 3 juv Sunampithoë pelagica: 10 adults, 31 juv Latreutes fucorum: 3 juv Planes minutus: 1 adult w/eggs, 8 juv Anoplodactylus petiolatus: 2 juv Litiopa melanostoma: 1 adult, 22 juv Histrio histrio: 1 juv
	6	Platynereis dumerillii: 1 adult, 16 juv Bagatus minutus: 6 adult (1 w/eggs), 38 juv Hippolyte coerulescens Planes minutus: 8 juv Litiopa melanostoma: 9 juv Unidentified nudibranch C: 3
383	1	Platynereis dumerillii: 4 adults, 7 juv Litiopa melanostoma: 3 large, 2 medium, 1 small
	3	Amphiscolops sargassi: 19 Platynereis dumerillii: 12 adults, 15 juv Latreutes fucorum: 2 with eggs, 1 small Litiopa melanostoma: 7 large, 4 med, 3 small Doto pygmaea: 3
385	1	Unidentified (mangled) shrimp: 2
	2	Platynereis dumerillii: 10 adults, 6 juv Litiopa melanostoma: 2 large, 3 med, 1 small Unidentified (immature) fish: 1
	3	Platynereis dumerillii: 5 adults, 4 juv Litiopa melanostoma: 11 large, 1 small
	4	Platynereis dumerillii: 5 adults, 1 juv Leander tenuicornis: 1 w/eggs Planes minutus: 1 very small Litiopa melanostoma: 23 large, 6 med, 10 small Unidentified nudibranch D: 1

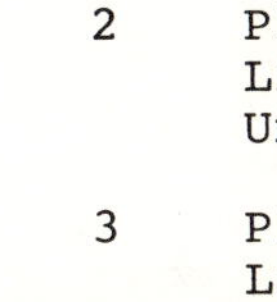

(because of the difficulty of handling the preserved nudibranchs, it was not possible to tell whether this was the same as Debbie Mogelberg's A, B, or C. It was not Doto pygmaea or any of the catalogued species. Its main characteristic was two 'ears' as in the sketch below --Jill Cadwallader)

See Eastward E22324-3, Hunt cruise 1-2

Station	Sample	Observations
385	5	Latreutes fucorum: 2 incl 1 w/eggs Litiopa melanostoma: 21 large, 2 med, 2 small Planes minutus: 1 very small Unidentifiable shrimp: 1
	6	Planes minutus: 2 very small Litiopa melanostoma: 14 large, 2 med, 20 small
	7	Amphiscolops sargassi: 7 Platynereis dumerillii: 6 adults, 6 juv Litiopa melanostoma: 3 large, 9 small Doto pygmaea: 23
	8	Platynereis dumerillii: 10 juv Litiopa melanostoma: 6 large, 1 small
386	1	Platynereis dumerillii: 6 adults, 7 juv Hoploplana grubei: 1 small (cf. Hunt samples) Litiopa melanostoma: 2 large Doto pygmaea: 1
	2	Latreutes fucorum: 8 incl 2 w/eggs, 1 with isopod eggs, 3 small.
	3	Amphiscolops sargassi: 3 Platynereis dumerillii: 36 adults, 2 juv Latreutes fucorum: 8 incl 1 w/eggs, 2 with isopod eggs, 2 small Litiopa melanostoma: 2 large
	4	Hoploplana grubei: 1 very large, 1 very small Parasitic isopod: 1 Latreutes fucorum: 10 incl 1 w/eggs, 2 small Planes minutus: 1 small Litiopa melanostoma: 1 large
	5	Latreutes fucorum: 18 incl 2 w/eggs, 1 with isopod eggs, 2 small
	6	Leander tenuicornis: 2 incl 1 w/eggs Litiopa melanostoma: 1 large
387	1	Platynereis dumerillii: 1 juv Parasitic isopod: 1 Latreutes fucorum: 3 incl 1 w/eggs Leander tenuicornis: 2 incl 1 w/eggs Litiopa melanostoma: 3 med, 4 small
	2	Platynereis dumerillii: 5 adults, 15 juv Planes minutus: 3 small Unidentified nudibranch D: 1

Station	Sample	Observations
387	3	Hoploplana grubei: 12, many of the small type Platynereis dumerillii: 6 adults, 2 juv Unidentified isopod: 2 Latreutes fucorum: 3 w/eggs, 2 others, 2 small Portunus sayi: 1 small Litiopa melanostoma: 1 large,4 med,2 small Unidentified nudibranch D: 9, many larger than usual
	4	Hoploplana grubei:1 small Platynereis dumerillii: 11 adults, 9 juv Hemiaegina minuta: 6 small Planes minutus: 1 small Litiopa melanostoma: 1 large, 2 med, 1 small Unidentified nudibranch D: 2 Unidentified creature (see sketch): 1 
	5	Platynereis dumerillii: 8 adults, 2 juv Latreutes fucorum: 2 w/eggs, 2 very small Planes minutus:3 small Unidentified nudibranch D: 1 Unidentified creature (see sketch): 1 
	6	Hoploplana grubei: 1 very small Unidentified nudibranch D: 3
388	1	Many 'fruiting bodies' on S. natans Hoploplana grubei: 1 large, 1 med, 1 very small Amphiscolops sargassi: 1 Platynereis dumerillii: 65 adults, 36 juv Bagatus minutus: 64 incl 4 w/eggs Latreutes fucorum: 6 incl 4 w/eggs, 1 with isopod eggs Leander tenuicornis: 4 incl 2 w/eggs Portunus sayi: 1 large, w/eggs Litiopa melanostoma: 1 large, 2 small Doto pygmaea: 2 Unidentified nudibranch D: 4 Unidentified creature(see sketch): 48  (Very similar to Unidentified nudibranch D in size, shape, texture, and color, only in place of 'tentacles' there are 2 eye spots, dark and relatively large. Somebody's eggs? - Jill Cadwallader)
	2	Hoploplana grubei: 2 large, 3 very small Platynereis dumerillii: 11 adults, 7 juv Latreutes fucorum: 6 incl 2 w/eggs, 1 very small Planes minutus: 12, all small Unidentified nudibranch D: 4 Unidentified creature (see sample 1 above): 1 Unidentified flatworm (circular with 2 central eyespots and mottled appearance): 1 very small Doto pygmaea: 5

Station Sample Observations

388 3 Hoploplana grubei: 3 large, 1 very small
Platynereis dumerillii: 14 adults, 4 juv
Parasitic isopod larva: 1
Planes minutus: 4 small
Anoplodactylus petiolatus: 22 incl 1 with eggs
Litiopa melanostoma: 1 med 2 small
Unidentified nudibranch D: 7
Unidentified creature (cf. 388-1): 4

4 Platynereis dumerillii: 25 adults, 15 juv
Bagatus minutus: 9 incl 1 with eggs
Parasitic isopod larva: 3
Latreutes fucorum: 10 (1 with eggs, 1 very small)
Leander tenuicornis: 1 with eggs
Doto pygmaea: 8
Anoplodactylus petiolatus: 23 incl 4 w/eggs
Litiopa melanostoma: 1 large, 1 med, 5 small
Unidentified nudibranch D: 6
Unidentified creature (cf. 388-1): 1

5 Hoploplana gruebei: 5 very small, 1 larger
Platynereis dumerllii: 3 adults, 2 juv
Doto pygmaea: 8
Litiopa melanostoma: 1 large, 4 small
Anoplodactylus petiolatus: 8 incl 3 w/eggs
Unidentified nudibrandh D: 10
Unidentified creature (see 388-1): 4

6 Platynereis dumerillii: 4 adults, 8 juv
Litiopa melanostoma: 2 small
Doto pygmaea: 5 (larger than usual)
Finona pinnata (?): 2
Unidentified nudibranch D: 6

389 1 Platynereis dumerillii: 3 adults, 18 juv
Latreutes fucorum: 24 incl 4 with eggs,
1 with isopod, 5 very small
Planes minutus: 14 small
Anoplodactylus petiolatus: 52 (8 w/eggs)
Doto pygmaea: 1
Unidentified nudibranch D: 3
Unidentified creature (see 388-1): 2
Parasitic isopod larva: 1

2 Platynereis dumerillii: 6 adults, 10 juv
Planes minutus: 12 small
Latreutes fucorum: 9 (2 very small)
Anoplodactylus petiolatus: 275 (10 w/eggs)
Unidentified nudibranch D: 14
Unidentified creature (see 388-1): 6

Station	Sample	Observations
389	3	Platynereis dumerillii: 17 adults, 6 juv
		Bagatus minutus: 18 (1 w/eggs)
		Hippolyte coerulescens: 1
		Latreutes fucorum: 3 (1 with eggs, 1 with isopod)
		Litiopa melanostoma: 1 small
		Unidentified nudibranch D: 3 
		Unidentified creature (see 388-1): 1 
	4	Platynereis dumerillii: 3 adults, 1 juv
		Latreutes fucorum: 10 (4 w/eggs)
		Leander tenuicornis: 4 (1 w/eggs)
		Planes minutus: 34 (1 w/eggs, 31 small)
		Anoplodactylus petiolatus: 31 (6 w/eggs)
		Litiopa melanostoma: 2 large 1 med
		Doto pygmaea: 2
		Fiona pinnata: 1
		Shrimp (mangled): 1
		Parasitic isopod: 1
		Parasitic isopod larva: 2
		Unidentified nudibranch D: 9
		Unidentified creature (see 388-1): 18
	5	Bagatus minutus: 13 (3 w/eggs)
		Sunampithoë pelagica: 13 (2 w/eggs)
		Latreutes fucorum: 15 (3 w/eggs, 3 very small)
		Shrimp (mangled): 1
		Parasitic isopod: 1
		Parasitic isopod larva: 1
	6	Platynereis dumerillii: 4 adults, 15 juv
		Bagatus minutus: 46 (1 w/eggs)
		Latreutes fucorum: 8 (1 w/eggs, 7 very small)
		Planes minutus: 16 small
		Anoplodactylus petiolatus: 53 (3 w/eggs)
		Litiopa melanostoma: 2 large
		Doto pygmaea: 2
		Fiona pinnata (?): 1
		Parasitic isopod larva: 1
		Unidentified nudibranch D: 8
		Unidentified creature (see 388-1): 19
390	1	Platynereis dumerillii: 1 adult, 6 small
		Latreutes fucorum: 13 (1 w/eggs, 6 very small)
		Litiopa melanostoma: 5 large, 2 small
		Parasitic isopod larva: 2
		Doto pygmaea: 1
	2	Hoploplana grubei: 1 small
		Gnescioceros sargassicola: 27 ("There seem to be 2 distinct types with the same body form. The plainer ones are usually larger than the more stippled ones."- J. Cadwallader) 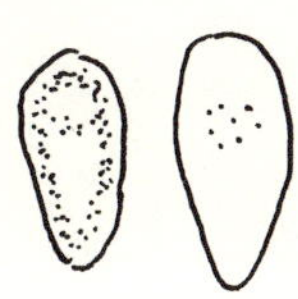

Station	Sample	Observations
390	2 cont.	Platynereis dumerillii: 2 adults, 12 juv Planes minutus: 2 medium, 12 small Latreutes fucorum: 6 (1 w/eggs, 2 small) Litiopa melanostoma: 2 large, 2 small Doto pygmaea: 1 Unidentifiable shrimp: 1 Unidentifed creature (see 388-1): 1
	3	Platynereis dumerillii: 9 adults, 10 juv Bagatus minutus: 53 (5 w/eggs) Hemiaegina minuta: 1 very small Hoploplana grubei: 1 small Latreutes fucorum: 5 (3 w/eggs) Leander tenuicornis: 5 (3 w/eggs, 1 small) Planes minutus: 6 (1 w/eggs) Anoplodactylus petiolatus: 5 (1 w/eggs) Litiopa melanostoma: 4 large Scyllaea pelagica: 2 large Unidentified nudibranch D: 1 Unidentified creature (see 388-1): 8
	4	Platynereis dumerillii: 19 adults, 2 juv Bagatus minutus: 176 (3 w/eggs, most small) Latreutes fucorum: 16 (2 w/eggs) Spurilla neapolitana: 1 juv Unidentified nudibranch D: 1 Spionid worms (planktonic): 9
	5	Platynereis dumerillii: 1 juv Bagatus minutus: 12 (2 w/eggs) Latreutes fucorum: 7 (1 w/eggs) Planes minutus: 4 small Litiopa melanostoma: 1 large Unidentified nudibranch D: 1 Spurilla neapolitana: 1
	6	Latreutes fucorum: 5 (1 w/eggs, 4 small) Planes minutus: 8 small Unidentified nudibranch D: 1
391	1	Hoploplana grubei: 2 small Platynereis dumerillii: 4 juv Latreutes fucorum: 5 (2 w/eggs, 1 small) Litiopa melanostoma: 1 large 1 med, 95 small
	2	Platynereis dumerillii: 5 adults, 5 juv Latreutes fucorum: 4 (1 w/eggs) Litiopa melanostoma: 14 small
	3	Platynereis dumerillii: 2 juv Latreutes fucorum: 1 small Litiopa melanostoma: 1 large 38 small Anoplodactylus petiolatus: 4 (1 w/eggs) Scyllaea pelagica: 1 juv

Station	Sample	Observations
391	4	Platynereis dumerillii: 1 very large, 13 adults, 12 juv Planes minutus: 2 (1 w/eggs) Anoplodactylus petiolatus: 13 (1 w/eggs, 2 very small) Litiopa melanostoma: 8 small Unidentified nudibranch D: 1
	5	Platynereis dumerillii: 4 adults (1 large), 13 juv Bagatus minutus: 16 large, 16 small (2 w/eggs) Latreutes fucorum: 1 small Planes minutus: 1 small Litiopa melanostoma: 7 small Unidentified nudibranch D: 1
	6	Hoploplana grubei: 2 small Platynereis dumerillii: 29 adults (2 large) 33 juv Bagatus minutus: 23 large (5 w/eggs), 11 juv Portunus sayi: 1 juv Litiopa melanostoma: 9 small
392	1	Hoploplana grubei: 1 small Platynereis dumerillii: 1 adult, 2 juv Bagatus minutus: 25 (1 w/eggs) Parasitic isopod larvae: 2 Latreutes fucorum: 9 (3 w/eggs, 1 small) Planes minutus: 4 (1 w/eggs) Litiopa melanostoma: 1 large, 2 small
	2	Hoploplana grubei: 1 small Latreutes fucorum: 8 (1 w/eggs, 1 very small) Anoplodactylus petiolatus: 26 (1 w/eggs) Litiopa melanostoma: 1 small Unidentified nudibranch D: 5 Sipunculid larva (planktonic) (?): 1
	3	Platynereis dumerillii: 19 adults, 4 juv Bagatus minutus: 14 large (1 w/eggs) 40 small Hemiaegina minuta: 12 large, 24 small Planes minutus: 1 small Litiopa melanostoma: 3 small Doto pygmaea: 1 small Unidentified creature (See 388-1): 5
	4	Latreutes fucorum: 2 small Hemiaegina minuta: 2 small Litiopa melanostoma: 1 large 1 med, 4 small Many loose eggs
	5	Sunampithoë pelagica: 1 small Planes minutus: 2 small Anoplodactylus petiolatus: 2 small Many loose eggs "The vial with the rest of the sample was broken but apparently little was in it."- J. Cadwallader

Station	Sample	Observations
392	6	Platynereis dumerillii: 2 juv Bagatus minutus: 17 large, 43 small Hemiaegina minuta: 2 very small Latreutes fucorum: 1 small Planes minutus: 1 small Spurilla neapolitana: 1 very small Unidentified nudibranch D: 2 (among the largest found yet)
393	1	Platynereis dumerillii: 12 adults, 3 juv Bagatus minutus: 263 (28 w/eggs) Latreutes fucorum: 7 (3 w/eggs) Leander tenuicornis: 1 w/eggs Planes minutus: 1 large Litiopa melanostoma: 3 med, 6 small Scyllaea pelagica: 1 small Unidentified nudibranch D: 6 Parasitic isopod larvae: 3
	2	Hoploplana grubei: 1 small Sunampithoë pelagica: 36 (1 w/eggs, 1 with "brood pouch" containing 8 young, miniatures of the adult) - J. Cadwallader Latreutes fucorum: 7 (2 w/eggs, 1 small Anoplodactylus petiolatus: 23 (3 w/eggs) Litiopa melanostoma: 33 large, 53 med, 26 small Doto pygmaea: 2 Scyllaea pelagica: 6 med, 1 small, 1 very small Unidentified nudibranch D: 11
	3	Hoploplana grubei: 2 small Platynereis dumerillii: 3 adults, 1 juv Bagatus minutus: 49 adults (5 w/eggs), 9 juv Sunampithoë pelagica: 25 adults, 2 juv Latreutes fucorum: 5 (2 w/eggs, 1 very small) Litiopa melanostoma: 16 large, 9 medium, 14 small Doto pygmaea: 3 Unidentified nudibranch D: 5
	4	Hoploplana grubei: 2 small Platynereis dumerillii: 12 adults, 11 juv Bagatus minutus: 64 adults (13 w/eggs) 44 juv Sunampithoë pelagica: 27 adults (1 w/eggs), 21 juv Latreutes fucorum: 2 (1 w/eggs) Anoplodactylus petiolatus: 6 (1 w/eggs) Litiopa melanostoma: 23 large, 5 med, 34 small Unidentified nudibranch D: 5
	5	Platynereis dumerillii: 2 adults, 1 juv Bagatus minutus: 106 adults (10 w/eggs), 26 juv Sunampithoe pelagica: 40 (2 w/eggs, 2 with brood pouch), 14 juv (64 total) Latreutes fucorum: 2 w/eggs Anoplodactylus petiolatus: 19 (2 w/eggs)

tation	Sample	Observations
393	5 cont.	Litiopa melanostoma: 33 large, 37 med, 66 small Doto pygmaea: 3 small Scyllaea pelagica: 2 very small Unidentified nudibranch D: 2 Unidentified creature (see 388-1): 1
	6	Hoploplana grubei: 1 small Bagatus minutus: 23 adutls (4 with eggs), 32 juv Sunampithoë pelagica: 9 adults (1 with brood pouch, 3 juv Latreutes fucorum: 4 (1 w/eggs, 1 with isopod eggs, 1 very small) Litiopa melanostoma: 22 large, 16 med, 36 small Doto pygmaea: 10 Unidentified nudibranch D: 4
	7	Hoploplana grubei: 2 small Litiopa melanostoma: 24 large, 26 med, 56 small Doto pygmaea: 195 Scyllaea pelagica: 1 very small Unidentified nudibranchs resembling very small Fiona pinnata: 2 Parasitic isopod larva: 1
	8	Hoploplana grubei: 2 small Bagatus minutus: 15 adults (3 w/eggs), 4 juv Sunampithoë pelagica: 21 adults (2 with brood pouch) 20 juv Litiopa melanostoma: 3 large, 1 med, 16 small Doto pygmaea: 1 Unidentified (mashed) shrimp: 1 Unidentified nudibranch D: 1 small
	9	Platynereis dumerilli: 17 adults, 9 juv Bagatus minutus: 41 adults (8 w/eggs) 27 juv Latreutes fucorum: 2 w/eggs Anoplodactylus petiolatus: 5 (1 w/eggs) Litiopa melanostoma: 4 large, 2 med, 4 small Scyllaea pelagica: 1 med, 1 small Unidentified nudibranch D: 1 small
	10	Bagatus minutus: 1 adult, 8 juv Litiopa melanostoma: 1 large, 4 small Scyllaea pelagica: 1 small Unidentified nudibranch D: 2 Parasitic isopod larva: 1
	11	Hoploplana grubei: 1 small Bagatus minutus: 1 juv Litiopa melanostoma: 3 large, 3 med, 17 small Unidentified nudibranch D: 1

Station	Sample	Observations
393	12	Hoploplana grubei: 1 small Sunampithoë pelagica: 3 adults, 1 juv Litiopa melanostoma: 2 med, 10 small
394	1	Acerotisa notulata: 14 (4 w/eggs) Hoploplana grubei: 1 large, 3 small Platynereis dumerillii: 11 adults, 3 juv Bagatus minutus: 253 adults (24 w/eggs), 154 juv Sunampithoë pelagica: 162 adults (9 w/eggs, 5 with brood pouch), 220 juv Planes minutus: 5 adults (2 w/eggs), 1 juv Latreutes fucorum: 28 (5 w/eggs, 7 very small) Litiopa melanostoma: 26 large, 11 med, 9 small Doto pygmaea: 16 Scyllaea pelagica: 1 med, 4 small, 3 very small Unidentified shrimp (head missing): 1 Parasitic isopod: 3 Unidentified nudibranch: 1 very small Unidentified nudibranch D: 6 ("including the best specimen I've seen"- J.C.)
	2	Sample of "budding algae" to W.R. Taylor Hoploplana grubei: 1 normal, 2 very small Platynereis dumerillii: 23 adults, 5 juv Bagatus minutus: 303 adults (22 w/eggs), 232 juv Sunampithoë pelagica: 89 adults (7 w/eggs, 3 with brood pouch), 70 juv Hippolyte coerulescens: 2 Latreutes fucorum: 18 (1 with isopod eggs, 4 small Doto pygmaea: 53 Litiopa melanostoma: 13 large, 10 med, 14 small Parasitic isopod: 1 Scyllaea pelagica: 1 large, 5 med, 1 small, 2 very small Unidentified nudibranch D: 5
	3	Algae growing on tar lumps saved to be identified Platynereis dumerillii: 8 adults, 2 juv Bagatus minutus: 82 adults (4 w/eggs), 28 juv Sunampithoë pelagica: 78 adults (12 w/eggs, 4 with brood pouch), 72 juv Latreutes fucorum: 7 (2 w/eggs, 1 with isopod, Doto pygmaea: 4 Planes minutus: 3 (1 w/eggs) Anoplodactylus petiolatus: 7 (2 w/eggs) Litiopa melanostoma: 13 large, 2 med, 3 small Unidentified nudibranch D: 4
	4	Bagatus minutus: 21 adults (1 w/eggs) 12 juv Sunampithoë pelagica: 23 adults (4 with eggs, 1 with brood pouch), 23 juv Planes minutus: 2 w/eggs Litiopa melanostoma: 1 med, 1 small Doto pygmaea: 1 Scyllaea pelagica: 1 med, 1 very small

Station	Sample	Observations
398	1	Samples marked "11 Feb 75", which is the date of station S-398. Different location? Litiopa melanostoma: 1 med, 1 large Doto pygmaea: 5
	2	Platynereis dumerillii: 7 adults, 8 juv Doto pygmaea: 4
	3	Doto pygmaea: 15
	5	Platynereis dumerillii: 15 adults, 4 juv Latreutes fucorum: 2 with isopods Doto pygmaea: 4
	6	Platynereis dumerillii: 11 adults, 3 juv Hippolyte coerulescens: 1 Unidentified shrimp: 1 Doto pygmaea: 3
	7	Platynereis dumerillii: 5 adults, 3 juv Bagatus minutus: 22 (1 w/eggs) Doto pygmaea: 19
399	1	Doto pygmaea: 3 Unidentified nudibranch D: 10 (some good ones)
	2	Unidentifiable shrimps: 3
	3	Gnescioceros sargassicola: 35 small, 14 med, 1 large (second entry below: 3) = 53 total Latreutes fucorum: 6 (4 w/eggs) Unidentifiable shrimp: 1 Litiopa melanostoma:25 small, 15 med Doto pygmaea: 8
	4	Litiopa melanostoma: 5 small, 11 med
	5	Litiopa melanostoma: 120 small, 30 med, 25 large Doto pygmaea: 80
	6	Latreutes fucorum: 5 w/eggs, 3 med Spurilla neapolitana: 2 Doto pygmaea: 3

APPENDIX IV

Motile Macrofauna on Samples from Various Cruises

CARIBBEAN CRUISE

MOTILE MACROFAUNA ON SARGASSUM

Station	1	2	2a	3	4	5
Date	6 Jan 71	7 Jan 71	7 Jan 71	7 Jan 71	9 Jan 71	14 Jan 71
Sample No.	1*	2*	2a*	3*	4*	5*
Location:	33°28.9'N 69°48.9W	30°10.4'N 68°55.0'W		28°52'N 68°35.2W	24°16'N 67°10'W	21°22'N 58°40'W
Wet weight, g						
Displacment vol. ml.	430	340	420	290	230	270
Sargassum species	mixed	mixed	mixed	N	mixed	mixed

Station	6	7
Date	17 Jan 71	18 Jan 71
Sample No.	6*	7*
Location:	18°40.2N 63°28'W	17°33'N 63°34.2'W
Wet weight, g		
Displacement vol. ml.	87	
Sargassum species	F	mixed

*Fauna not counted. See notes for observations.

CARIBBEAN (January 1971)

Station	Sample	Observations
1		33°29'N, 69°49'W. 6 January 1971. 1012 hr Surface temp 19.7°C. Half meter net tow. Mixed sargassum, displacement volume 430 ml Hoploplana grubei:"Rolled up into a ball, it looked like a bladder" -MLF Planes minutus: "Most had a white oval on the forward part of the carapace." Fauna not counted.
2		30°10'N, 68°55'W. 7 Jan 1971. 1015 hr. ½ meter net. Surface temp 21.3°C. Mixed sargassum, 340 ml
	2A	Mixed sargassum, 420 ml, same location as 2
3		28°52'N, 68°35'W. 7 January 1971. 2010 hr. Dipnetting at night. Surface temp. 21.4°C Sargassum natans: 290 ml Fauna not counted.
4		24°16'N, 67°10'W. 9 Jan 1971. 1600 hr "First longline station. Dipnetted from dense windrow. Sargassum loaded with congealed oil, for the most part small and ratty clumps." MLF Mixed sargassum, 230 ml. (a sample 4a, taken for hydroids, was totally dried out and discarded)
5		21°22'N, 58°40'W. 14 Jan 1971. Temp: 25.3°C "Sargassum rare since last stations. Windrows have sparse small clumps. Sargassum more abundant further north" - MLF. Mixed sargassum, 270 ml. Fauna not counted.
6		18°40'N, 63°28'W. 17 January 1971. 1200 hr Temp. 26.3 °C. "Near Anagada passage. Dipnet during longlining. Old and new pieces around. some S. hystrix in water." - MLF Sargassum fluitans (small amount of natans):87 ml
7		17°33'N, 63°34'W. 18 January 1971. Temp. 26.3°C "An unusual Sargassum species is in a little bag to key out. S. hystrix is around too, and pelagic species." -MLF Sample lost.

(Samples collected by Michael L. Fine)

BERMUDA
Sir Horace Lamb
21 Jan. — 19 Feb. 1972
ATLANTIC OCEAN
60°
20°
DOMINICAN
REPUBLIC
PUERTO
RICO
BARBUDA
ANTIGUA
GUADELOUPE
DOMINICA
MARTINIQUE
ST. LUCIA
BARBADOS
GRENADA
TOBAGO
1
2
3
4
5
6
7
8
9
10
11

MOTILE MACROFAUNA ON SARGASSUM

XBT Station No.	#3			#4		
Station	1			2		
Date	21 Jan 72			21 Jan 72		
Sample No.	1	2	3	1	2	3
Location:	29°10'N 63°45'W			28°13'N 63°30'W		
Wet weight, g	42.4	33.3	62.8	58.7	50.7	31.3
Displacement vol. ml.	50	29	60	60	49	29
Sargassum species	N	N	N	N	N	N
Fauna:						
44 Anemonia sargassensis						
46 Acerotisa notulata			1			5
47 Hoploplana grubei						
48 Gnescioceros sargassicola			3	7	5	5
51 Platynereis dumerillii		1	1	1	22	
70 Bagatus minutus		5			28	
73 Sunampithoë pelagica				1		
74 Biancolina sp.			1		1	1
76 Hemiaegina minuta				1	16	4
80 Latreutes fucorum						
82 Leander tenuicornis						
83 Planes minutus						
84 Portunus sayi						
87 Anoplodactylus petiolatus	6		21	4		3
89 Litiopa melanostoma	2	1	18	1		6
95 Scyllaea pelagica				2	1	
100 Histrio histrio						
101 Syngnathus pelagicus						
Other Species:			13	15		2
Total Number of Specimens:	8	7	58	32	73	26

Note: N = *S. natans*

SIR HORACE LAMB

MOTILE MACROFAUNA ON SARGASSUM

XBT Station No.		#11			#13		
Station		3			4		
Date		22 Jan 72			22 Jan 72		
Sample No.		1	2	3	1	2	3
Location:		24°43'N 62°29"W			23°00'N 62°46'W		
Wet weight, g		61.4	47.2	25.5	70	27.6	
Displacement vol. ml		60	45	22	87	23	74
Sargassum species		N	N	N	N	N	N
	Fauna:						
44	Anemonia sargassensis						
46	Acerotisa notulata	?	17	22	9		2
47	Hoploplana grubei	1?					
48	Gnescioceros sargassicola	5	1	1		1	1
51	Platynereis dumerillii	2	1				3
70	Bagatus minutus	1					2
73	Sunampithoë pelagica						
74	Biancolina sp.						3
76	Hemiaegina minuta	1					
80	Latreutes fucorum	1					
82	Leander tenuicornis						
83	Planes minutus				1	2	
84	Portunus sayi						
87	Anoplodactylus petiolatus	5	7	2	7	1	8
89	Litiopa melanostoma	200	17	29	8		6
95	Scyllaea pelagica	1			1		
100	Histrio histrio						
101	Syngnathus pelagicus						
	Other Species:	59	24	20		1	
Total Number of Specimens:		276	67	74	26	5	25

Note: N = *S. natans*

MOTILE MACROFAUNA ON SARGASSUM

XBT Station No.	#15			#19		
Station	5			6		
Date	23 Jan 72			23 Jan 72		
Sample No.	1	2	3	1	2	3
Location:	21°50'N 61°48'W			20°55'N 61"33"W		
Wet weight, g						
Displacement vol. ml.	67	48	38	85	120	48
Sargassum species	N	N	N	N	N	N
Fauna:						
44 Anemonia sargassensis						
46 Acerotisa notulata	25	5		1		
47 Hoploplana grubei	2	5	3		1	
48 Gnescioceros sargassicola	3	3	18	11	13	4
51 Platynereis dumerillii	2		1			
70 Bagatus minutus						
73 Sunampithoë pelagica						
74 Biancolina sp.						
76 Hemiaegina minuta						
80 Latreutes fucorum						
82 Leander tenuicornis						
83 Planes minutus					1	
84 Portunus sayi						
87 Anoplodactylus petiolatus	5	11	3	1	2	1
89 Litiopa melanostoma	117	23	42	23	16	9
95 Scyllaea pelagica	1		2			
100 Histrio histrio						
101 Syngnathus pelagicus						
Other Species:	85	22	29	15	24	5
Total Number of Specimens:	240	69	98	51	57	19

Note: N = S. natans

SIR HORACE LAMB

MOTILE MACROFAUNA ON SARGASSUM

XBT Station No.	#25			#26		
Station	7			8		
Date	19 Feb 72			18 Feb 72		
Sample No.	1	2	3	1	2	3
Location:	18°14.2N 62°52.5W			18°6.3'N 62°44.8'W		
Wet weight, g						
Displacement vol. ml.	58	44	50	60	40	79
Sargassum species	F	N	F	F+?	N	N
Fauna:						
44 Anemonia sargassensis						
46 Acerotisa notulata				1		2
47 Hoploplana grubei						
48 Gnescioceros sargassicola	4	3	3	1	2	
51 Platynereis dumerillii				1	2	
70 Bagatus minutus				3		
73 Sunampithoë pelagica						
74 Biancolina sp.			1	1		
76 Hemiaegina minuta						
80 Latreutes fucorum	3	3	3		2	1
82 Leander tenuicornis	1			1		
83 Planes minutus						
84 Portunus sayi						
87 Anoplodactylus petiolatus		7	4	4		2
89 Litiopa melanostoma	9	4	12	11	4	10
95 Scyllaea pelagica	1					
100 Histrio histrio						
101 Syngnathus pelagicus						
Other Species:	4	7	6	33	20	6
Total Number of Specimens:	22	24	29	56	30	21

Note: F = S. fluitans; N = S. natans

MOTILE MACROFAUNA ON SARGASSUM

No.							
	XBT Station No.	#27			#28	#34	
	Station	9			10	11	
	Date	24 Jan 72			17 Feb 72	25 Jan 72	
	Sample No.	1	2	3	1	1	2
	Location:	18°2'N 60°47'W			17°46'N 62°51'W	11°57'N 61°46'W	
	Wet weight, g					32.5	
	Displacement, vol. ml.	55	36	30	26	31	25
	Sargassum species	N	N	N	F	H	H
	Fauna:						
44	Anemonia sargassensis						
46	Acerotisa notulata		1				
47	Hoploplana grubei						
48	Gnescioceros sargassicola	1	1				
51	Platynereis dumerillii				1		
70	Bagatus minutus				1		
73	Sunampithoë pelagica				1		
74	Biancolina sp.						
76	Hemiaegina minuta						
80	Latreutes fucorum			2			
82	Leander tenuicornis			1			
83	Planes minutus						
84	Portunus sayi			2	1		
87	Anoplodactylus petiolatus	1		1			
89	Litiopa melanostoma	6	6	9	7		
95	Scyllaea pelagica						
100	Histrio histrio						
101	Syngnathus pelagicus						
	Other Species:	13	2	2	12	4	1?
	Total Number of Specimens:	21	10	17	23	4	1

Note: F = *S. fluitans*; N = *S. natans*; H = *S. hystrix*

Station	Sample	Observations
Sir Horace Lamb cruise, 21 Jan to 19 Feb 1972		
1		21 Jan 72; 29°10'N, 63°45'W
	1	Anoplodactylus petiolatus: 6 juv
	2	Bagatus minutus: 3 adults (1 w/eggs), 2 juv
	3	unidentified isopod "crescent moon shape": 1 Platynereis dumerilli: 1 juv Litiopa melanostoma: 9 adults, 9 juv unidentified nudibranch "A": 7 unidentified nudibranch "C": 5
2		21 Jan 72; 28°13'N, 63°30'W
	1	Platynereis dumerilli: 1 juv Spirorbis sp. (without shell) Scutellidium longicauda: 27 Dactylopodia tisboides?: 1 Scyllaea pelagica: 1 adult, 1 juv unidentified nudibranch "A": 15 hydroids
	2	hydroids Platynereis dumerilli: 22 juv Spirorbis sp. (with and without shells) polychaetes, juv copepods, not sorted Bagatus minutus: 13 adults (1 w/eggs), 15 juv Hemiaegina minuta: 15 adults, 1 juv
	3	Hemiaegina minuta: 1 adult, 3 juv Litiopa melanostoma: 1 adult, 5 juv unidentified nudibranch "A": 1 (Doto? See S-375-2) Endeis spinosa: 1
3		22 Jan 72; 24°43'N, 62°29'W
	1	Platynereis dumerilli: 2 juv Latreutes fucorum: 1 juv Scyllaea pelagica: 1 juv unidentified nudibranch "A": 53 (adult and juv) unidentified nudibranch "C": 6
	2	Platynereis dumerilli: 1 juv unidentified nudibranch "A": 20 unidentified nudibranch "B": 4
	3	unidentified nudibranch "A": 19 unidentified nudibranch "B": 1
4		22 Jan 72; 23°00'N, 62°46'W
	1	Acerotisa notulata: 9 juv Anoplodactylus petiolatus: 7 juv Planes minutus: 1 juv Litiopa melanostoma: 2 adults, 6 juv

Station	Sample	Observations
SHL 4	2	Planes minutus: 2 juv unidentified nudibranch "A": 1
	3	Platynereis dumerilli: 1 adult, 2 juv Bagatus minutus: 1 adult, 1 juv Biancolina sp.: 1 adult, 2 juv Litiopa melanostoma: 6 juv
5		23 Jan 72; 21°50'N, 61°48'W
	1	Platynereis dumerilli: 2 juv Litiopa melanostoma: 117 juv Scyllaea pelagica: 1 juv unidentified nudibranch "A": 5 adults, 79 juv unidentified nudibranch "B": 1 juv
	2	Anoplodactylus petiolatus: 11 juv Litiopa melanostoma: 23 juv unidentified nudibranch "A": 2 adults, 20 juv
	3	Holoplana grubei: 3 juv Platynereis dumerilli: 1 juv Anoplodactylus petiolatus: 1 adult, 2 juv Litiopa melanostoma: 4 adults, 38 juv Scyllaea pelagica: 2 juv unidentified nudibranch "A": 29 juv Note: some S. fluitans in sample.
6		23 Jan 72; 20°55'N, 61°33'W
	1	Gnesciocers sargassicola: 2 adults, 9 juv Anoplodactylus petiolatus: 1 juv Litiopa melanostoma: 23 juv unidentified nudibranch "A": 4 adults, 11 juv
	2	Gnesciocers sargassicola: 13 juv Anoplodactylus petiolatus: 2 juv Litiopa melanostoma: 2 adults, 14 juv unidentified nudibranch "A" 9 adults, 15 juv
	3	Gnesciocers sargassicola: 1 adult, 3 juv Anoplodactylus petiolatus: 1 juv Litiopa melanostoma: 9 juv unidentified nudibranch "A": 4 adults, 1 juv
7		19 Feb 72; 18°14.2'N, 62°52.5'W
	1	Latreutes fucorum: 3 (1 adult w/eggs) Litiopa melanostoma: 9 juv Scyllaea pelagica: 1 juv unidentified nudibranch "A": 3 juv unidentified nudibranch "B": 1

Station	Sample	Observations
SHL 7	2	Latreutes fucorum: 3 (1 adult w/eggs) Anoplodactylus petiolatus: 7 juv unidentified nudibranch "A": 2 adults, 5 juv
	3	Latreutes fucorum: 3 with eggs Litiopa melanostoma: 12 juv unidentified nudibranch "A": 2 adults, 4 juv
8		18 Feb 72; 18° 6.3' N, 62°44.8'W
	1	Looks like S. fluitans, but might be S. polyceratium Platynereis dumerilli: 1 juv Bagatus minutus: 1 adult, 2 juv unidentified nudibranch "A": 6 adults, 27 juv
	2	Latreutes fucorum: 2 (1 adult w/eggs) unidentified nudibranch "A": 1 adult, 18 juv unidentified nudibranch "B": 1
	3	Latreutes fucorum: 1 adult w/eggs Anoplodactylus petiolatus: 2 juv Litiopa melanostoma: 10 juv unidentified nudibranch "A": 5 juv unidentified nudibranch "B": 1
9		24 Jan 72; 18° 2'N, 60° 47'W
	1	Gnesciooeros sargassicola: 1 juv Litiopa melanostoma: 6 juv unidentified nudibranch "A": 13 juv
	2	Litiopa melanostoma: 6 juv unidentified nudibranch "A": 2 juv
	3	Latreutes fucorum: 2 (1 adult w/eggs) Leander tenuicornis: 1 juv Anoplodactylus petiolatus: 1 juv Litiopa melanostoma: 9 juv unidentified nudibranch "A": 2 juv
10		17 Feb 72; 17°46'N, 62°51'W
	1	Amphiscolops sargassi?: 6 ("uncommon") Platynereis dumerilli: 1 juv four different unidentified polychaetes Bagatus minutus: 1 juv unidentified isopod "crescent moon-shape": 1 Portunus sayi: 1 juv Litiopa melanostoma: 7 juv unidentified nudibranch: 1

Station	Sample	Observations
SHL 11		25 Jan 72; 11°57' N, 61°46'W (Grenada)
	1	Sargassum hystrix (Hystrix hystrix?) unidentified flatworm: 1 →  unidentified polychaete: 1 unidentified amphipod resembling Sunampithoë pelagica: 1 Lepas anatifera: 1 ("uncommon")
	2	Hystrix buxifolium (Sargassum hystrix v. buxifolium) unidentified isopod(crescent shaped) unidentified nudibranch remainder of motile fauna are copepods

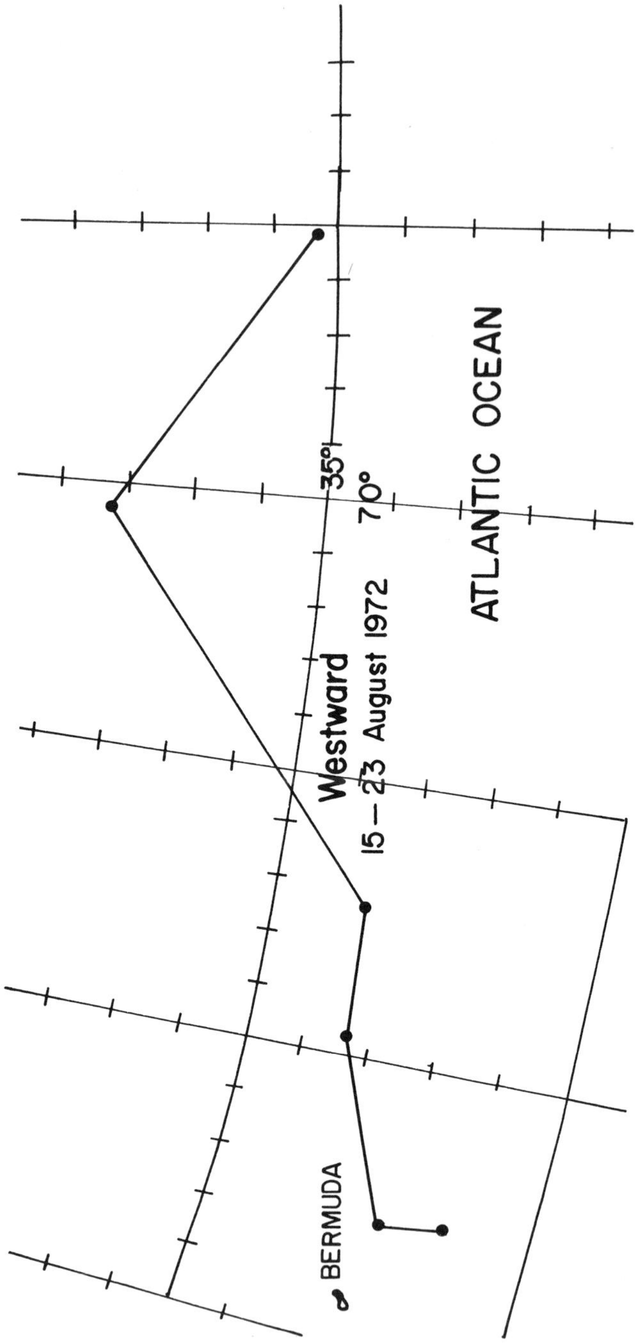
ATLANTIC OCEAN
Westward
15 — 23 August 1972
35°
70°
BERMUDA

MOTILE MACROFAUNA ON SARGASSUM

Cruise 2													
Station		#1			#2			#3			#4		#5
Date		15 Aug 72			16 Aug 72			17 Aug 72			21 Aug 72		23 Aug 72
Sample No.		1	2	3	4	5	6	7	8	9	10	11	12
Location:		32°40'N 63°0'W			33°25'N 59°45'W			33°35'N 57°20'W			38°25'N 50°15'W		35°36'N 45°3'W
Wet weight, g													
Displacement vol. ml.		100	28	52	90	60	80	55	40	20	70	110	100
Sargassum species:		N	N	N	N	N	N	F	N	N	N	N	N
	Fauna:												
44	Anemonia sargassensis							4			45	5	2
46	Acerotisa notulata				2	1							11
47	Hoploplana grubei						9						32
48	Gnescioceros sargassicola	2	1	4	4	2	11	14	8		19	18	47
51	Platynereis dumerillii	29	16	12	5	21	79	24	27	4	18	21	46
70	Bagatus minutus	13	1	3			28	11			23	2	12
73	Sunampithoë pelagica							3			8		
74	Biancolina sp.	1		1		1	16						1
76	Hemiaegina minuta			18	10			260	2			1	
80	Latreutes fucorum						1		1	1	6	4	8
82	Leander tenuicornis					1	1						2
83	Planes minutus	1		3				3	1		11	10	1
84	Portunus sayi												
87	Anoplodactylus petiolatus	24	1			40	1		3		42	8	22
89	Litiopa melanostoma	5	6			5	5	1	3	1	5	24	77
95	Scyllaea pelagica										1		1
100	Histrio histrio												
101	Syngnathus pelagicus												
	Other Species:				3	3	17		1	1			7
Total Number of Specimens:		75	25	41	24	74	168	320	46	7	178	93	269

Note: F = *S. fluitans*; N = *S. natans*;

WESTWARD CRUISE 1: June 1972

One station only: 29 June 1972. 37°40'N, 70°10' W
Sargassum natans I, Sargassum natans II,
Sargassum V, Sargassum natanx IX.
Displacement volume 95 ml total.
Motile fauna not sorted or counted.

WESTWARD CRUISE 2: August 1972

Station	Sample	Observations
1	1	At 32°40'N, 63°00'W. Time 0900. 15 Aug 72 Platynereis dumerillii: 13 adults, 16 juv Planes minutus: 1 adult, 2 juv Anoplodactylus petiolatus: 10 adults (3 pregnant), 11 juv
	2	Platynereis dumerillii: 10 adults, 6 juv
	3	Platynereis dumerillii: 9 adults, 3 juv Planes minutus: 3 juv
2	4	33°25'N, 59°45'W. Time 1500, 16 Aug 72 Platynereis dumerillii: 5juv Amphiscolops sargassi: 1 Doto pygmaea: 2
	5	Platynereis dumerillii: 5 adults, 16 juv Anoplodactylus petiolatus: 30 adults, 10 juv Spurilla neapolitana: 1 juv Doto pygmaea: 2
	6	Amphiscolops sargassii: 1 Platynereis dumerillii: 74 adults, 5 juv Doto pygmaea: 16
3	7	33°35'N, 57°20'W. Time 1300 17 Aug 72 Platynereis dumerillii: 21 adults, 3 juv
	8	Platynereis dumerillii: 23 adults, 4 juv Planes minutus: 1 very small juv Anoplodactylus petiolatus: 3 (1 w/eggs) Doto pygmaea: 1
	9	Platynereis dumerillii: 2 adults, 2 juv Hippolyte coerulescens: 1
4	10	38°25'N, 50°15'W. Time 1700 21 Aug 72 Platynereis dumerillii: 16 adults, 2 juv Planes minutus: 3 adults, 8 juv Anoplodactylus petiolatus: 42 (3 w/eggs)
	11	Planes minutus: 10 all small

5 12 35°36'N, 45°03'W. Time 1500 23 Aug 72
Platynereis dumerillii: 28 adults, 18 juv
Latreutes fucorum: 8 (and 4 ? with broken-off rostrum)
Planes minutus: 1 small
Anoplodactylus petiolatus: 20 adults, 2 juv
Litiopa melanostoma: 54 adults, 23 juv
Doto pygmaea: 6
Triggerfish (Balistes carolinensis?): 1 juv

HALIFAX CRUISE

(approx. $62^{o}30'W$, $44^{o}30'N$)

MOTILE MACROFAUNA ON SARGASSUM

	Station	D-22	D-22	D-22	D-22	D-22	D-22
	Date	8 Nov 72	8 Nov 72	8 Nov 72			
	Sample No.	1X	1A	1B	1C	1D	1E
	Wet weight, g.						
	Displacement vol. ml.	440	80	40	50	52	70
	Sargassum species	mixed	mixed	N	mixed	mixed	N
	Fauna:						
44	Anemonia sargassensis		7	24	1	5	
46	Acerotisa notulata		1		1		
47	Hoploplana grubei		2			1	
48	Gnescioceros sargassicola		78	15	61	59	16
51	Platynereis dumerillii		76	2	54	98	1
70	Bagatus minutus		8	2	23	28	
73	Sunampithoë pelagica		4				
74	Biancolina sp.						
76	Hemiaegina minuta		4		1	10	
80	Latreutes fucorum		17	4	7	1	14
82	Leander tenuicornis		3	2			2
83	Planes minutus	3					
84	Portunus sayi	1					
87	Anoplodactylus petiolatus		5	6			2
89	Litiopa melanostoma			2	3		
95	Scyllaea pelagica						
100	Histrio histrio						1
101	Syngnathus pelagicus						
	Other species:		8			1	4
	Total Number of Specimens:	4	213	57	151	203	40

Note: F = S. *fluitans*; N = *S. natans*;

HALIFAX CRUISE

MOTILE MACROFAUNA ON SARGASSUM

	Station	D-23	D-23	D-23	D-24
	Date	8 Nov 72			8 Nov 72
	Sample No.	2A	2B	2C	3A
	Wet weight, g				
	Displacement vol. ml.	65	78	120	14
	Sargassum species	N	N	N	N
	Fauna:				
44	Anemonia sargassensis	45	216	8	
46	Acerotisa notulata				
47	Hoploplana grubei			1	
48	Gnescioceros sargassicola	58	60	81	2
51	Platynereis dumerillii	3	2	38	
70	Bagatus minutus	5	2		6
73	Sunampithoë pelagica				
74	Biancolina sp.				
76	Hemiaegina minuta	2	2		
80	Latreutes fucorum	20	21	21	
82	Leander tenuicornis	4		4	6
83	Planes minutus				
84	Portunus sayi				
87	Anoplodactylus petiolatus	9	2	23	3
89	Litiopa melanostoma		1		7
95	Scyllaea pelagica				1
100	Histrio histrio		1?		
101	Syngnathus pelagicus				
	Other Species		1	16	10
	Total Number of Specimens:	146	308	192	35

Note: F = *S. fluitans*; N = *S. natans*;

HALIFAX CRUISE (approx 60°W, 44°N)
8 November 1972

Station	Sample	Observations
D-22	1x	Trawl. Begin 0900. Surface temp. 73.7 °F Sea state 3, waves 10 ft, wind 8 knots. Tow one nautical mile. Mouth of net under surface about 25% of the time. Rear mouth not completely under. Air forced into plankton net. Missed about 3 good clumps of Sargassum. Sargassum 440 mg; tar 0.795g.
	1A	Dip net. No tar. Platynereis dumerillii: 4 adults, 72 juv Bagatus minutus: 5 adults (2 w/eggs) 3 juv Sunampithoe pelagica: 3 adults, 1 juv Hemiaegina minuta: 4 juv Hippolyte coerulescens: 1 juv Latreutes fucorum: 17 juv Leander tenuicornis: 3 juv Unidentified nudibranch C: 7
	1B	No tar. Platynereis dumerillii: 2 juv Bagatus minutus: 1 adult, 1 juv Latreutes fucorum: 4 juv Leander tenuicornis: 1 adult, 1juv Anoplodactylus petiolatus: 6 incl 2 w/eggs
	1C	No tar Platynereis dumerillii: 8 adults, 46 juv Bagatus minutus: 6 adults (2 w/eggs), 17 juv Latreutes fucorum: 7 juv Litiopa melanostoma: 1 adult, 2 juv
	1D	No tar Platynereis dumerillii: 98 juv Bagatus minutus: 10 adults (2 w/eggs), 18 juv Latreutes fucorum: 1 juv Unidentified nudibranch C: 1
	1E	No tar Platynereis dumerillii: 1 juv Latreutes fucorum: 8 adults(5 w/eggs), 6 juv Leander tenuicornis: 1 adult, 1 juv Histrio histrio: 1 juv "A. sargassicola" (<u>not</u> Gnescioceros sargassicola, which is already on the list. Possibly intended to be Amphiscolops sargassii or Anemonia sargassensis. Listed as "other"): 4
D-23	2x	Trawl. Time 1500. Surface temp. 73.0 °F Sargassum natans. Displaced volume 180 ml Tar 409 mg. No fauna counted.

Station	Sample	Observations
D-23	2A	Dip net. No tar. Platynereis dumerillii: 3 juv Bagatus minutus: 1 adult 4 juv Hemiaegina minuta: 2 juv Latreutes fucorum: 11 adults (9 w/eggs), 9 juv Leander tenuicornis: 2 adults (1 w/eggs), 2 juv Anoplodactylus petiolatus: 9 (3 w/eggs)
	2B	No tar. Platynereis dumerillii: 2 juv Bagatus minutus: 2 juv Latreutes fucorum: 7 adults (6 w/eggs) 14 juv Anoplodactylus petiolatus: 2 (1 w/eggs) Litiopa melanostoma: 1 juv Unidentified nudibranch C: 1 Histrio histrio: 1 juv
	2C	Platynereis dumerillii: 38 juv Latreutes fucorum: 5 adults w/eggs, 16 juv Leander tenuicornis: 3 adults (2 w/eggs), 1 juv Anoplodactylus petiolatus: 23 (7 w/eggs) Unidentified nudibranch C: 16
D-24	3x	Time 2100 hrs. Surface temp. 69.8 °F. No tow.
	3A	Dip net. No tar. Bagatus minutus: 3 adults (1 w/eggs), 3 juv Leander tenuicornis: 1 adult w/eggs, 5 juv Endeis spinosa: 2 adults (1 w/eggs), 1 juv Spurilla neapolitana: 1 Unidentified nudibranch A: 6 (Doto pygmaea?)

(Samples collected by Byron F. Morris)

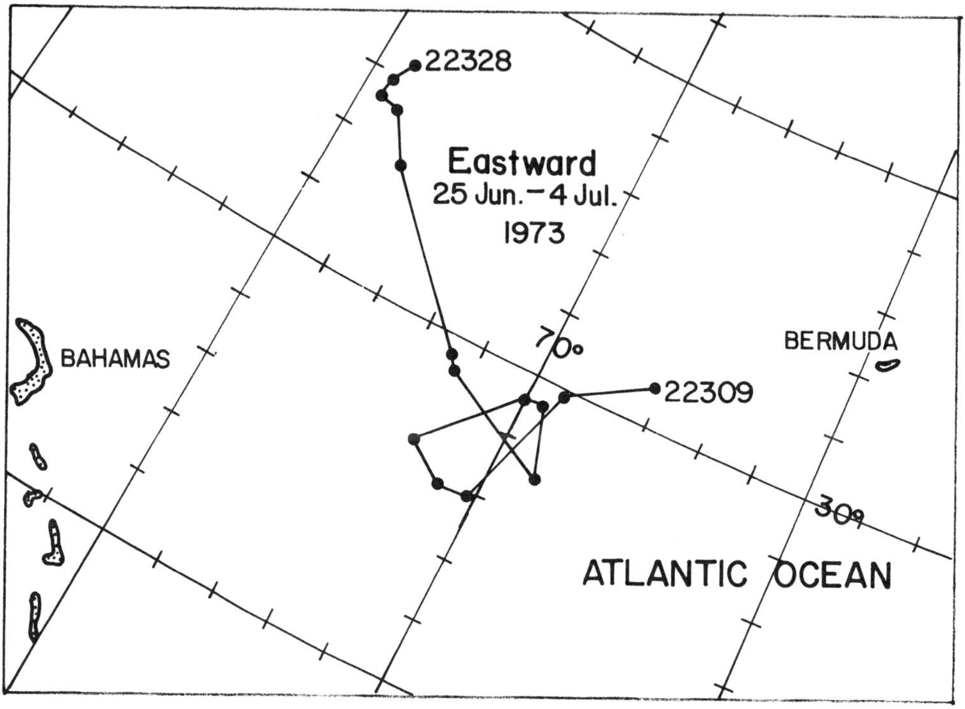
22328
Eastward
25 Jun. – 4 Jul.
1973
70°
BAHAMAS
BERMUDA
22309
30°
ATLANTIC OCEAN

MOTILE MACROFAUNA ON SARGASSUM

	Station	E-22309				E-22310				E-22310A
	Date	25 Jun 73				25 Jun 73				25 Jun 73
	Sample No.	1	2	3	4	1	2	3	4	1
	Location:	30°32'N 68°14.5'W				29°52'N 69°26'W				29°52'N 69°26'W
	Wet weight, g									
	Displacement vol. ml.	105	50	20	50	100	75	68	28	120
	Sargassum species	----------All S. Natans----------								
	Fauna:									
44	Anemonia sargassensis									
46	Acerotisa notulata									
47	Hoploplana grubei	2	11	5	6	5	1			
48	Gnescioceros sargassicola	9	7	2	4	22	8	7	13	14
51	Platynereis dumerillii	53	2	16	35	141	105	29	11	104
70	Bagatus minutus	1	1		3	6	4	9	6	10
73	Sunampithoë pelagica	1	9	4	3	35	17	14	9	39
74	Biancolina sp.	31	13	5	6	27	36	16	25	40
76	Hemiaegina minuta									
80	Latreutes fucorum	2	3	2	6	1		2		3
82	Leander tenuicornis		2		1					1
83	Planes minutus	1		1	2	1	1	1		8
84	Portunus sayi									
87	Anoplodactylus petiolatus									
89	Litiopa melanostoma	3	3		3	1	4	2		3
95	Scyllaea pelagica									
100	Histrio histrio									
101	Syngnathus pelagicus									
	Other Species:					1				1
	Total Number of Specimens:	103	51	35	69	240	176	80	64	223

RV EASTWARD

MOTILE MACROFAUNA ON SARGASSUM

	Station	E-22312					E-22313			
	Date	26 Jun 73					26 Jun 73			
	Sample No.	1	2	3	4	5	1	2	3	4
	Location:	27°56'N 70°02'W					27°50.5N 70°38'W			
	Wet weight, g									
	Displacement vol. ml.	150	160	30	30	45	175	65	40	40
	Sargassum-species	----------	----------	----------	---all	*S.*	*natans*---	----------	----------	----------
	Fauna:									
44	Anemonia sargassensis									
46	Acerotisa notulata		2		1			1		
47	Hoploplana grubei	2	5	15	1	3		1		1
48	Gnescioceros sargassicola	23	12	7	3	6	15	6	4	9
51	Platynereis dumerillii	206	178	23	49	44	214	34	57	14
70	Bagatus minutus	20	11		2		12	2	4	2
73	Sunampithoë pelagica	48	11		11	1	29	3	2	3
74	Biancolina sp.	35	50	7	12		18	9	6	19
76	Hemiaegina minuta						4			
80	Latreutes fucorum		1					1		
82	Leander tenuicornis	2	5		1	1	1			
83	Planes minutus	8	9		2	1	10			
84	Portunus sayi									
87	Anoplodactylus petiolatus						1			
89	Litiopa melanostoma	6	7	4	3	5	5	1		
95	Scyllaea pelagica									
100	Histrio histrio									
101	Syngnathus pelagicus									
	Other Species:					1		1		
	Total Number of Specimens:	350	291	56	85	62	309	59	73	48

RV EASTWARD

MOTILE MACROFAUNA ON SARGASSUM

		E-22317					E-22319			
Station		E-22317					E-22319			
Date		27 Jun 73					28 Jun 73			
Sample No.		1	2	3	4	5	1	2	3	4
Location:		28°25'N 71°23'W					29°38.5N 69°42'W			
Wet weight, g										
Displacement vol. ml.		40	35	30	40	100	55	55	50	50
<u>Sargassum</u> Species		------------------All <u>S</u>. <u>natans</u>-------------------								
	Fauna:									
44	Anemonia sargassensis									
46	Acerotisa notulata					1				
47	Hoploplana grubei	4	3?	1	1	1	4	3		6
48	Gnesioceros sargassicola	3	4	2	2			4	2	3
51	Platynereis dumerillii	47		26	10	28	25	75	4	33
70	Bagatus minutus				4	2		2		
73	Sunampithoë pelagica		1?	3	9	1?	7	1		3
74	Biancolina sp.	2		1			11	9	10	12
76	Hemiaegina minuta									
80	Latreutes fucorum	2	1		7	5	2	2	1	2
82	Leander tenuicornis					2			1	
83	Planes minutus		1		1	3	1	1	4	2
84	Portunus sayi									
87	Anoplodactylus petiolatus									
89	Litiopa melanostoma	1		3		4	2	1	1	
95	Scyllaea pelagica									
100	Histrio histrio									
101	Syngnathus pelagicus									
	Other Species:			1	1				1	
Total Number of Specimens:		59	10	37	35	47	52	98	24	61

RV EASTWARD

MOTILE MACROFAUNA ON SARGASSUM

	Station	E-22321					E-22324				
	Date	29 Jun 73					2 July 73				
	Sample No.	1	2	3	4	5	1	2	3	4	5
	Location:	28°31'N 69°10'W					29°49'N 71°25'W				
	Wet weight, g										
	Displacement vol. ml.	55	55	120	50	45	200	70	70	30	50
	Sargassum species	N	N+F	N	N	N	N	N	N	N	N
	Fauna:										
44	Anemonia sargassensis										
46	Acerotisa notulata						15		6	5	5
47	Hoploplana grubei	1	5	3	6	1	3	1			
48	Gnescioceros sargassicola	2	2	14	5	3	15	6	15	4	5
51	Platynereis dumerillii	10	92	70	46	34	105	1	109	111	145
70	Bagatus minutus	2	5	3	2	1	33	3	13	6	8
73	Sunampithoë pelagica	7	34	1	3	1	12	12	11	3	5
74	Biancolina sp.	17	26	32	39	17	17	3	10	7	3
76	Hemiaegina minuta		1								3
80	Latreutes fucorum	1	4	2	2	1	5			2	1
82	Leander tenuicornis	1?		1			4	1	1	2	
83	Planes minutus	1		2	2	1	1		1		2
84	Portunus sayi						1	1			
87	Anoplodactylus petiolatus										
89	Litiopa melanostoma	3		17	2	7	16	5	1	1	7
95	Scyllaea pelagica										
100	Histrio histrio						1				
101	Syngnathus pelagicus										
	Other Species:			1		1	2		1		1
	Total Number of Specimens:	45	169	146	107	67	230	33	168	141	185

Note: F = S. fluitans; N = S. natans

MOTILE MACROFAUNA ON SARGASSUM

Station		E-22326				E-22327				E-22328		
Date		3 July 73				3 July 73				3 July 73		
Sample No.		1*	2*	3	4*	1*	2	3	4	1	2	3
Location:		31°51'N 73°42'W				32°26'N 74°17'W				32°46'N 74°40'W		
Wet weight, g												
Displacement, vol. ml.		170	60	30	75	80	300	70	45	70	60	40
Sargassum species		N	N	N	N	N	N+F	N	N	F	N	N
	Fauna:											
44	Anemonia sargassensis						?	27?	49?			
46	Acerotisa notulata											
47	Hoploplana grubei									2?		
48	Gnescioceros sargassicola			2								2?
51	Platynereis dumerillii			15			146	76	63	42	13	18
70	Bagatus minutus						173	51	65	7	1	4
73	Sunampithoë pelagica						65	6	20	3		3
74	Biancolina sp.			2			28	13	5	12	4	2
76	Hemiaegina minuta						18	2	30			
80	Latreutes fucorum			4			8	10		5		8
82	Leander tenuicornis			1			2			3	1	
83	Planes minutus						11		1			1
84	Portunus sayi											
87	Anoplodactylus petiolatus						2	1	1			
89	Litiopa melanostoma			2			6	3	2		4	
95	Scyllaea pelagica											
100	Histrio histrio											
101	Syngnathus pelagicus											
	Other Species:			2			33+	41	11	2		6
Total Number of Specimens:				28			492	230	247	76	23	44

Note: F = S. fluitans; N = S. natans

*Fauna not counted. See notes.

EASTWARD CRUISE E-6B: June-July 1973

Station	Sample	Observations
E-22309		30°32'N, 68°15'W. Neuston tow 1031 to 1046 hrs @ 3 knots, 25 June 1973. Air temp. 80.5°F, Sea state: 3. 4 Sarg. clumps. 14.06 g = 32.0 mg/m^2 ta (Tar collections described by Sleeter and Butler, Deep-Sea Research 1974, 21: 773-5)
	1	Platynereis dumerillii: 42 adults, 11 juv Latreutes fucorum: 2 (1 with parasite)
	2	Platynereis dumerillii: 2 juv Litiopa melanostoma: 2 adults, 1 juv
	3	Platynereis dumerillii: 9 adults, 7 juv
	4	Platynereis dumerillii: 22 adults, 13 juv Latreutes fucorum: 6 incl 1 with parasite
E-22310		29°52'N, 69°26'W. Neuston tow 1836 to 1851 hrs. 25 June 1973. Surface temp. 26.5°C, sea state 3. One of the five bags of Sargassum contains 2 clumps but they were entangled together and packaged as E-22310-A, all others labeled E-22310 with formalin. 18.97 g net tar = 28.1 mg/m^2.
	1	Amphiscolops sargassii: 1 Platynereis dumerillii: 123 adults, 18 juv
	2	Platynereis dumerillii: 98 adults, 7 juv
	3	Platynereis dumerillii: 17 adults, 12 juv
	4	Platynereis dumerillii: 6 adults, 5 juv
E-22310-A		Platynereis dumerillii: 79 adults 25 juv Planes minutus: 2 med, 5 small, 1 tiny Parasitic isopod: 1
E-22312		27°56'N, 70°02'W. Neuston tow @3 knots 1511 to 1526 hrs 26 June 1973. Air Temp 87.1°F, Surface temp. 27.2°C. Sea state 3 126.6 g net tar = 90.6 mg/m^2.
	1	Platynereis dumerillii: 197 adults, 9 juv Planes minutus: 4 adults, 3 juv, 1 very small Litiopa melanostoma: 3 large, 3 small
	2	Platynereis dumerillii: 92 adults, 86 juv Planes minutus: 6 adults, 3 juv
	3	Platynereis dumerillii: 14 adults, 9 juv Litiopa melanostoma: 2 large, 2 small

Station	Sample	Observations
E-22312	4	Platynereis dumerillii: 32 adults, 17 juv Planes minutus: 1 adult, 1 large juv. form
	5	Platynereis dumerillii: 26 adults, 18 juv Litiopa melanostoma: 3 large, 2 small Parasitic isopod: 1
E-22313		27°50'N, 70°38'W. Neuston tow @3 knots 1836 to 1851 hrs 26 June 1973 Air temp. 84°F, Surface 27.3°C, sea state 3 22.79 g net tar = 16.4 mg/m^2. Many large crabs observed at this station.
	1	Platynereis dumerillii: 175 adults, 39 juv Planes minutus: 6 adults, 4 smaller
	2	Platynereis dumerillii: 28 adults, 6 juv Hippolyte coerulescens: 1 (no dorsal spine on rostrum)
	3	Platynereis dumerillii: 36 adults, 21 juv
	4	Platynereis dumerilllii: 5 adults, 9 juv
E-22317		28°25'N, 71°23'W. Neuston tow @ 3 knots 1311 to 1326 hours 27 June 1973 Air temp. 85°F, suface temp 27.3°C Sea state 3, wind 10 knots 110° (raining?) 89.96 g net tar = 64.7 mg/m^2
	1	Platynereis dumerillii: 37 adults, 10 juv
	2	Planes minutus: 1 small Hoploplana grubei: 3 mangled
	3	Platynereis dumerillii: 22 adults, 4 juv Unidentified (mangled) shrimp): 1
	4	Platynereis dumreillii: 9 adults, 1 juv Planes minutus: 1 small Parasitic isopod: 1
	5	Platynereis dumerillii: 9 adults, 19 juv Planes minutus: 2 small, 1 juv Litiopa melanostoma: 3 large, 1 small
E-22319		29°38'N, 69°42'W. Neuston tow @ 3 knots 1538 to 1553 hrs 28 June 1973 Surface temp 26.5°C, wind 8 knots 045° Sea state 2, hard rain, high swells 13.64 g net tar = 9.81 mg/m^2
	1	Platynereis dumerillii: 15 adults, 10 juv

Station	Sample	Observations
E-22319	2	Platynereis dumerillii: 51 adults, 24 juv Litiopa melanostoma: 1 extremely small Planes minutus: 1 small
	3	Platynereis dumerillii: 1 adult 3 juv Planes minutus: 3 small, 1 juv Latreutes fucorum. 1 with larval isopod on gill (Cryptoniscan larva, order isopoda, suborder epicaridea)
	4	Platynereis dumerillii: 11 adults, 21 juv
E-22320		29°22'N, 69°36'W. 1538 to 1556 hrs 28 June 1973 Surface Temp. 26.5°C, Wind 8 knots 045° Sea state 2: hard rains, high swells. "During the 15 minute tows on the starboard side, Sargassum clumps were gathered with dip nets on the port side. All the clumps were put in plastic bags and preserved in formalin. All organisms were gathered from the net and placed in the bags. From the neuston tows, only the tar was retrieved. The weed was pulled out clump by clump and the tar removed by hand and placed in plastic bags, then frozen."- T.Sleeter No motile macrofauna counts made on these samples
E-22321		28°31, 69°10'W. Neuston tow @ 3 knots 1005 to 1022 hrs 29 June 1973 Surface Temp. 26.9°C, wind 17 knots 010° Sea state 2. 44.53 g net tar = 28.27 mg/m^2
	1	Platynereis dumerillii: 4 adults, 6 juv Litiopa melanostoma: 3 small
	2	Platynereis dumerillii: 81 adults, 11 juv
	3	Platynereis dumerillii: 47 adults, 23 juv Litiopa melanostoama: 1 large, 2 med, 14 small Amphiscolops sargassi: 1 Planes minutus: 1 small
	4	Platynereis dumerillii: 39 adults, 7 juv Litiopa melanostoma: 1 large, 1 small
	5	Platynereis dumerillii: 17 adults, 17 juv Cryptoniscid isopod larva: 1 (Cryptoniscan larva, order isopoda, suborder epicaridea) Litiopa melanostoma: 2 large 5 small

Station	Sample	Observations
E-22324		29°49'N, 71°25'W. Neuston tow at 3 knots 1810 to 1825 2 July 1973 Surface Temp. 26.4°C, wind 22 knots 050° Sea state 2. 5.77 g net tar = 4.15 mg/m^2
	1	Platynereis dumerillii: 74 adults, 31 juv Parasitic isopod:2 Portunus sayi: 1 used in gut analyses 18 June 1974 by T.D. Sleeter
	2	Portunus sayi: 1 used in gut analyses
	3	Platynereis dumerillii: 73 adults, 36 juv 2 halves of unidentifiable creature (see sketch) in with flatworms. (Later designated'Unidentified nudibranch D') See Hunt 1-2 and S-385-4 to 399-1.
	4	Platynereis dumerillii: 66 adults, 45 juv Litiopa melanostoma: 1 very small
	5	Platynereis dumerillii: 62 adults, 83 juv Parasitic isopod: 1 Planes minutus: 1 very small Litiopa melanostoma: 5 large, 2 small
E-22325		29°52'N, 71°28'W. 1845 to 1900 on 2 July 1973 12.37 g net tar = 8.9 mg/m^2.
E-22326		31°51'N, 73°42'W. 1113 to 1128 on 3 July 1973 24.86 g= 17.9 g/m^2 tar. 4 bags Sargassum frozen; formalin not added until 5 July 73 in Cambridge.
	1	5 jars of washings are all clouded with reddish junk. Fauna not counted.
	2	"Like above, with reddish-brown junk. There are few animals, and most of those are in various stages of deterioration. Forget it!"-J.Cadwallader
	3	Specimens are all in bad shape, similar to above but counted anyway. Unidentifiable flatworm: 1 Unidentified shrimp (mangled): 1
	4	Sample deteriorated. Fauna not counted.
E-22327		32°27'N, 74°18'W. 1600 to 1615 on 3 July 73 15.69 g= 11.3 g/m^2 tar. 4 bags Sargassum frozen; formalin added 5 July 73 in Cambridge.
	1	Deteriorated sample. Fauna not counted
	2	This sample was somewhat deteriorated, but for the most part distinguishable. Platynereis dumerillii: 125 adults, 21 juv

Station	Sample	Observations
E-22327	2 Cont.	Flatworms (all to crummy to be identified): 29 Anemones look different from the usual Anemonenia sargassensis. Could this be a fixation artifact? Number not given. Unidentifiable shrimp: 4 Planes minutus: 11 incl 6 with eggs
	3	Flatworms (mostly all too raggedy to be identified): 37 Anemones (see above): 27 Platyneries dumerillii: 51 adults, 25 juv Latreutes fucorum: 10 incl 1 w/eggs Unidentified (mangled) shrimp: 4
	4	Anemones: 49 Flatworms (unidentifiable): 9 Platynereis dumerillii: 52 adults, 11 juv Litiopa melanostoma: 1 large 1 small Shrimp (too badly destroyed to identify): 2 incl 1 with eggs
E-22328		32°46'N, 74°40'W 1839 to 1856, 3 July 1973 Sea state 1, little wind. Tow at 3 knots 10.85 g net tar = 7.80 mg/m². 3 bags of Sargassum frozen. Sargassum very sparse in this area. Found the bottom and 1 side of a plastic bucket in the net with 3 tar lumps greater than 1 cm diameter and serveral smaller ones. Also about 11 small barnacles. The fouling on the bucket looked as if it had been the water for about a month.- T. Sleeter
	1	Platynereis dumerillii: 35 adults, 7 juv Flatworms (H grubei?): 2 Unidentifiable shrimp: 2
	2	Platynereis dumerillii: 12 adults, 1 juv
	3	Gnescioceros sargassicola (?): 2 Unidentifiable flatworms: 3 Platynereis dumerillii: 16 adults, 2 juv Unidentifiable shrimp: 3
E-22329		34°21'N, 76°29'W. 0806 to 0821 on 4 July 1973 9.27 g net tar = 6.67 mg/m². Wind 17 knots 325° No sargassum. (West of the Gulf stream)
E-22330		34°33'N, 76°39'W. 1012 to 1027 on 4 July 1973 1.24 g net tar = 0.89 mg/m². Wind 9.5 knots 270° No sargassum. Hundreds of 5 cm long white tubes that look like polychaete tubes but feel like plant material. Possibly a marsh grass remnant encountered here and in a nearby estuary in May of this year.-T. Sleeter

(Samples collected by Thomas D. Sleeter)

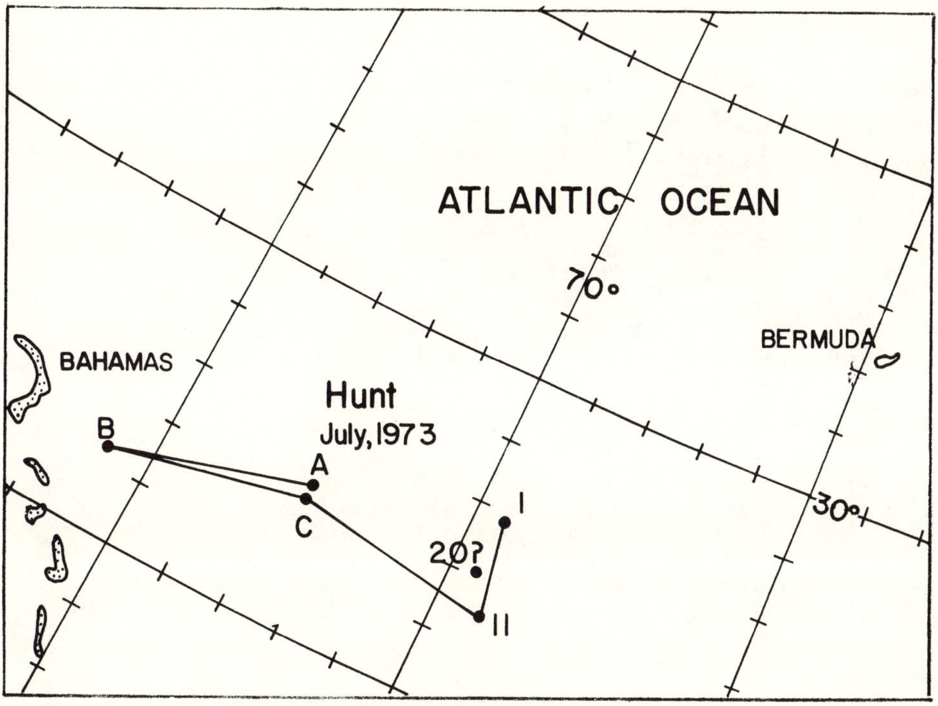
ATLANTIC OCEAN
70°
BERMUDA
BAHAMAS
Hunt
July, 1973
B
A
C
I
20?
II
30°

HUNT CRUISE

MOTILE MACROFAUNA ON SARGASSUM

Station	1				11			Central Mooring		
Date	8 July 73				7 July 73			4 July 73		
Sample No.	1	2	3	4	1	2	3	1	2	3
Location:	27°58'N 69°39.2'W				26°27.8'N 69°19.6'W			27°58'N 69° 39'W		
Wet weight, g										
Displacement vol. ml.	120	50	70	110	69	24	19	18	35	51
<u>Sargassum</u> Species	-------------------All <u>S</u>. <u>natans</u>-------------------------									
Fauna:										
44 Anemonia sargassensis										
46 Acerotisa notulata	5	7		2	13		2			3
47 Hoploplana grubei	13	10	18	17	27	11	5	17	17	21
48 Gnescioceros sargassicola	1	10	21	5	34	8	9	4	4	4
51 Platynereis dumerillii	52	31	15	27	63	13	43	3	32	82
70 Bagatus minutus	1	2			4				9	2
73 Sunampithoë pelagica	3	3	3	4	1	3	1		7	1
74 Biancolina sp.	6	6	7	3	3				6	1
76 Hemiaegina minuta										
80 Latreutes fucorum	5	1	1		1	1	1			
82 Leander tenuicornis				1	1	1				
83 Planes minutus										
84 Portunus sayi				1		1				
87 Anoplodactylus petiolatus										
89 Litiopa melanostoma	31	29	25	25	13	3	4	4	13	13
95 Scyllaea pelagica										
100 Histrio histrio										
101 Syngnathus pelagicus										
Other Species:	1	4	3	5	6		4	3	1	
Total Number of Specimens:	118	103	93	90	166	41	69	31	89	127

HUNT CRUISE

MOTILE MACROFAUNA ON SARGASSUM

Station	A	B				C		
Date								
Sample No.	1	1	2	3	4	1	2	3
Location:	27°04'N 72°40'W	26°11'N 75°47'W				27°07'N 72°40'W		
Wet weight, g								
Displacement vol. ml.	130	25	10	83	98	40	27	55
Sargassum species	---------------All *S. natans*---------------							
Fauna								
44 Anemonia sargassensis								
46 Acerotisa notulata	4		1	7	12			
47 Hoploplana grubei	4			5	1	6	8	
48 Gnescioceros sargassicola	4	3	1	4	25	1	4	
51 Platynereis dumerillii	252	32	12	106	295	57	45	74
70 Bagatus minutus	11	2	1	7	24	6		5
73 Sunampithoë pelagica	3	1		1?	19	6		11
74 Biancolina sp.	4	1		12	3	6		4
76 Hemiaegina minuta								
80 Latreutes fucorum	1	2	1	1		4	1	3
82 Leander tenuicornis	2	2	2	2			1	2
83 Planes minutus			1		2	1		4
84 Portunus sayi								
87 Anoplodactylus petiolatus								
89 Litiopa melanostoma	121	5	2	124	32	11	9	33
95 Scyllaea pelagica								
100 Histrio histrio								
101 Syngnathus pelagicus								
Other Species:	7+	1		7	18	9	4	1
Total Number of Specimens:	413	49	21	276	431	107	72	137

HUNT CRUISE (July 1973)

Station Sample Observations

1 — $27^{\circ}58'$N, $69^{\circ}39'$W, 8 July 1973

1 Amphiscolops sargassii: 1 (greenish tinged?)
Platynereis dumerillii: 50 adults, 2 juv
Litiopa melanostoma: 11 large, 17 med, 3 small

2 Amphiscolops sargassii: 2
Platynereis dumerillii: 18 adults, 13 juv
Litiopa melanostoma: 9 large, 8 med, 12 small
Unidentified nudibranch D: 2 (See sketch below, also Eastward E22324-3 and Station S-385-4 to 399-1)

length 1-2 mm

3 Hoploplana grubei: 18 (2 may be parasitized?)
Platynereis dumerillii: 7 adults, 8 juv
Litiopa melanostoma: 7 large, 13 med, 5 small
Unidentified nudibranch D: 3

4 Gnescioceros sargassicola: 5 (1 with parasite)
Platynereis dumerillii: 20 adults, 7 juv
Portunus sayi: 1 juv
Litiopa melanostoma: 8 large, 10 med, 7 small
Unidentified nudibranch D: 5

11 — 26° $28'$N, $69^{\circ}20'$W, 7 July 1973 (noted $60^{\circ}32.7$W in error)

1 Hoploplana grubei: 27 small
Gnescioceros sargassicola: 34 small
Platynereis dumerillii: 25 adults, 38 juv
Latreutes fucorum: 1 juv
Litiopa melanostoma: 6 large, 2 med, 5 small
Doto pygmaea:3
Unidentified nudibranch D: 3

2 Platynereis dumerillii: 6 adults, 7 juv
Latreutes fucorum: 1 with eggs
Portunus sayi: 1 juv
Litiopa melanostoma: 1 large 2 small

3 Platynereis dumerillii: 16 adults, 27 juv
Litiopa melanostoma: 4 med
Doto pygmaea: 1
Unidentified nudibranch D: 3

Central Mooring — $27^{\circ}58'$N, $69^{\circ}39'$W. 4 July 1973

1 Hoploplana grubei: 17, all small, mostly $<$ 1 mm ("Is it possible that some of these smaller specimens of H. grubei could be a new species? They don't seem to be identical to the usual type, in that they are 'patterned' and some-

Station	Sample	Observations
Central Mooring	1 cont.	times a large oval sucker (?) is present ventrally (like Gnescioceros sargassicola). However, unlike G. sargassicola, they are thick, oval, and don't have eyespots on the tentacles. Representatives of this sort of flatworm are in many of the Hunt samples." - Jill Cadwallader Platynereis dumerillii: 2 adults, 1 juv Litiopa melanostoma: 1 large, 1 med, 2 small Unidentified nudibranch D: 3 (good specimens)
	2	Hoploplana grubei: 17, mostly all <1 mm Platynereis dumerillii: 26 adults, 6 juv Litiopa melanostoma: 1 large 12 small Bagatus minutus: 9 (2 with eggs) Unidentified nudibranch D: 1
	3	Hoploplana grubei: 21, many very small Acerotisa notulata: 3 large Platynereis dumerillii: 48 adults, 34 juv Bagatus minutus: 2 (1 w/eggs) Litiopa melanostoma: 6 large, 7 small
A		$27^{\circ}04'$N, $72^{\circ}40'$W
	1	Hoploplana grubei: 4 small Platynereis dumerillii: 206 adults, 46 juv Bagatus minutus: 11 (1 w/eggs) Litiopa melanostoma: 28 large, 31 med, 62 small Unidentifiable shrimp: 1 Unidentified nudibranch D: 5 Oyster?: 1 juv "flies" (?): several small
B		$26^{\circ}11'$N, $75^{\circ}47'$W
	1	Platynereis dumerillii: 16 adults, 16 juv Litiopa melanostoma:2 large, 1 med, 2 small Parasitic isopod: 1
	2	Platynereis dumerillii: 8 adults, 4 juv Leander tenuicornis: 2 (1 w/eggs) Litiopa melanostoma: 2 small
	3	Hoploplana grubei: 5, all small type Amphiscolops sargassii: 2 Platynereis dumerillii: 77 adults, 29 juv Litiopa melanostoma: 10 large, 15 med, 99 small Unidentified nudibranch D: 5

Station	Sample	Observations
B	4	Amphiscolops sargassi: 7 Platynereis dumerillii: 97 adults, 198 juv Litiopa melanostoma:2 large, 7 med, 23 small Hoploplana grubei: 1 small type Unidentified Nudibranch D: 11 (some very good specimens)
C		27°07'N, 72°40'W
	1	Hoploplana grubei: 5 small type, 1 normal Platynereis dumerillii: 36 adults, 21 juv Latreutes fucorum: 4 (1 w/eggs, 1 with isopod eggs?) Litiopa melanostoma: 1 large 2 med, 8 small Parasitic isopod larva: 1 Unidentified undibranch D: 8
	2	Hoploplana grubei: 6 small, 2 normal Platynereis dumerillii: 39 adults, 6 juv Litiopa melanostoma: 9, all small Unidentified nudibranch D: 4
	3	Planes minutus: 4 (2 w/eggs) Platynereis dumerillii: 62 adults, 12 juv Litiopa melanostoma: 5 large 13 med, 15 small Shrimp (mangled): 1

(Samples collected by Laura Brown)

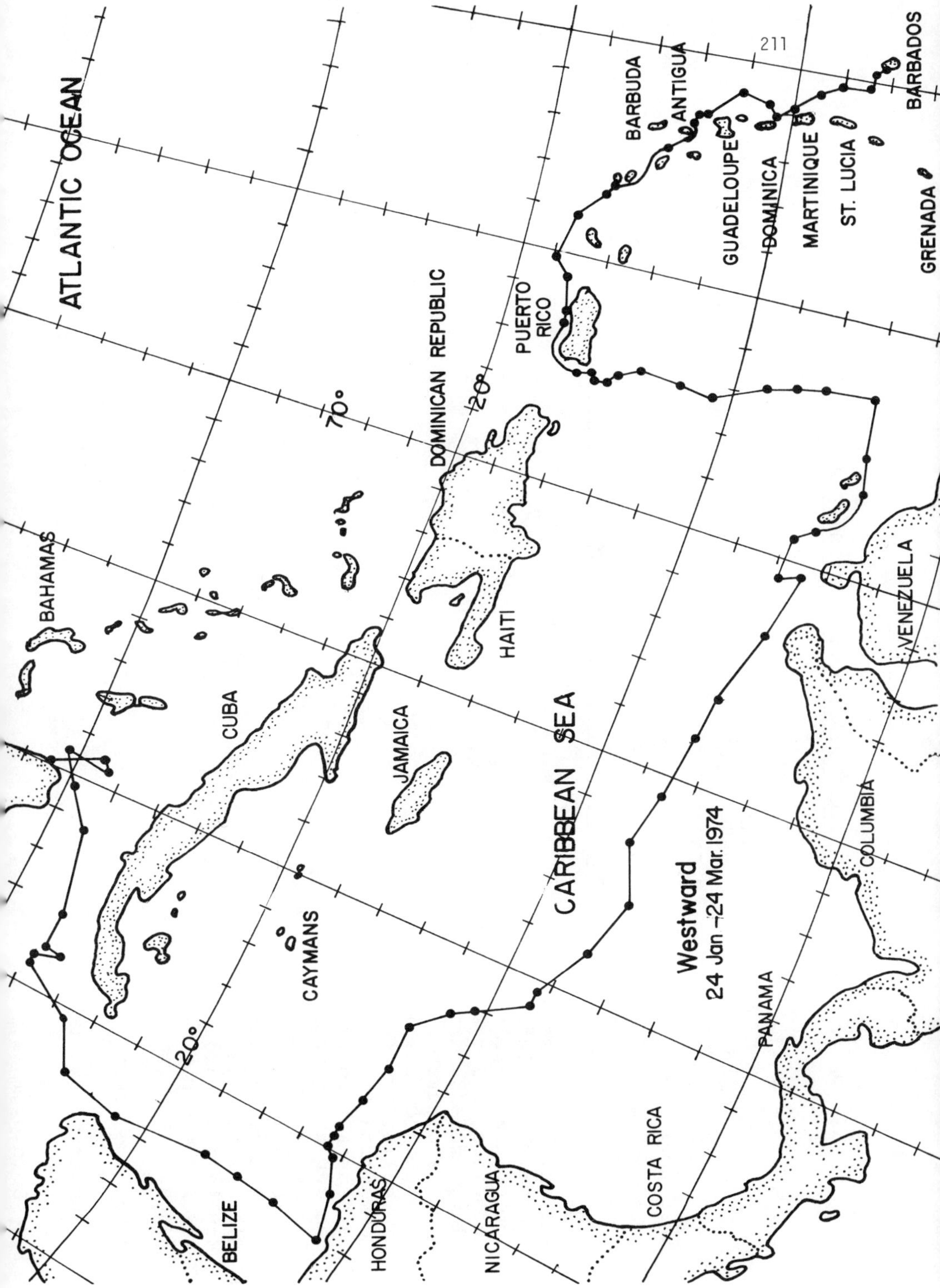
ATLANTIC OCEAN
BAHAMAS
CUBA
CAYMANS
JAMAICA
HAITI
DOMINICAN REPUBLIC
PUERTO RICO
BARBUDA
ANTIGUA
GUADELOUPE
DOMINICA
MARTINIQUE
ST. LUCIA
BARBADOS
GRENADA
CARIBBEAN SEA
Westward
24 Jan -24 Mar. 1974
VENEZUELA
COLUMBIA
PANAMA
COSTA RICA
NICARAGUA
HONDURAS
BELIZE
70°
20°
20°

RV WESTWARD

MOTILE MACROFAUNA ON SARGASSUM

Station							
Date	24 Jan 74					25 Jan 74	26 Jan 74
Time	0954	1002	1004	1005	1007	1720	1325
Location:	16°40'N 61°22'W					17°20'N 62°20'W	18°23'N 63°39'W
Wet weight, g							
Displacement vol. ml.	5	14	18	5	5	80	5
Sargassum species	N	N	N	N	N	N	N
	All *S. natans*						
Fauna:							
44 Anemonia sargassensis							
46 Acerotisa notulata			1				
47 Hoploplana grubei	2	2	3	3	3	11	1
48 Gnesioceros sargassicola		1					
51 Platynereis dumerillii			1				
70 Bagatus minutus	3				1		
73 Sunampithoë pelagica							
74 Biancolina sp.							
76 Hemiaegina minuta							
80 Latreutes fucorum			2	2			
82 Leander tenuicornis			2			10	1
83 Planes minutus						2	2
84 Portunus sayi						2	
87 Anoplodactylus petiolatus	4	4	2		2	4	
89 Litiopa melanostoma	5	8	19	5	17	24	3
95 Scyllaea pelagica							
100 Histrio histrio							
101 Syngnathus pelagicus							
Other Species:*	10	13	21	6	15	44	14
Total Number of Specimens:	24	28	51	16	38	97	21

*Mostly Doto pygmaea. See notes.

MOTILE MACROFAUNA ON SARGASSUM

Station	1							
Date	26 Jan 74		8 Feb 74					
Time	1330	1337	1400	1407	1409	1500*	1502	1514
Location:	18°23'N 63°39'W		18°00'N 67°30'W					
Wet weight, g								
Displacement vol. ml.	18	5	80	80	92	10	5	70
Sargassum species	All *S. natans*							
Fauna:								
44 Anemonia sargassensis								
46 Acerotisa notulata								
47 Hoploplana grubei	1		13	1	13		7?	
48 Gnesioceros sargassicola		1	5	1	2			5
51 Platynereis dumerillii			4	2	6			
70 Bagatus minutus			3					1
73 Sunampithoë pelagica								
74 Biancolina sp.								
76 Hemiaegina minuta								
80 Latreutes fucorum	1		5	7	3		5	3
82 Leander tenuicornis			2	1			1	
83 Planes minutus								
84 Portunus sayi								
87 Anoplodactylus petiolatus	4	8	20	2	29		3	13
89 Litiopa melanostoma	3		11	8	8		8	11
95 Scyllaea pelagica								
100 Histrio histrio								
101 Syngnathus pelagicus								
Other Species:**	8	15	25	34	41		8	27
Total Number of Specimens:	17	24	88	56	102		32	60

*No motile fauna.

**Mostly *Doto pygmaea*, some *Spurilla neapolitana*. See notes.

RV WESTWARD

MOTILE MACROFAUNA ON SARGASSUM

Station								
Date	9 Feb 74			1 Mar 74				
Time	1750	1810	1820	1410	1420*	1430	1440	1450
Location:	17°19'N 67°26'W			13°48'N 78°47'W				
Wet weight, g								
Displacement vol. ml.	5	5	10	5	5	5	5	2
Sargassum species	N	N	N	N	N	N	N	N
Fauna:								
44 Anemonia sargassensis		1		27?		11	112	57
46 Acerotisa notulata								
47 Hoploplana grubei								
48 Gnescioceros sargassicola				10		5	10	4
51 Platynereis dumerillii								
70 Bagatus minutus								
73 Sunampithoë pelagica								
74 Biancolina sp.	1					1		
76 Hemiaegina minuta								
80 Latreutes fucorum	2	1	5			3		1
82 Leander tenuicornis	1							
83 Planes minutus								
84 Portunus sayi								
87 Anoplodactylus petiolatus								
89 Litiopa melanostoma	3?		1			1		
95 Scyllaea pelagica								
100 Histrio histrio								
101 Syngnathus pelagicus								
Other Species			2					
Total Number of Specimens:	7	2	8	37	*	21	122	62

Note: F = *S. fluitans*; N = *S. natans*;

MOTILE MACROFAUNA ON SARGASSUM

Station				A	B	C	D
Date	1 Mar 74			8 Mar 74			
Time	1500	1520	1530	0800	0800	0800	0800
Location:	13°48'N 78°47'W			16°39'N 84°38'W			
Wet weight, g							
Displacement, vol. ml.	5	5	5	40	42	50	60
<u>Sargassum</u> species	N	N	N	?	?	?	?
Fauna							
44 Anemonia sargassensis	105	62					
46 Acerotisa notulata					1		1
47 Hoploplana grubei							
48 Gnescioceros sargassicola	5	2		4	13	13	7
51 Platynereis dumerillii							
70 Bagatus minutus				3	3		9
73 Sunampithoë pelagica				1		1	
74 Biancolina sp.							1
76 Hemiaegina minuta							
80 Latreutes fucorum		1		9	13	2	12
82 Leander tenuicornis					1		1
83 Planes minutus							
84 Portunus sayi							
87 Anoplodactylus petiolatus							
89 Litiopa melanostoma			11				
95 Scyllaea pelagica							
100 Histrio histrio							
101 Syngnathus pelagicus							
Other Species:				4			
Total Number of Specimens:	110	65	11	21	31	16	31

Note: F = <u>S</u>. <u>fluitans</u>; N = <u>S</u>. <u>natans</u>

RV WESTWARD

MOTILE MACROFAUNA ON SARGASSUM

Station							
Date	8 Mar 74						17 Mar 74
Time	0930	0930	0930	1100	1100	1100	1350
Location:	---------16°39'N------ 84°47'W			--------16°40'N------ 84°56'W			23°59'N 84°29'W
Wet weight, g							
Displacement vol. ml.	110	5	45	50	120	180	70
Sargassum species	T	?	?	N	N?	?	?
Fauna:							
44 Anemonia sargassensis	5				2	7	13
46 Acerotisa notulata			1			1	
47 Hoploplana grubei	2					2	
48 Gnescioceros sargassicola	68	4	7	5	1	2	22
51 Platynereis dumerillii	47	1				1	21
70 Bagatus minutus	104	1	6	19	5	43	29
73 Sunampithoë pelagica							
74 Biancolina sp.	2			2	1		2
76 Hemiaegina minuta	8						
80 Latreutes fucorum	45	2	7	5	14	10	6
82 Leander tenuicornis	1		6		1	1	2
83 Planes minutus							
84 Portunus sayi	2						
87 Anoplodactylus petiolatus	3					14	
89 Litiopa melanostoma	1			5			
95 Scyllaea pelagica							
100 Histrio histrio							
101 Syngnathus pelagicus							
Other Species:	18	3	1	6	20	31	1
Total Number of Specimens:	306	11	28	42	44	112	96

Note: F = S. fluitans; N = S. natans; T = Turbinaria turbinata

RV WESTWARD

MOTILE MACROFAUNA ON SARGASSUM

	Station	
	Date	24 Mar 74
	Time	1740
	Location:	24°11'N 79°48'W
	Wet weight, g	
	Displacement vol. ml.	150
	Sargassum species	S. natans
	Fauna:	
44	Anemonia sargassensis	
46	Acerotisa notulata	1
47	Hoploplana grubei	
48	Gnescioceros sargassicola	8
51	Platynereis dumerillii	43
70	Bagatus minutus	
73	Sunampithoë pelagica	1
74	Biancolina sp.	1
76	Hemiaegina minuta	1
80	Latreutes fucorum	9?
82	Leander tenuicornis	2
83	Planes minutus	
84	Portunus sayi	6
87	Anoplodactylus petiolatus	18
89	Litiopa melanostoma	7
95	Scyllaea pelagica	
100	Histrio histrio	
101	Syngnathus pelagicus	
	Other Species:	
	Total Number of Specimens:	97

WESTWARD CRUISES 13 and 14 (January-March 1974)

Date	Time	Observations
24 Jan 74	0950	Hoploplana grubei: 1 small Latreutes fucorum: 1 w/eggs Litiopa melanostoma: 6 med Doto pygmaea: 8
	0954	Hoploplana grubei: 2 (1 very small) Anoplodactylus petiolatus: 4 (1 w/eggs, 1 very small) Litiopa melanostoma: 2 large, 3 med Doto pygmaea: 9 Unidentified nudibranch D: 1
	1000	Position interpolated: 16°40'N, 61°22'W between Stations W-13-8 (0210 hrs @ 16°05'N,60°47'W) Station W-13-9 (1200 hrs at 16°49'N, 61°31'W)
	1002	Anoplodactylus petiolatus: 4 (1 w/eggs) Litiopa melanostoma: 4 med, 4 small Doto pygmaea: 13
	1004	Hoploplana grubei: 3 (1 very small) Leander tenuicornis: 2 (1 w/eggs) Anoplodactylus petiolatus: 2 w/eggs Litiopa melanostoma: 2 large, 7 med, 10 small Doto pygmaea: 18 Fiona pinnata (?):2 small Unidentified nudibranch D: 1
	1005	Hoploplana grubei: 3 (1 very small) Latreutes fucorum: 2 (1 w/eggs) Litiopa melanostoma: 3 med, 2 small Doto pygmaea: 6 (1 very small)
	1007	Hoploplana grubei: 3 (2 very small) Litiopa melanostoma: 4 large, 5 med, 2 small Doto pygmaea: 11 Unidentified nudibranch D: 4
25 Jan 74	1720	Position interpolated: 17°20'N, 62°20'W between Stations W-13-9 (1200 hrs 24 Jan) and W-13-10 (0030 hrs 26 Jan at 17°28'N, 62°32'W) Latreutes fucorum: 10 (4 w/eggs) Portunus sayi: 2 small Litiopa melanostoma: 5 large, 16 med, 3 small Doto pygmaea: 43 Unidentified nudibranch D: 1

Date	Time	Observations
26 Jan 74	1325	Position interpolated: 18°23'N, 63°39'W between Stations W-13-11 (1130 h @ 18°12'N,63°26'W W-13-12 (1900 hrs at 18°55'N, 64°15'W Leander tenuicornis: 2 large (1 w/eggs) Doto pygmaea: 14
	1330	Doto pygmaea: 8
	1337	Doto pygmaea: 11  Unidentified creature (see sketch): 4 (Apparently the same as Unidentified nudibranch D. See 8 March 0930 sample)
8 Feb 74	1400	Position approx. 18°N, 67°30'W (just west of Puerto Rico), preceeding station W-14-1 (9 Feb 74, 0716 hrs) Hoploplana grubei: 13 small Gnescioceros sargassicola: 5 large Platynereis dumerillii: 1 large, 3 small Bagatus minutus: 3 (1 w/eggs) Tozeuma carolinense ("rare"): 1 small Latreutes fucorum: 1 small, 3 med w/eggs, 1 large Spurilla neapolitana: 5 Doto pygmaea: 18 Unidentified nudibranch: 1
	1407	Latreutes fucorum: 7 (3 w/eggs) Spurilla neapolitana:1 Doto pygmaea: 31 Flatworm (very small H. grubei?): 2
	1409	Platynereis dumerillii: 3 adults, 3 juv Anoplodactylus petiolatus: 29 (4 w/eggs) Spurilla neapolitana: 6 Doto pygmaea: 29 Unidentified nudibranch D: 6
	1500	No motile fauna
	1502	Hoploplana grubei: 1 normal, 6 very small (possibly another flatworm species?) Spurilla neapolitana: 1 Doto pygmaea: 7
	1514	Latreutes fucorum: 3 (1 w/eggs) Spurilla neapolitana: 9 Doto pygmaea: 18

Date	Time	Observations
9 Feb 74	1750	Position interpolated: 17°19'N, 67°26'W between stations W-14-2 (9 Feb, 1245 hrs) and W-14-3 (10 Feb, 0024 hrs) Notebook entry "Liddioca" interpreted as "Litiopa melanostoma"
	1810	Position interpolated: 17°17'N, 67°26'W Latreutes fucorum: 1 small adult
	1820	Latreutes fucorum: 5 (2 w/eggs, 1 very small) Doto pygmaea: 2
1 March 74	1410	Position interpolated: 13°48'N, 78°47'W between W-14-21 (1300 hrs 27 Feb at 13°33'N, 77°58'W) and W-14-22 (2000 hrs 5 March at 14°18'N, 80°27'W) Vessel speed assumed constant for that six-day period in the absence of other data (approx 1 knot). Notebook entry "Anemone" interpreted as "Anemonia sargassensis: but might be an unidentified species.
	1420	No specimens? No animals counted
	1430	"Anemone" interpreted "Anemonia sargassensis":11 Latreutes fucorum: 3 juv Litiopa melanostoma: 1 very small
	1440	Anemonia sargassensis: 112
	1450	"Anenome" interpreted "Anemonia sargassensis":5
1 March 74	1500	"Anemone" interpreted "Anemonia sargassensis":10
	1520	"Anemone": 62 Latreutes fucorum: 1 very small
	1530	Litiopa melanostoma: 11 (No other animals counted)
8 March 74	0800	Position interpolated: 16°39'N, 84°38'W between W-14-27 (0030 hrs at 16°36'N, 83°53'W) and W-14-28 (1230 hrs at 16°40'N, 85°05'W)
	0800A	Latreutes fucorum: 4 adults, 5 juv Unidentified nudibranch D: 4

Date	Time	Observations
8 Mar 74	0800B	Latreutes fucorum: 13 (2 w/eggs) Leander tenuicornis: 1 juv
	0800C	Latreutes fucorum: 2 (1 w/eggs)
	0800D	Latreutes fucorum: 12 (5 w/eggs) Leander tenuicornis: 1 juv
	0930	Hoploplana grubei:2 tiny Platynereis dumerillii: 40 adults, 7 juv Bagatus minutus: 104 (20 w/eggs, 22 juv) Latreutes fucorum: 16 adults (7 w/eggs), 29 juv Leander tenuicornis: 1 juv Hemiaegina minuta: 5 adults, 3 juv Portunus sayi: 2 juv Anoplodactylus petiolatus: 3 (1 w/eggs) Parasitic isopod: 1 Unidentified nudibranch D: 9 Unidentified creature (see sketch): 2 "Fairly large"- J.C. Doto pygmaea: 6
	0930	Position interpolated: $16^{\circ}39'$N, $84^{\circ}47'$W between W-14-27 (0030 hrs) and W-14-28 (1230 hrs)
	0930	Unidentified nudibranch D: 2 (same as Unidentified creature in sample of 26 Jan 74 1337) Triggerfish (Balistes carolinensis?):1 juv
	0930	Latreutes fucorum: 6 adults (2 w/eggs) 1 juv Leander tenuicornis: 1 small adult, 2 juv Triggerfish (Balistes carolinensis?): 1 juv
	1100	Position interpolated: $16^{\circ}40'$N, $84^{\circ}56'$W between W-14-27 (0030 hrs) and W-14-28 (1230 hrs)
	1100	Latreutes fucorum: 5 (1 w/eggs) Unidentified nudibranch D: 6
	1100	Latreutes fucorum: 14 (4 w/eggs) Unidentified nudibranch D: 20 Leander tenuicornis: 1 juv
	1100	Hoploplana grubei (?): 1 very small Platynereis dumerillii: 1 juv Bagatus minutus: 43 (1 w/eggs) Latreutes fucorum: 10 (1 w/eggs, 3 very small) Unidentified nudibranch D: 28 Triggerfish (Balistes carolinensis?): 2 very small

Date	Time	Observations
17 Mar 74	1350	Position interpolated: 23°59' N, 84°29'W between stations W-14-37 (1335 hrs at 24°00'N,84°30'W) and W-14-38 (0000 hrs at 23°21'N,84°05'W)
	1350	*Platynereis dumerillii*: 20 adults, 1 juv *Bagatus minutus*: 29 (2 w/eggs) *Latreutes fucorum*: 6(3 w/eggs) Parasitic isopod: 1
24 Mar 74	1740	Position interpolated: 24°11'N, 79°48'W between W-14-45 (1720 hrs, 24 March at 24°08'N,79°48'W)and W-14-46 (0525 hrs March 25 at 25°30'N,80°02'W) *Gnescioceros sargassicola*: 8 (1 small) *Latreutes fucorum*: 9 (4 w/eggs, 3 small adults)

Note: These samples were collected by Byron F. Morris, Jan Tuttleman, and Paul Deslauriers.

Most *Sargassum* collections were made *en route* and did not correspond to the stations for neuston and tar collections. These were described by T.D. Sleeter, B.F. Morris and J.N. Butler in *Deep-Sea Research* 1976, *23*: 467-474. Table 2 of that paper is reproduced below:

Date	Station number	Location	Tar (mg/m²)
21 Jan. 74	1 (W-13-1)	13°35'N, 59°50'W	8.37
22 Jan. 74	2 (W-13-2)	13°34'N, 60°05'W	1.27
	3 (W-13-3)	14°10'N, 60°10'W	1.24
	4 (W-13-4)	14°31'N, 60°30'W	0.17
	5 (W-13-5)	14°55'N, 60°50'W	0.19
23 Jan. 74	6 (W-13-6)	15°13'N, 61°11'W	0.03
	7 (W-13-7)	15°37'N, 60°56'W	0.32
24 Jan. 74	8 (W-13-8)	16°05'N, 60°47'W	0.30
	9 (W-13-9)	16°49'N, 61°31'W	0.27
26 Jan. 74	10 (W-13-10)	17°28'N, 62°32'W	0.37
	11 (W-13-11)	18°12'N, 63°23'W	0.02
	12 (W-13-12)	18°55'N, 64°15'W	2.87
27 Jan. 74	13 (W-13-13)	19°07'N, 65°05'W	0.23
	14 (W-13-14)	18°50'N, 65°28'W	0.07
9 Feb. 74	15 (W-14-1)	18°01'N, 67°28'W	13.4
	16 (W-14-2)	17°45'N, 67°33'W	0.31
10 Feb. 74	17 (W-14-3)	16°44'N, 67°18'W	0.34
	18 (W-14-4)	15°57'N, 67°19'W	1.9
	19 (W-14-5)	15°22'N, 67°18'W	0.96

Date	Station number	Location	Tar (mg/m^2)
11 Feb. 74	20 (W-14-6)	14°23'N, 66°50'W	0.39
	21 (W-14-7)	13°42'N, 66°41'W	1.62
	22 (W-14-8)	13°03'N, 66°39'W	0.23
12 Feb. 74	23 (W-14-9)	12°08'N, 66°36'W	0.004
19 Feb. 74	24 (W-14-10)	12°02'N, 67°47'W	0.023
20 Feb. 74	25 (W-14-11)	11°55'N, 68°32'W	0.09
22 Feb. 74	26 (W-14-12)	11°18'N, 69°18'W	0.043
23 Feb. 74	27 (W-14-13)	11°38'N, 69°46'W	0.002
	28 (W-14-14)	13°00'N, 70°32'W	0.0
24 Feb. 74	29 (W-14-15)	12°32'N, 70°22'W	0.0
25 Feb. 74	30 (W-14-16)	12°52'N, 71°48'W	0.002
	31 (W-14-17)	13°23'N, 73°13'W	0.36
26 Feb. 74	32 (W-14-18)	13°29'N, 74°13'W	0.25
	33 (W-14-19)	13°39'N, 75°33'W	3.63
27 Feb. 74	34 (W-14-20)	13°49'N, 76°40'W	1.7
	35 (W-14-21)	13°33'N, 77°58'W	10.5
28 Feb. 74	36 (W-14-21A)	14°20'N, 80°15'W	2.43
5 Mar. 74	37 (W-14-22)	14°18'N, 80°27'W	4.9
6 Mar. 74	38 (W-14-23)	15°11'N, 81°05'W	0.67
	39 (W-14-24)	15°40'N, 81°23'W	0.06
7 Mar. 74	40 (W-14-25)	16°13'N, 82°00'W	0.64
	41 (W-14-26)	16°28'N, 83°04'W	0.37
8 Mar. 74	42 (W-14-27)	16°36'N, 83°53'W	8.5
	43 (W-14-28)	16°40'N, 85°05'W	1.43
	44 (W-14-29)	16°17'N, 85°55'W	0.01
12 Mar. 74	45 (W-14-30)	16°03'N, 86°46'W	0.014
13 Mar. 74	46 (W-14-31)	17°06'N, 86°30'W	0.16
	47 (W-14-32)	18°18'N, 86°30'W	0.61
14 Mar. 74	48 (W-14-33)	18°55'N, 86°27'W	0.45
15 Mar. 74	49 (W-14-34)	20°48'N, 86°49'W	0.41
16 Mar. 74	50 (W-14-35)	22°17'N, 86°19'W	0.18
	51 (W-14-36)	22°48'N, 85°17'W	0.06
17 Mar. 74	52 (W-14-37)	24°00'N, 84°30'W	6.0
18 Mar. 74	53 (W-14-38)	23°21'N, 84°05'W	3.0
	54 (W-14-39)	23°45'N, 84°14'W	0.11
19 Mar. 74	55 (W-14-40)	23°45'N, 83°11'W	0.043
	56 (W-14-41)	24°15'N, 81°13'W	0.023
20 Mar. 74	57 (W-14-42)	24°48'N, 80°22'W	2.82
	58 (W-14-43)	25°25'N, 79°33'W	0.03
22 Mar. 74	59 (W-14-44)	24°14'N, 79°59'W	0.009
24 Mar. 74	60 (W-14-45)	24°09'N, 79°48'W	0.14
25 Mar. 74	61 (W-14-46)	25°30'N, 80°02'W	1.02

WESTWARD CRUISE 22 (June 1975)

Date	Time	Sample	Observations
22 June 75	2050		29°52'N, 60°29'W
		059	Sargassum natans, 53 ml. Fauna not counted.
		060	Sargassum fluitans, 32 ml. Platynereis dumerillii: 5 Biancolina sp.: 1 Planes minutus: 2 Litiopa melanostoma: 5 med, 14 large
		061	Sargassum fluitans, 20 ml Biancolina sp.: 1 Litiopa melanostoma: 2 large
24 June 75	2000	068	Sargassum fluitans, 170 ml. Not counted 30°42'N, 61°14'W
		069	Sargassum natans, 80 ml. Fauna not counted.
28 June 75	1130	077	29°00'N,64°30'W Sargassum natans, 50 ml. Not counted
		078	Sargassum natans, 70 ml. Not counted
		079	Sargassum natans, 20 ml. Not counted
29 June 75	1400	081	30°20'N, 64°52'W Sargassum natans, 60 ml. Not counted
		082	Sargassum natans, 60 ml. Not counted
		083	Sargassum natans, 70 ml. Not counted

(Samples collected by E.H. Wheeler)

APPENDIX V

SESSILE FAUNA AND FLORA:
Observations by

Deborah D. Mogelberg
Byron F. Morris
Jill Cadwallader

1972-73

SESSILE FAUNA AND FLORA

Station	Sample	Observations
346	1	Clytia noliformis: abundant (90%) new growth, healthy, few gonophores. Clytia johnstoni (= Clytia hemisphaerica): rare (2%) one dense colony only, new growth, healthy, with gonophores. Occurs with Obelia hyalina. Obelia hyalina: rare (5%), middle region and new regions, no gonangium. Plumularia halecioides: rare(1%), healthy, in washings Sertuleria cornicina: rare(1%), washings only Membranipora tuberculata: healthy, approx 5% of surface Spirorbis violaceous (= corrugatus; see Morris and Mogelberg, 1973)
346	2	(see chart) Membranipora tuberculata: a large number of new colonies Clytia noliformis: a few gonophores on 2^{O}F Eggs found over large area Encrusting calcareous spp. present on some stems & leaves(2^{O}S)*

(In each cell: entry above the diagonal / entry below the diagonal)

		2^{O}S	2^{O}L	2^{O}F	3°S	3°L	3°F
Zanclea costata	rare	10 / 1	5 / 1	– / –	– / –	– / –	– / –
Clytia noliformis	common	1 / 1	8 / 3	4 / 2	3 / 10	10 / 10	5 / 10
Obelia dichotoma	v. rare	– / –	+ / +	– / –	– / –	– / –	– / –
Obelia hyalina	v. rare	– / –	– / 1	– / –	– / –	– / –	– / –
Plumularia setaceoides	v. rare	+ / +	+ / +	– / –	– / –	– / –	– / –
Spirorbis sp.	rare	– / –	– / –	10 / 2	+ / +	– / –	– / –
Membranipora tuberculata	common	6 / 10	10 / 10	4 / 10	2 / 2	1 / 1	– / –
Algae "A"	rare	+ / +	– / –	+ / +	– / –	+ / +	+ / +
Algae "B"	common	1 / 1	10 / 4	3 / 4	+ / +	– / –	– / –

These Sargassum samples contain no primary stem (1^{O}S). Secondary stems (2^{O}S), leaves (2^{O}L), floats (2^{O}F) and tertiary stems, leaves, floats were sampled. Entries above and below the diagonal lines are replicate observations.

---

*Spindler (1980) identified these white encrusting organisms (noted as "Flustra?" in the tables below) as foraminifera: Planorbulina acervalis and Rosalina globularis. See his paper for details. As of this writing, no quantitative survey of their population has been carried out.

ation	Sample	Observations
346	3	sessile fauna not counted
3	4	Clytia noliformis: a few gonophores on 2°L and 3°L. Circular pattern of eggs found on 2°S, 3°S, 3°L:

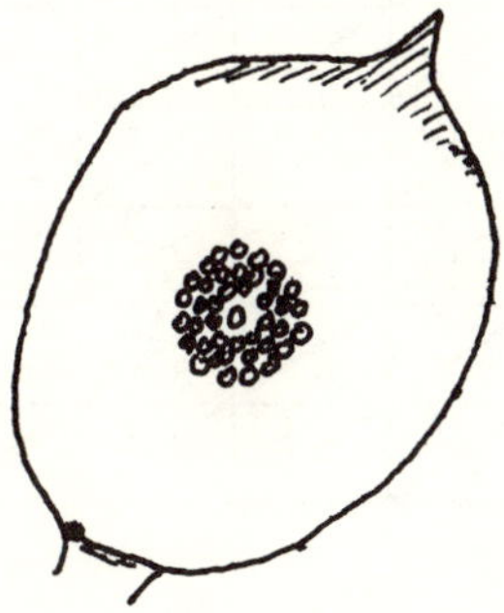

		2°S	2°L	2°F	3°S	3°L	3°F
anclea costata	rare	+ / +	− / −	− / −	− / −	− / −	− / −
lytia noliformis	com.	1 / 2	7 / 5	− / −	2 / 6	10 / 10	5 / 10
belia dichotoma	rare	+ / +	10 / 4	5 / 10	− / −	1 / 2	− / −
lumularia strictocarpa	rare	− / −	+ / +	10 / 5	− / −	− / −	− / −
lustra sp.?	common	4 / 10	10 / 10	1 / 1	5 / 10	− / −	+ / +
lgae "A"	rare	− / −	+ / +	+ / +	− / −	+ / +	− / −
lgae "B"	less rare	3 / 1	− / −	10 / 3	+ / +	− / −	− / −

The encrustation tentatively identified as the bryozoan Flustra sp. may be a foram (Spindler 1977) Plumularia strictocarpa is called P. sargassi in the lab notebook. These species are the same (Morris and Mogelberg 73).

346	5	No observations of sessile fauna
347	1	Tertiary floats (3°F) practically bare See chart next page. Note on Flustra sp. as above.

Station Sample Observations

347 1 (cont.)

		2^oS	2^oL	2^oF	3^oS	3^oL	3^oF
Zanclea costata	rare	+ / +	- / -	10 / 1	- / -	- / -	- / -
Clytia noliformis		- / -	2 / 1	+ / +	2 / 10	10 / 10	- / 2
Plumularia strictocarpa		- / -	10 / 2	- / -	- / -	2 / 2	- / -
Spirorbis sp.	rare	- / -	- / -	+ / +	- / -	- / -	- / -
Membranipora tuberculata		10 / 4	- / -	+ / +	- / -	- / -	- / -
Flustra sp.?	common	6 / 10	10 / 10	9 / 10	- / -	- / -	- / -
Algae "A"	common	7 / 3	4 / 2	10 / 8	1 / 4	+ / 1	1 / 10
Algae "B"	common	10 / 5	6 / 5	5 / 3	- / -	+ / +	- / -

347 2 Sample unusually bare.
Egg sac found on 3^oL, covering edge area
Plumularia strictocarpa: only one colony found (2^oS)

		2^oS	2^oL	2^oF	3^oS	3^oL	3^oF
Clytia noliformis	common	4 / 10	10 / 10	3 / 10	6 / 10	8 / 10	5 / 10
Plumularia strictocarpa		+ / +	- / -	- / -	- / -	- / -	- / -
Flustra sp.?	rare	+ / +	10 / 1	- / -	- / -	- / -	- / -
Algae "A"	rare	+ / +	- / -	- / -	- / -	- / -	- / -
Algae "B"	rare	10 / 5	- / -	1 / 2	- / -	- / -	- / -

347 3

		2^oS	2^oL	2^oF	3^oS	3^oL	3^oF
Clytia noliformis	common	+ / 3	+ / +	+ / 2	10 / 10	4 / 10	3 / 10
Obelia dichotoma		- / -	6 / 10	10 / 10	- / -	- / -	- / -
Plumularia strictocarpa		10 / 10	- / -	- / -	- / -	- / -	- / -
Algae "A"	rare	10 / 5	2 / 5	3 / 5	4 / 3	1 / 1	1 / 1

Station	Sample	Observations
347	4 to 7	No observations of sessile organisms
347	8	Membranipora tuberculata unusually abundant Plumularia strictocarpa: new colonies on 3°L Spirorbis sp.: only one specimen on 3°L Sketch of Algae "C":

This sample is unusual in having a primary stem.

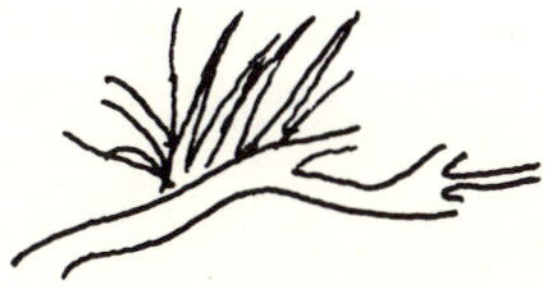

(Each cell: upper-right value / lower-left value)

		1°S	2°S	2°L	2°F	3°S	3°L	3°F
Zanclea costata	rare	10 / 1	5 / 1	– / –	2 / 1	– / –	– / –	– / –
Obelia dichotoma	common	– / –	3 / 1	+ / +	4 / 1	9 / 10	10 / 10	6 / 8
Plumularia strictocarpa		– / –	2 / 2	+ / +	+ / 1	10 / 7	8 / 5	– / –
Spirorbis sp.	rare	– / –	– / –	– / –	– / –	– / –	+ / +	– / –
Membranipora tuberculata		6 / 9	10 / 10	9 / 10	8 / 10	5 / 4	2 / 6	1 / 10
Algae "A"	common	10 / 10	5 / 6	3 / 2	4 / 5	+ / +	1 / 2	1 / 1
Algae "B"	common	2 / 5	9 / 5	10 / 8	5 / 4	1 / 1	– / –	– / –
Algae "C"	rare	10 / 2	2 / 3	1 / 1	+ / +	– / –	– / –	– / –

347 9

		2°S	2°L	2°F	3°S	3°L	3°F
Clytia noliformis	common	1 / 10	– / –	2 / 2	8 / 10	10 / 10	7 / 10
Plumularia strictocarpa	com.	+ / +	5 / 10	2 / 4	– / –	10 / 8	/ 1
Flustra sp.?	few	2 / 2	1 / 2	10 / 10	– / –	– / –	– / –
Algae "A"	few	2 / 1	10 / 6	4 / 5	– / –	– / –	+ / +
Algae "B"	rare	– / –	+ / +	+ / +	– / –	– / –	– / –

Station	Sample	Observations
348	1	Aglaophenia latecanaria: common (50%), no gn. Plumularia setaceoides: few (5%), no gnph. Spirorbis sp. : few Membranipora tuberculata: rare

Station	Sample			Observations					
348	2								
				2°S	2°L	2°F	3°S	3°L	3°F
Clytia noliformis		common		3 / 8	6 / 10	- / -	10 / 10	9 / 10	2 / 10
Algae "A"		f. common		10 / 2	3 / 2	2 / 10	+ / +	+ / +	+ / +
Flustra?		rare		- / -	+ / 1	- / -	+ / 2	+ / 2	+ / +
Plumularia setaceoides		rare		10 / 4	- / -	- / -	- / -	- / -	- / -
Zanclea costata		rare		+ / +	- / -	- / -	+ / +	- / -	- / -
Algae "B"		rare		10 / 1	- / -	5 / 1	2 / 1	- / -	- / -

348 5

		2°S	2°L	2°F	3°S	3°L	3°F
Membranipora tuberculata	common	5 / 10	10 / 10	6 / 10	- / -	- / -	+ / 10
Algae "A"	few	10 / 3	+ / +	6 / 4	+ / +	+ / +	+ / +
Algae "B"	common	4 / 6	8 / 6	10 / 8	- / -	- / -	- / -
Obelia dicholoma	rare	10 / 1	- / -	2 / 1	- / -	- / -	- / -
Algae "C"	few	- / -	10 / 3	3 / 2	- / -	- / -	- / -
Spirorbis violaceus*	v. rare	- / -	- / -	+ / +	- / -	- / -	- / -
Zanclea costata*	v. rare	+ / +	- / -	- / -	- / -	- / -	- / -

*Only one found.

348 6

		2°S	2°L	2°F	3°S	3°L	3°F
Plumularia strictocarpa	common	10 / 10	9 / 10	3 / 8	2 / 4	5 / 3	8 / 10
Clytia johnstoni*	common	2 / 5	- / -	- / -	6 / 10	10 / 10	4 / 6
Zanclea Costata	rare	+ / 1	- / -	+ / 1	- / -	- / -	- / -
Algae "A"	few	+ / +	7 / 1	10 / 10	1 / 1	+ / +	- / -
Flustra?	few	10 / 3	- / -	5 / 6	- / -	- / -	- / -
Spirorbis violaceus	rare	- / -	- / -	+ / +	- / -	- / -	- / -

*not present on 2°L, 2°F; re-appears 3°S, 3°L, 3°F

Station Sample Observations

349 2

		1°S	2°S	2°L	2°F	3°S	3°L	3°F
Algae "A"	few	1 / 10	10 / 6	6 / 5	3 / 5	- / -	+ / 10	+ / 10
Algae "B"	few	1 / 5	10 / 5	7 / 4	5 / 4	- / -	- / -	- / -
Clytia noliformis	rare	1 / 1	+ / +	- / -	10 / 1	- / -	+ / +	- / -
Membranipora tuberculata	few	- / -	10 / 10	6 / 10	8 / 10	- / -	- / -	- / -
Obelia dichotoma	rare	- / -	- / -	10 / 1	- / -	- / -	- / -	- / -
Spirorbis violaceus	rare	- / -	+ / +	- / -	10 / 2	- / -	- / -	- / -
Zanclea costata	rare	- / -	- / -	- / -	- / -	+ / +	- / -	- / -

Only hydroid stems found in sample. Remainder possibly in washings or eaten by organisms.

353 1

		1°S	2°S	2°L	2°F	3°S	3°L	3°F
Algae "A"	common	8 / 10	10 / 10	2 / 2	1 / 3	- / -	- / -	- / -
Aglaeophenia latecarinata*	common	1 / 2	2 / 2	9 / 10	6 / 10	1 / 10	10 / 10	5 / 10
Zanclea costata	rare	- / -	10 / 2	+ / 1	- / -	- / -	- / -	- / -
Membranipora tuberculata	rare	- / -	- / -	+ / +	+ / +	- / -	- / -	- / -
Flustra?	v. rare	- / -	- / -	+ / +	- / -	- / -	- / -	- / -
Plumularia setaceoides**	rare	- / -	- / -	- / -	1 / +	- / -	- / -	- / -

*Originally identified as <u>Aglaeophenia minuta</u>. Found several colonies with gonophores (2°F) (3°L) (3°F)
**Found only on oldest floats (2°)

353 2

		1°S	2°S	2°L	2°F	3°S	3°L	3°F
Algae "A"	rare	+ / +	+ / 1	- / -	10 / 2	+ / 1	+ / 1	+ / 1
Clytia noliformis	common	- / -	2 / 5	8 / 10	7 / 10	2 / 2	10 / 10	7 / 10
Obelia sp.*	rare	- / -	- / -	- / -	+ / 1	- / -	- / -	- / -
Zanclea costata	few	+ / +	3 / 10	4 / 5	+ / 1	10 / 10	2 / 7	1 / 2
Flustra sp?	common	- / -	2 / 5	10 / 10	2 / 3	+ / 1	+ / 1	3 / 10
Membranipora tuberculata	rare	- / -	- / -	- / -	+ / 1	- / -	- / -	- / -

*no gonophores. Only one colony found

Station	Sample		Observations
353	3		

Species	Abundance	1°S	2°S	2°L	2°F	3°S	3°L	3°F
Aglaeophenia latecarinata[1]	Abundant	2 / 10	3 / 10	6 / 10	6 / 10	8 / 10	10 / 10	1[illegible] / 10
Plumularia setaceoides[1]	common	1 / 2	10 / 4	2 / +	2 / 6	3 / +	1 / +	1 / +
Clytia noliformis[2]		- / -	- / -	- / -	- / -	- / +	- / -	- / -
Halecium nanum[3]	few	8 / 3	10 / 2	- / -	+ / 2	- / -	- / -	- / -
Zanclea costata[4]	few	1 / 1	/ 1	- / -	- / -	10 / 2	- / -	- / -
Spirorbis violaceus	rare	- / -	+ / +	+ / +	+ / +	- / -	- / -	- / -
Membranipora tuberculata[5]	few	- / -	4 / 8	10 / 2	10 / 5	- / -	2 / 1	1 / 1
Algae "A"[6]	v. few	+ / 1	+ / +	+ / +	+ / +	- / -	- / -	+ / 1
Zanclea Gemmaria[7]	rare	- / -	- / -	- / -	1 / +	- / -	- / -	- / -

1 entire growth, no gonangia
2 very few; washings mostly - stem 3 (possibly from S. natans found in sample
3 old age growth, no gonangia
4 middle & old aged growth, few with medusard buds
5 encrusting approx. 12 bladders & stems, older growth
6 old growth, occurring on 3-4 stems
7 secondary stems only, no med.

Station	Sample
353	4

Species	Abundance	1°S	2°S	2°L	2°F	3°S	3°L	3°F
Aglaeophenia latecarinata	common	- / -	3 / 10	9 / 10	7 / 10	4 / 10	10 / 10	8 / 10
Halecium nanum	few	- / -	10 / 6	- / -	2 / 2	6 / 2	- / -	- / -
Algae	rare*	- / -	+ / +	+ / +	+ / +	10 / 3	+ / +	+ / +
Zanclea costata	few	- / -	8 / 3	- / -	2 / 2	10 / 4	- / -	- /
Spirorbis violaceus	v. rare**	- / -	+ / +	- / -	+ / +	+ / +	- / -	- / -
Membranipora tuberculata	rare	- / -	1 / +	4 / 1	10 / 2	- / -	+ / +	/ -
Plumularia setaceoides	few	- / -	10 / 8	- / -	5 / 4	- / -	- / -	- / -

*only found in small patches
**only ∿ 12 in sample

ation Sample Observations

353 6

		1°S	2°S	2°L	2°F	3°S	3°L	3°F
Clytia noliformis*	common	1 / 1	4 / 10	9 / 10	6 / 10	3 / 10	10 / 10	5 / 10
Membranipora tuberculata	common	10 / 10	1 / 2	1 / 1	2 / 5	+ / +	+ / +	+ / +
Spirorbis violaceus	rare	10 / 2	- / -	+ / +	+ / +	- / -	- / -	- / -
Zanclea costata**	rare	- / -	- / -	- / -	- / -	- / -	- / -	+ / +

*gonophores found on several colonies (3°F)
**only found on one float

354 1

		1°S	2°S	2°L	2°F	3°S	3°L	3°F
Aglaeophenia latecarinata*	common	- / -	1 / 2	9 / 10	1 / 10	1 / 1	10 / 10	2 / 5
Plumularia setaceoides	common	- / -	5 / 10	2 / 1	6 / 5	10 / 10	7 / 6	9 / 10
Membranipora tuberculata	rare	- / -	+ / +	10 / 1	4 / 2	- / -	+ / +	2 / 1
Zanclea costata	few	- / -	10 / 6	- / -	1 / 1	9 / 5	2 / 1	1 / 1
Spirorbis violaceus	none	- / -	- / -	- / -	- / -	- / -	- / -	- / -
Halecium nanum **	v. rare	- / -	2 / 1	10 / 1	- / -	- / -	- / -	- / -

*few gonophores on (2°S) (2°L) (2°F) (3°S) (3°L)
**only oldest part of Sargassum, few leaves of (2°L)

354 4

		1°S	2°S	2°L	2°F	3°S	3°L	3°F
Algae "A" *	few	2 / 10	10 / 4	- / -	5 / 2	- / -	- / -	- / -
Aglaeophenia latecarinata**	common	- / -	6 / 10	9 / 10	7 / 10	4 / 10	10 / 10	8 / 10
Zanclea costata	rare	- / -	1 / +	10 / 2	- / -	+ / +	+ / +	2 / 1
Membranipora tuberculata	few	- / -	6 / 1	2 / 1	10 / 7	+ / +	- / -	- / -
Plumularia setaceoides***	rare	- / -	+ / 1	+ / 1	+ / 1	+ / 1	- / -	- / -
Spirorbis violaceus	rare	- / -	+ / +	+ / +	+ / +	- / -	- / -	- / -
Halecium nanum	v. rare	- / -	10 / 1	- / -	- / -	- / -	- / -	- / -

*fairly common in isolated areas
**few gonophores on (3°F) (3°L) (3°S) (2°S) (2°L) (2°F)
***not common enough for rank

Station Sample

354 5

Species	Abundance	1°S	2°S	2°L	2°F	2°S	3°L	3°F
Membranipora tuberculata	common	7 / 10	10 / 10	8 / 10	9 / 10	- / -	1 / 3	- / -
Algae "A"	common	10 / 6	7 / 6	3 / 2	1 / 1	- / -	- / -	- / -
Clytia noliformis	common	1 / 5	2 / 5	9 / 8	7 / 6	3 / 10	10 / 10	8 / 10
Halecium nanum	very rare	- / -	+ / +	- / -	- / -	- / -	- / -	- / -
Zanclea costata	very rare	- / -	+ / +	- / -	- / -	- / -	- / -	- / -
Plumularia setaceoides	rare	- / -	+ / 2	+ / 1	+ / +	- / -	- / -	- / -
Algae "B"	rare*	- / -	+ / 1	- / -	- / -	- / -	- / -	- / -
Spirorbis violaceus	rare	- / -	- / -	- / -	+ / +	- / -	- / -	- / -

(Observations: each cell shows upper / lower values.)

*isolated area of growths (2°S)

APPENDIX VI

11 Transfer of Petroleum and Biogenic Hydrocarbons in the Sargassum Community

B. F. Morris, J. Cadwallader, J. Geiselman and *J. N. Butler*

Introduction

Between 10^{12} and 10^{13} g of petroleum are annually introduced into the marine environment (NAS, 1975) and occur as floating oil slicks and tar lumps, as particles (10 to 300 μm sized oil droplets, tar specks, or hydrocarbons adsorbed on other kinds of particles) suspended in the water column, and as dissolved or colloidal material dispersed throughout the water column.

Studies in the Sargasso Sea have indicated the concentrations of some of these fractions throughout the water column. At the sea surface, tar lumps average about 10 mg/m^2 (Morris, 1971; Butler *et al.*, 1973) while the dispersed fraction in the surface microlayer contains about 10 μg/m^2 (Keizer and Gordon, 1973; Wade and Quinn, 1975). Within the water column, hydrocarbon concentrations (total particulate and dispersed) range from 10 μg/l near the surface to less than 1 μg/l at depths greater than 100 m (NAS, 1975; Gordon and Keizer, 1974). About 0.5 μg/l is in the form of tar particles (Morris *et al.*, 1976), and the remainder is in the form of colloidal micelles or adsorbed onto seston. Only the lightweight and highly polar hydrocarbons are truly dissolved (McAuliffe, 1974).

The incorporation of hydrocarbons from the environment into marine organisms has been shown to occur via adsorption of dissolved compounds from the water (Lee *et al.*, 1972a, b; Stegeman and Teal, 1973) as well as by ingestion (Conover, 1971; Parker, 1971). Whether these compounds are then metabolized by the organisms, directly released, or subsequently transferred along the food web is known in very few cases. There is some evidence that food web magnification does *not* occur (Burns and Teal, 1973; Lee *et al.*, 1972b; Corner, 1975; Corner *et al.*, 1975). Marine organisms can synthesize hydrocarbons (Blumer *et al.*, 1964; Clark, 1966; Blumer *et al.*, 1971) but it is often difficult to distinguish petroleum pollutants from the natural (biogenic) hydrocarbons produced by the organisms themselves or transferred along the food chain.

A number of criteria for distinguishing biogenic from petroleum hydrocarbons have been developed (NAS, 1975). We have employed those based on gas chromatography, which depend on measuring the relative amounts of the most abundant nonpolar compounds (usually normal paraffins, isoprenoids, or olefins). Specifically:

1. Distinctive, relatively large peaks (often olefinic) are almost certainly biogenic.

2. A series of normal alkanes from C_{16} to C_{33} where the adjacent peaks are of essentially equal size usually indicates petroleum. However, some special biogenic sources, such as bacterial cultures (Davis, 1968) or mixtures of phytoplankton (Clark and Blumer, 1967) also show this pattern.

3. A series of normal alkanes where the odd-numbered compounds are about 1.7 times the concentration of even-numbered compounds is typical of biogenic sources, particularly phytoplankton (Clark, 1966; Clark and Blumer, 1967).

4. The presence of pristane (C_{19} isoprenoid) but not phytane (C_{20} isoprenoid) indicates a biogenic source (Blumer *et al.*, 1964).

5. The presence of both pristane and phytane in a ratio of 1.5 to 2.5 is typical of petroleum (Blumer and Sass, 1972). Petroleum also may contain a C_{18} isoprenoid which is not biogenic.

6. A large envelope of unresolved compounds (alicyclic, branched aliphatic, etc.) is typical of petroleum, but its size relative to the alkane peaks depends on the column resolution. Hydrocarbons extracted from unpolluted organisms rarely, if ever, have so many compounds present in such small amounts. (Note that the extraction procedure and column chromatography on silica/alumina prior to gas chromatography removes aromatics, fatty acids, and other complex mixtures which might produce such a pattern.)

The pelagic *Sargassum* community contains a variety of organisms in a relatively simple ecological framework (Morris and Mogelberg, 1973). These organisms live in association with petroleum pollution, either in the form of tar lumps floating at the surface (Butler *et al.*, 1973) or as dispersed hydrocarbons in the water column or in the surface microlayer (Wade and Quinn, 1975). Clark (1966) found a range of hydrocarbons in *Sargassum;* Burns and Teal (1973) reported a significant level of petroleum hydrocarbon contamination of both the algae and the fauna.

This work was undertaken to educidate the transfer processes of hydrocarbons in the *Sargassum* community, in hopes that it would help us understand the larger and more inaccessible open ocean ecosystem. In particular, the questions asked were (1) What are the amounts of petroleum hydrocarbons in *Sargassum* and in the organisms of its associated community? (2) What is the ratio of biogenic to petroleum hydrocarbons in the various members of the community? (3) How are these hydrocarbons transferred through the food web? (4) Are hydrocarbons absorbed directly from the environment? (5) Does the association of a particular *Sargassum* community with pelagic tar increase the amount of petroleum hydrocarbons in its members? (6) Does association with pelagic tar alter the ecology of the community?

Not all these questions have yet been answered. The results presented in this paper represent partial answers to the first three questions, for *Sargassum* and the larger animals associated with it. For the motile and sessile microfauna, or the epiphytic algae, the only information available concerns their position in the food web, or in some cases their abundance, but much work remains to be done.

Materials and Methods

The procedures used in the sampling, extraction, and gas-chromatographic analysis of the samples will be reported elsewhere in detail (Morris, Cadwallader and Butler, 1976). Essentially, these procedures follow the recommendations of Grice *et al.* (1972) for avoiding sample contamination and of Burns and Teal (1973) and Zafiriou *et al.* (1972), with some modifications suggested by Clark and Finley (1974) for extraction.

The samples were dip-netted near Bermuda, taken to the laboratory and sorted the same day, and stored frozen until analysis. Extraction was by the Soxhlet procedure. All reagents were redistilled and purity checked, and chemicals, handling equipment etc. were cleaned in twice-distilled pentane before use. The nonpolar hydrocarbon fraction resulting from silica-alumina column chromatography was analyzed by gas-liquid chromatography on a nonpolar column (Supelco 2100). The temperature was programmed from 50 to 300° C, and *n*-alkanes were quantitatively determined in the C_{14} - C_{40} range, by comparison of peak height ratios with known standards. A standard and a blank were run with each day's sample for calibration. The unresolved envelope was measured by planimetry and quantified by comparison to the area of the *n*-alkane peak occurring at maximum envelope height. Typical chromatograms are shown in Figures 11-1 – 11-3.

Alkanes were identified by their retention indices. The quantities of the various hydrocarbons in each sample were determined from the calibration and the known amount of the total sample injected. These hydrocarbons were classified into four groups: "biogenic," "petroleum *n*-alkanes," "unresolved envelope," and "others." Biogenic peaks (normally near the retention index for a normal alkane) were distinguished by an abundance of greater than twice the adjacent *n*-alkane peaks. Petroleum *n*-alkanes were categorized as those whose abundance was similar along the series from C_{14} up to C_{40} (odd:even carbon ratio = 1). The unresolved envelope was measured between the base of the hydrocarbon peaks and the column bleed baseline as determined daily. Other measurable peaks with retention indices differing from the normal-alkanes occurred, usually in small quantities, and were probably either branched petroleum alkanes or biogenic alkenes (Blumer, Guillard and Chase, 1971; Clark and Blumer, 1967).

Sargassum samples were collected on 18 different occasions between 1972 and 1975, all from the vicinity of Bermuda. The individual species comprising the community were sorted and when enough material of one species was collected, it was extracted and analyzed for hydrocarbon content. To date, only

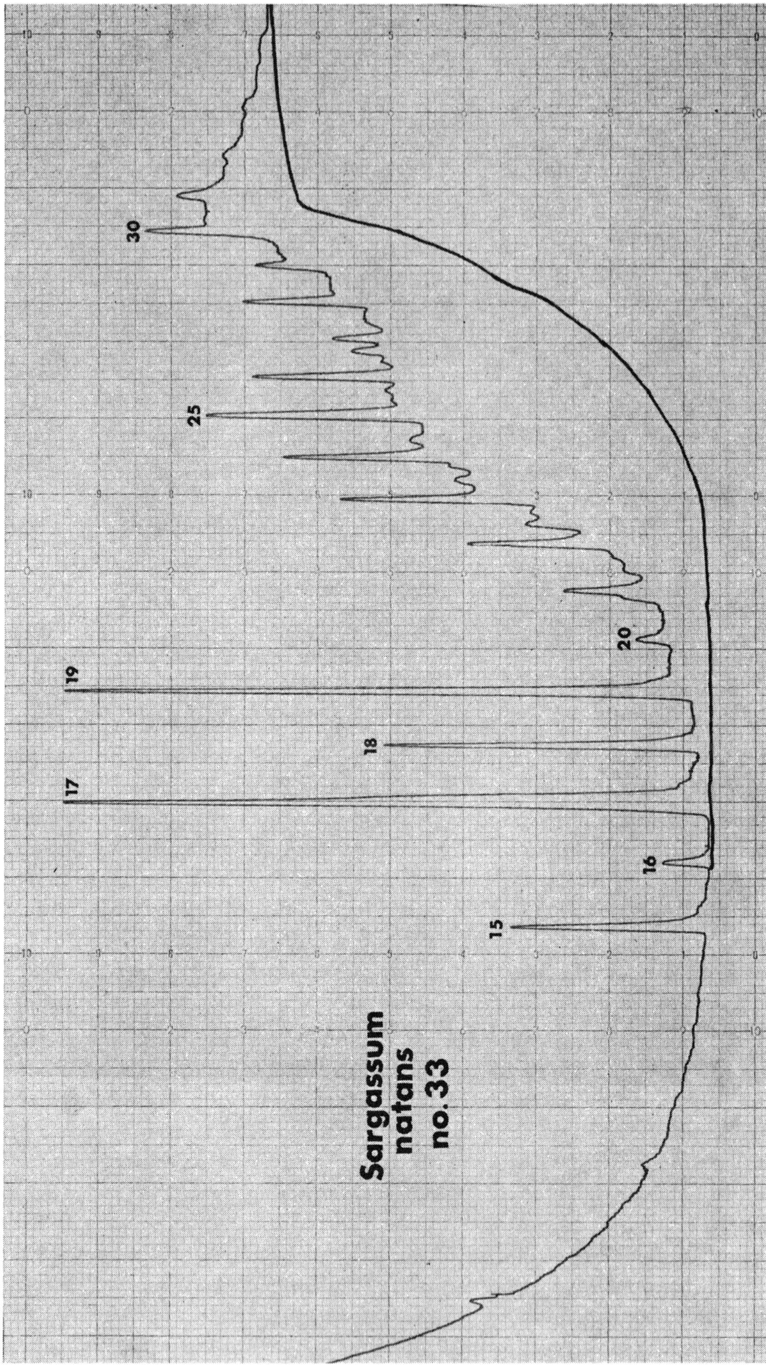

Figure 11-1. Typical gas chromatogram of extract from *Sargassum natans.*

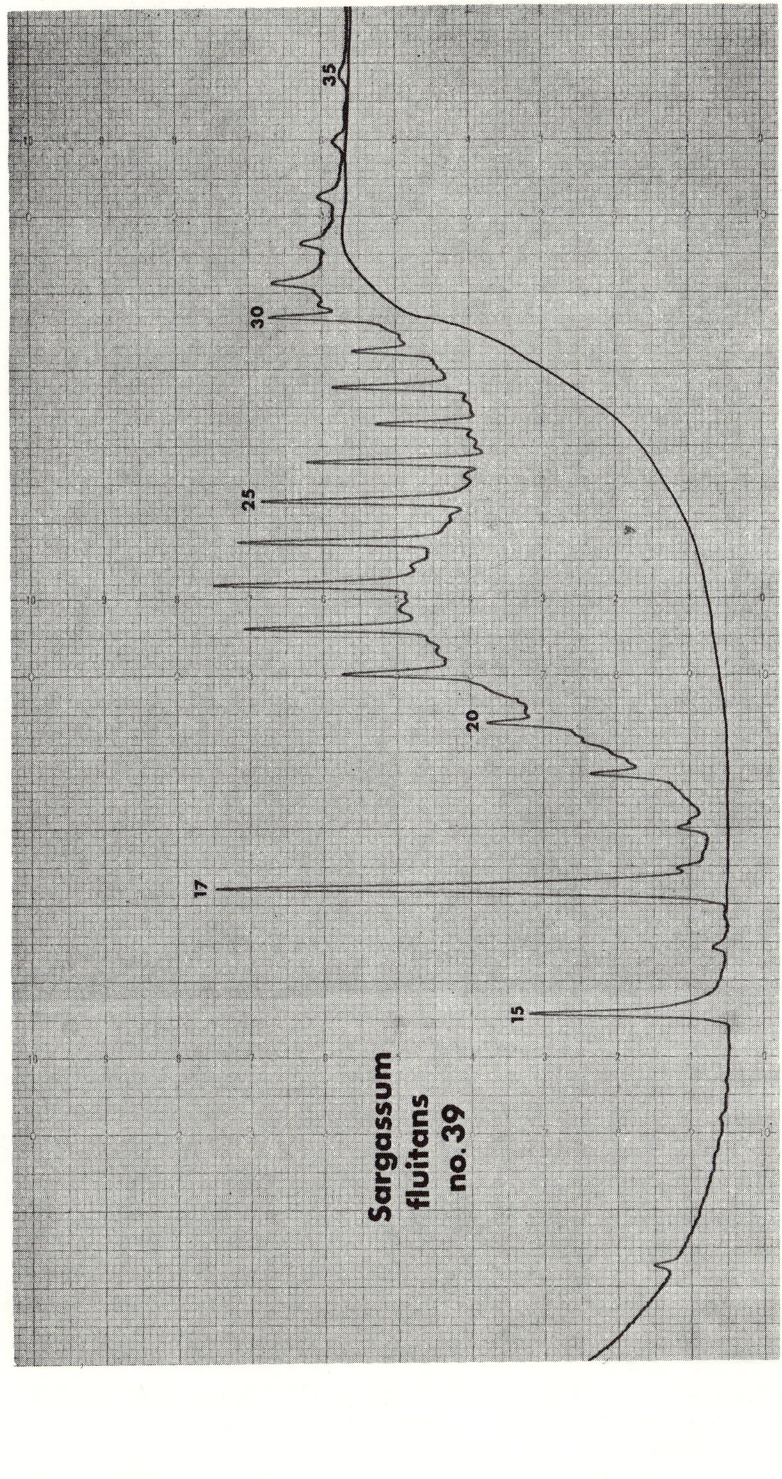

Figure 11-2: Typical gas chromatogram of extract from *Sargassum fluitans*.

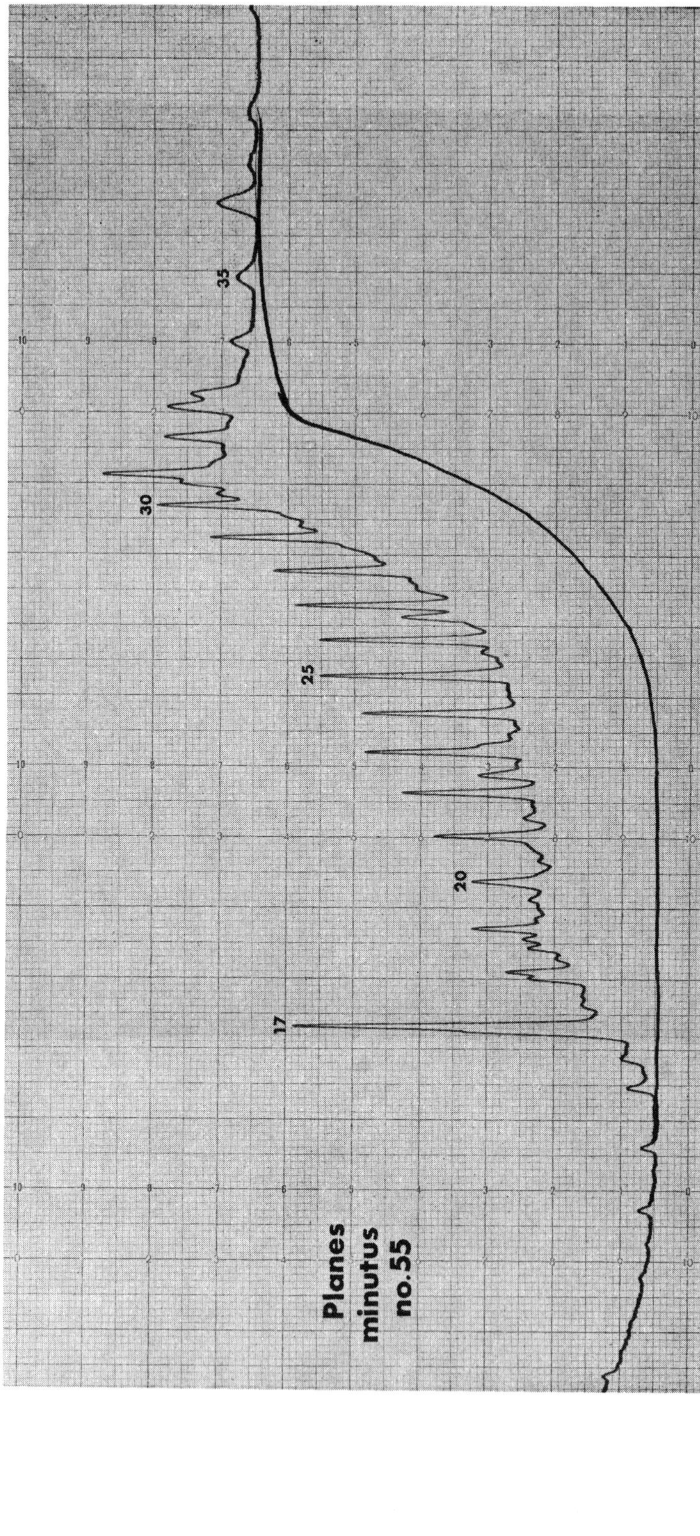

Figure 11-3. Typical gas chromatogram of extract from crab *Planes minutus*.

the larger organisms have been studied. Except for a few mixed species collections, the smaller organisms could not be obtained in sufficient quantity for a satisfactory analysis.

The pentane washings prior to the extraction procedure were often examined by GLC. These washings contained petroleum hydrocarbons, but the biogenic hydrocarbons were generally not abundant. This indicated that the surface of the *Sargassum* and organisms is often contaminated by tar particles and adsorbed hydrocarbons. But, since the amounts of hydrocarbons found in the washings averaged only about 2-3% (0.48 μg/l) of the total extractable hydrocarbons, we are confident that most of the hydrocarbons extracted were definitely contained within the sample tissues and that surface contamination did not affect our results. Burns and Teal (1973) also found low hydrocarbon levels in their pentane washings of the *Sargassum* algae.

Studies of the food web of the *Sargassum* community were performed both by gut analyses of the species and by direct observation of feeding behavior in the laboratory (Morris and Geiselman, 1976).

Results

Food Web

A preliminary summary diagram of the *Sargassum* community food web is given in Figures 11-4 and 11-5 (Morris and Geiselman, 1976). The community is supported by four major food sources: (1) the plankton and particulate organic matter in the surrounding water; (2) the *Sargassum* alga itself; (3) the epiphytes, including blue-green algae and diatoms, attached to Sargassum; and (4) the detritus produced within the community. At the first consumer level are the planktivores (filter feeding bryozoans, serpulid polychaetes, barnacles, hydroids), omnivores feeding on *Sargassum* and its epiphytes (amphipods, polychaetes, and other organisms with major alternative food sources), and detritus feeders (harpacticoid copepods, amphipods, and flatworms). Trophic pathways overlap even at the first trophic level, and above this level the distinction between levels is totally blurred.

For the higher level carnivores there are alternative food sources, and it is apparent that the limits of food availability within the *Sargassum* community have imposed a necessity for the largest organisms (crabs, shrimps, and fish) to have the greatest number of alternative food sources available to them. As an example, the common sargassum crab *Planes minutus,* seems to be an opportunistic feeder whose choice of food is determined primarily by its availability, success of capture, and the scarcity of preferred food sources. *Planes* has been observed to eat *Sargassum* and even tar lumps in the absence of other food sources.

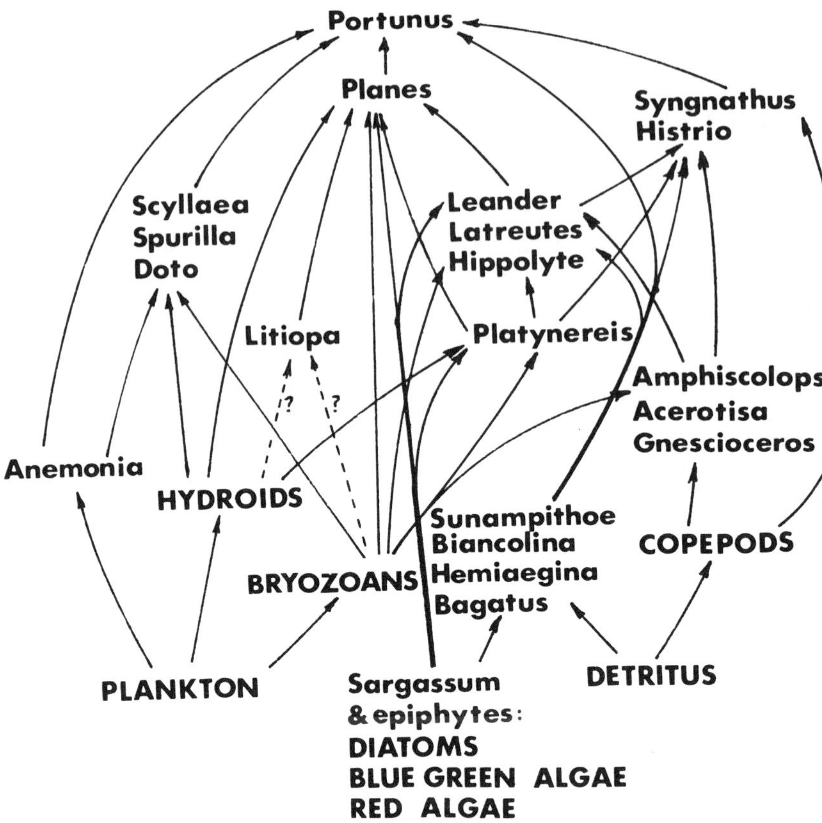

Figure 11–4. Schematic diagram of the food web of the Sargassum community.

These studies still require quantitative evaluation of the types and amounts of food consumed and of food selectivity. Further refinement of the food web is necessary, but a general indication of the complexity and hierarchy of the structure is apparent, and this is sufficient for a preliminary interpretation of the hydrocarbon analyses.

Hydrocarbon Levels

Analyses of various samples of ten species from the *Sargassum* community are summarized in Table 11-1. Only the means are reported in the table. The variability of the samples was quite large, and is indicated on some of the figures. Full details will be presented elsewhere (Morris, Cadwallader and Butler, 1976).

At the base of the *Sargassum* food web are the two algae with hydrocarbon

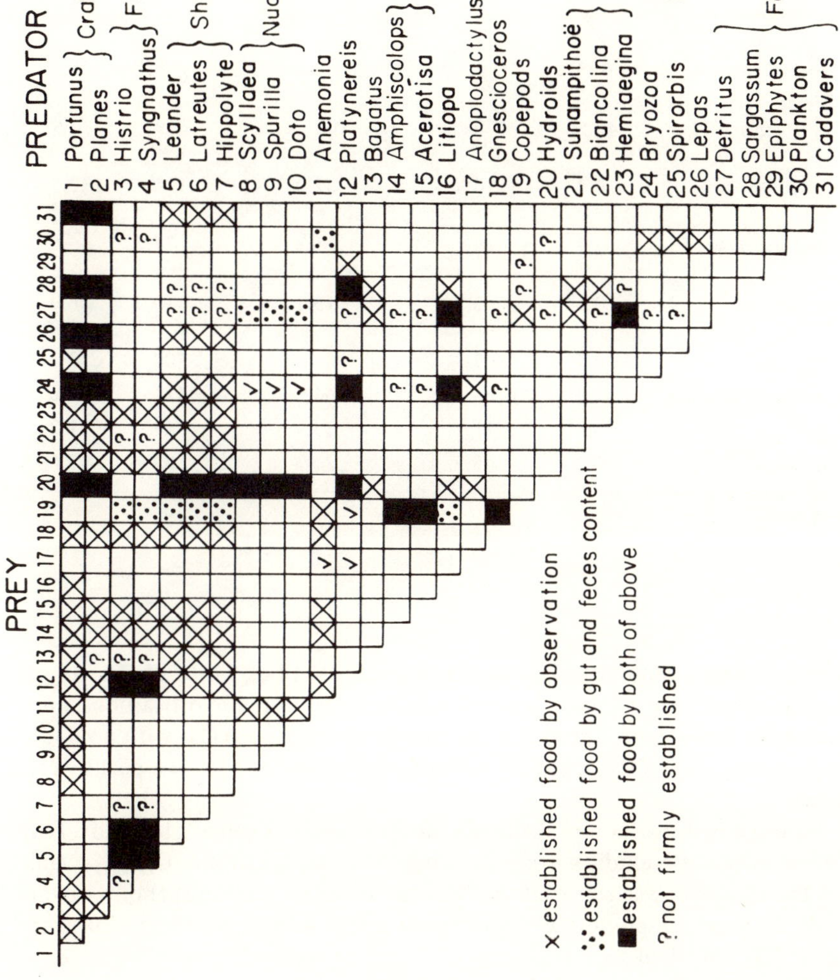

Figure 11-5. Summary diagram of the food sources of the Sargassum community fauna.

Table 11-1
Average Concentrations (μg/g – Dry Weight) of Nonpolar Hydrocarbons in Members of the Pelagic Sargassum Community

Organism (number of samples)	n-Alkanes Biogenic	n-Alkanes Petroleum	Unresolved Envelope	Total	Petroleum/ Biogenic
Sargassum natans (37)	1.4[a]	5.2	17.6	23.1	16.3
Sargassum fluitans (10)	0.8[b]	3.3	7.8	12.3	13.9
Litiopa melanostoma (4)	0.3[c]	4.4	3.2	8.1	25.3
Scyllaea pelagica (1)	40.5	420	104	619	12.9
Latreutes fucorum (4)	0.2[d]	50.3	82.0	135	132.5
Leander tenuicornis (2)	7.2[d]	18.1	51.2	77.7	44.6
Planes minutus (5)	0.6[d]	2.5	13.1	16.5	26.0
Portunus sayi	0.8[d]	109	381	500	612.1
Syngnathus pelagicus	34.9[d]	1466	1764	3411	92.6
Carangidae ("Jacks")	3.5	21.2	53.9	81.0	21.2
Mixed Fauna	2.3	45.9	31.3	83.1	33.6

[a] C_{15} 0.2, C_{17} 1.0, C_{19} 0.2.
[b] C_{15} 0.3, C_{17} 0.5.
[c] C_{15} 0.1, C_{17} 0.2.
[d] Principally C_{17}.

contents between about 10 to 20 μg/l (Figures 11-1, 11-2, 11-6, 11-7). Although biogenic hydrocarbons were noticeable, most of the hydrocarbons appeared to be petroleum: a series of *n*-alkanes between C_{20} to C_{36} with no odd/even preference and a C_{18}/phytane ratio near unity when these two peaks were resolved.

Biogenic hydrocarbon peaks for *Sargassum natans* (Figures 11-1 and 11-6) occurred at retention indices corresponding to normal alkanes of C_{15}, C_{17}, and C_{19}. The C_{17} peak was consistently the most abundant, and probably was comprised of several compounds which were not resolved by our column. Burns and Teal (1973) identified six compounds in the multiplet, including *n*-heptadecane, and four unidentified peaks with retention indices on Apiezon L of 1662, 1674, 1684, and 1696. These peaks could not be resolved on our column. A doublet was also found at C_{15} which consisted of *n*-pentadecane and pentadecene. We did not find the C_{21}:6 or C_{24} and higher biogenic compounds they also reported and assume that they were associated with attached organisms rather than with

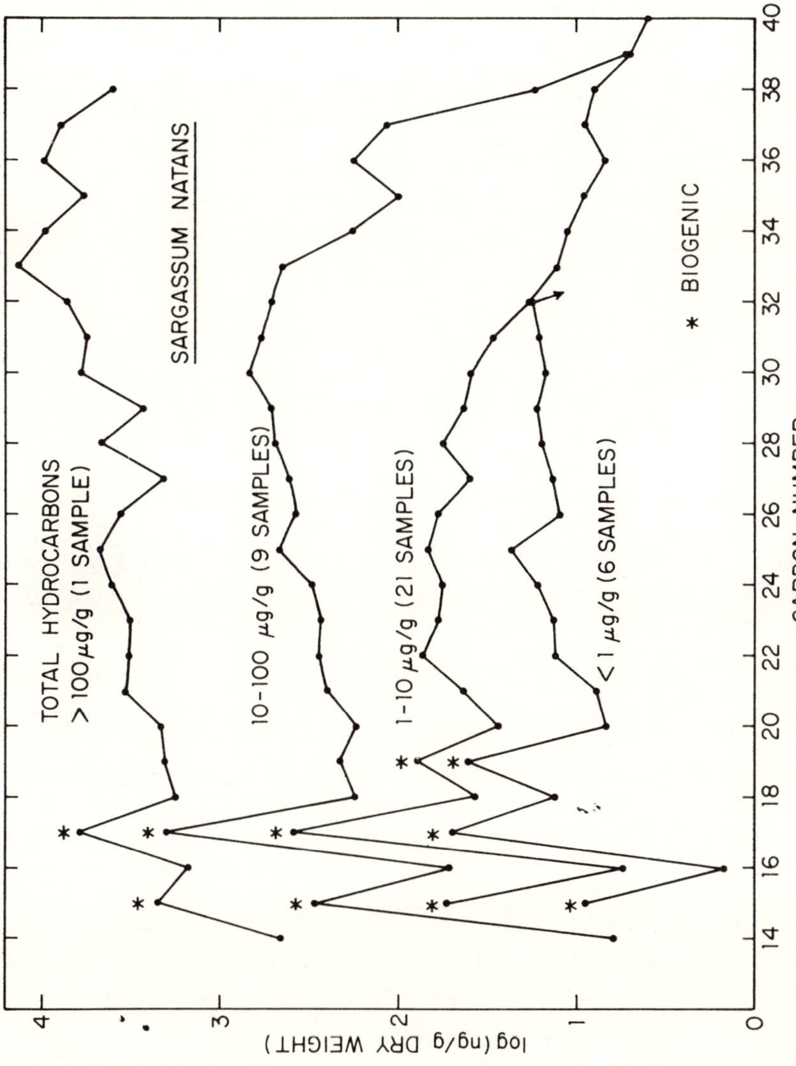

Figure 11-6. Hydrocarbon content of *Sargassum natans.*

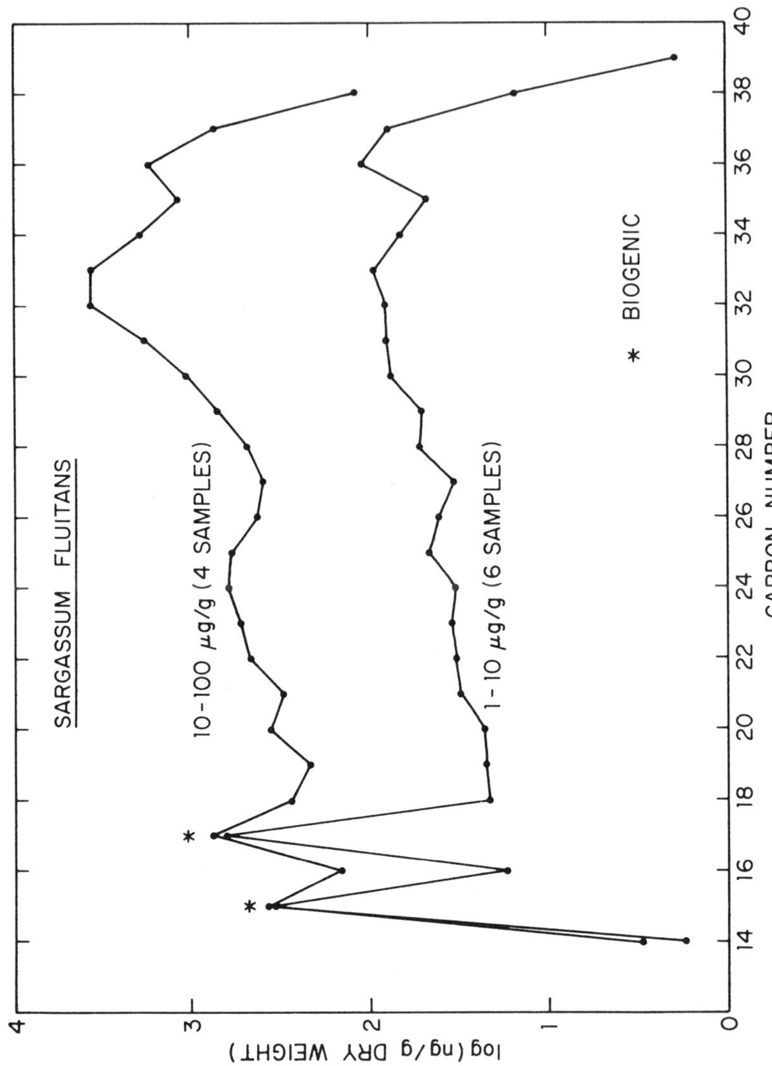

Figure 11-7. Hydrocarbon content of *Sargassum fluitans.*

Sargassum itself. *N*-nonadecane (C_{19}) was obvious as a biogenic hydrocarbon only in our samples which showed the lowest levels of petroleum contamination. At higher levels C_{19} abundance relative to adjacent *n*-alkanes was near unity; the biogenic component was masked by the C_{19} petroleum component.

The *Sargassum* community fauna all showed high levels of petroleum contamination. Biogenic hydrocarbons were at low levels and consisted primarily of C_{17} and in some cases C_{15}. The ratio of petroleum to biogenic compounds ranged from 13 to over 600. The average ratio for the mixed fauna samples was 34:1. In almost all cases a large unresolved envelope was present and phytane was often detectable (e.g., the C_{18} doublet in Figure 11-3).

The lowest contamination ratio occurred in the nudibranch *Scyllaea* (Figure 11-8) which feeds on anemones and hydroids. A distinct difference was found between the ratios for the two crabs. *Planes minutus* (Figure 11-9) feeds on a variety of foods, including *Sargassum,* and showed relatively low contamination (total 16 μg/g; ratio = 26). *Portunus sayi* (Figure 11-10) is the top predator in the community and had high hydrocarbon content (500 μg/g) and a petroleum biogenic hydrocarbon ratio of 612. The pipefish *Syngnathus* (Figure 11-11) had the highest hydrocarbon content (3300 μg/g). Hydrocarbon distribution for the snail *Litiopa* (Figure 11-12) and the shrimps *Leander* and *Latreutes* (Figures 11-13 and 11-14) are similar to the other animals.

We could not find any correlation between the hydrocarbon content of the various species and the amount of tar associated with the samples at the time of collection. This certainly indicated that our preparation procedures removed external tar, and this source of sample contamination was not a problem. But more importantly, it indicates that although the association between tar and *Sargassum* quantities fluctuates, the overall resulting petroleum load to the community is in response to the long-term general level of pollution, not to local concentration of tar.

Discussion

The only other study of hydrocarbons in the *Sargassum* community is that of Burns and Teal (1973). Table 11-2 compares their results with ours. This required us to correct their data to a dry weight basis, and this was done using wet/dry ratios determined in our laboratory. The agreement is within the range of variability encountered between individual samples for most species. The somewhat higher petroleum hydrocarbon contents we report probably reflect the higher level of pollution in the waters surrounding Bermuda than in the more northern Sargasso Sea where their samples were collected. Only the pipefish *Syngnathus* had very much higher petroleum levels in our samples. They may be due to our working with ungutted specimens, but we have to confirm whether this fish ingests tar. *Planes minutus* has been observed in the laboratory

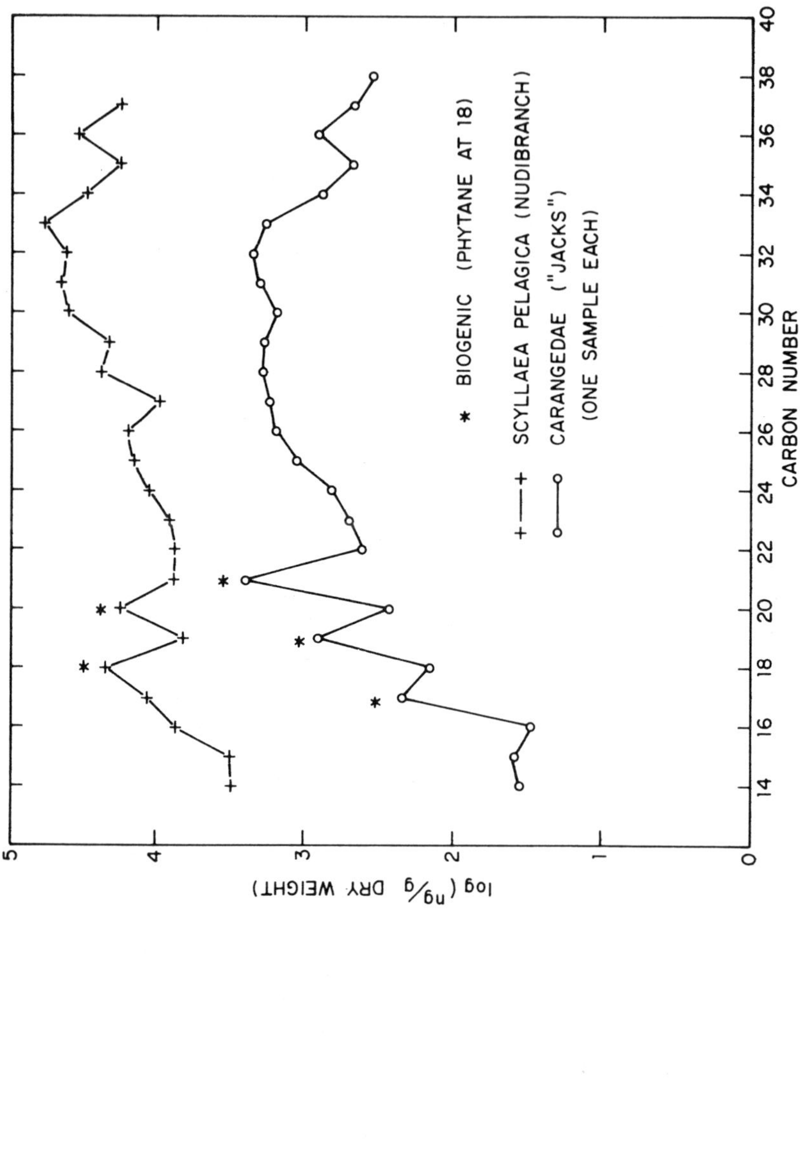

Figure 11-8. Hydrocarbon Content of *Scyllaea pelagica* and *Carangedae.*

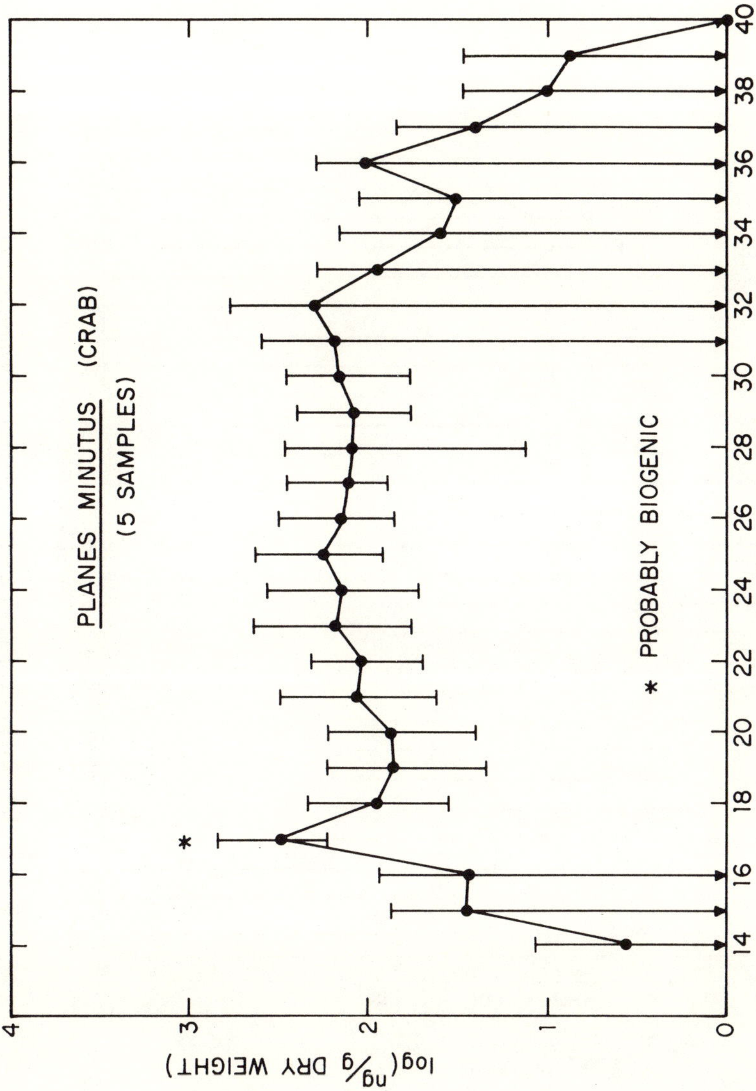

Figure 11-9. Hydrocarbon Content of *Planes minutus*.

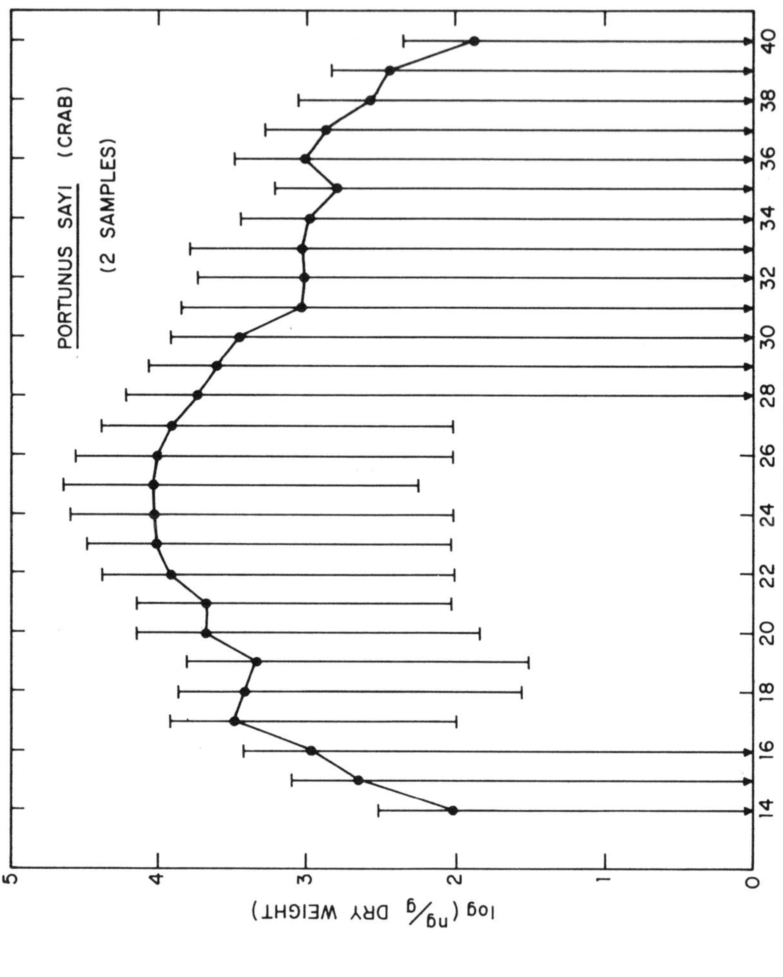

Figure 11-10. Hydrocarbon Content of *Porthunus sayi.*

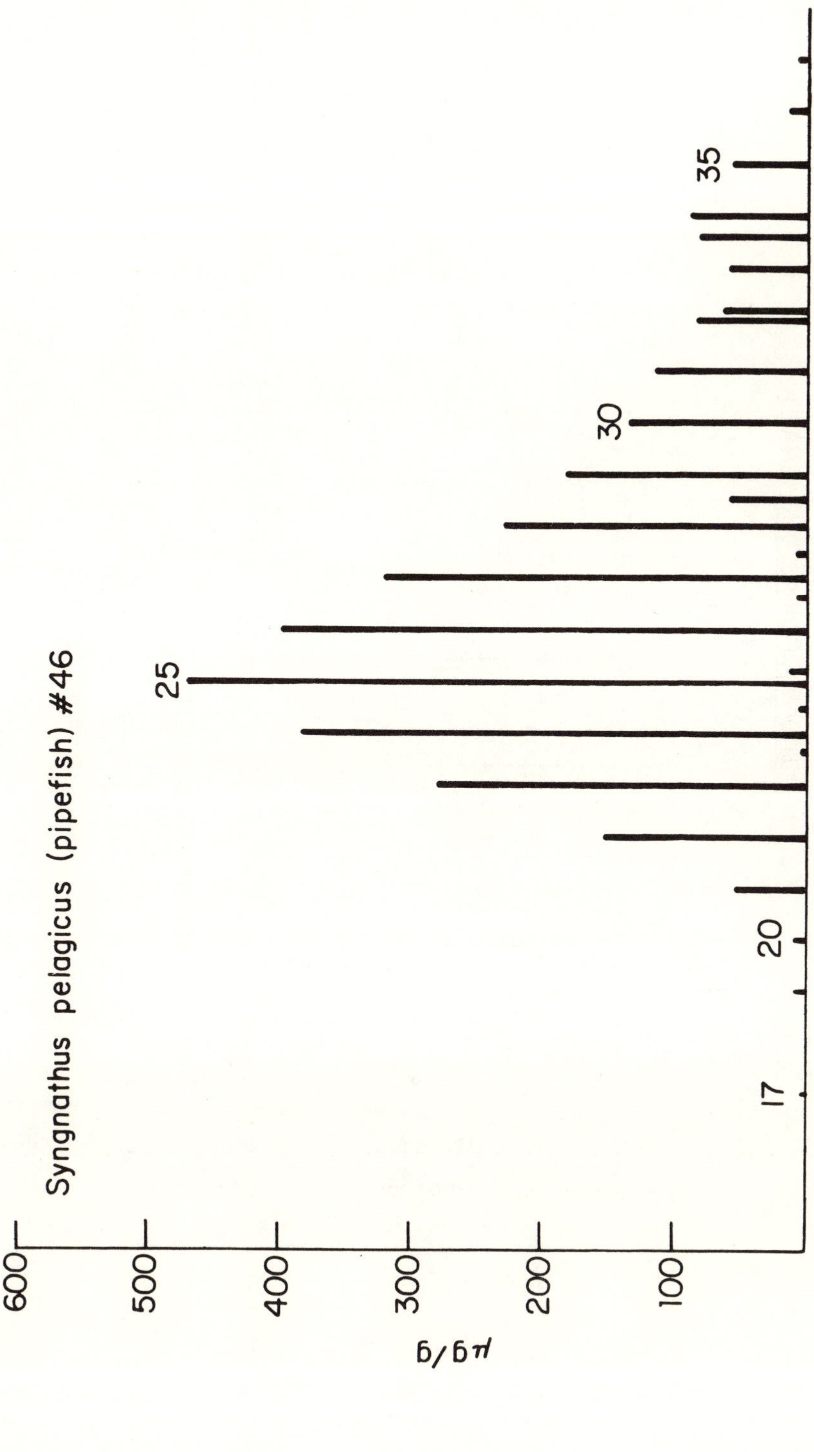

Figure 11-11. Reconstructed gas chromatogram #46, extract of 18 fish (*Syngnathus*) 5-10 cm long, 1.66 g dry weight.

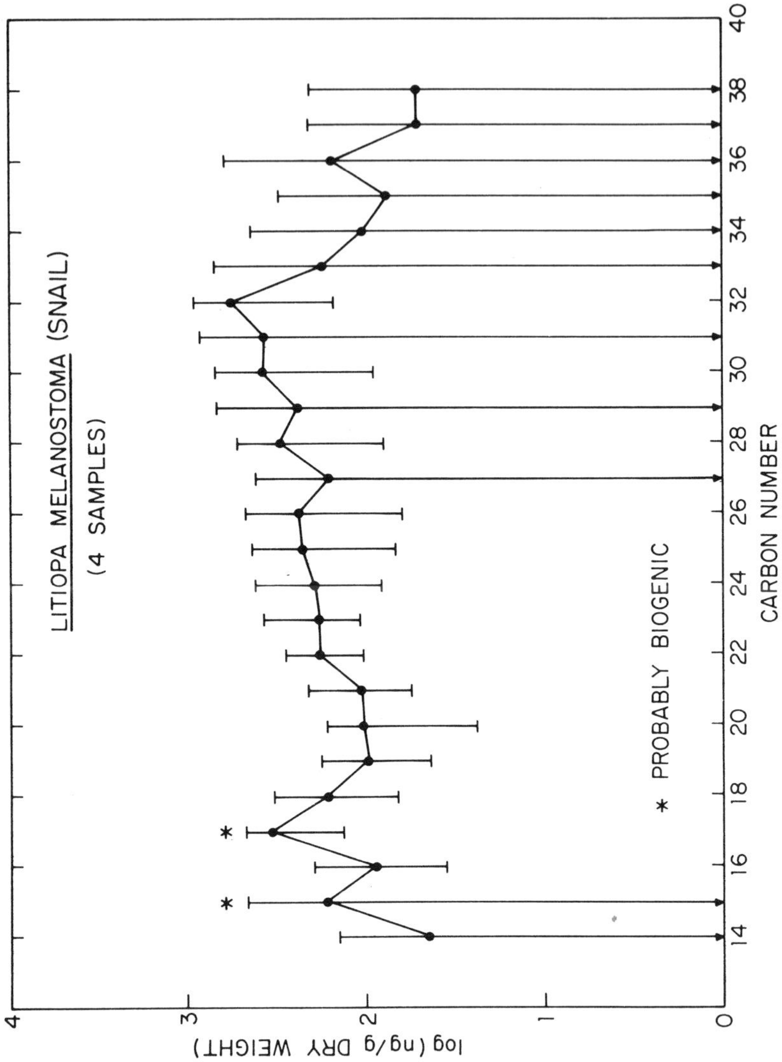

Figure 11-12. Hydrocarbon Content of *Litiopa melanostoma*.

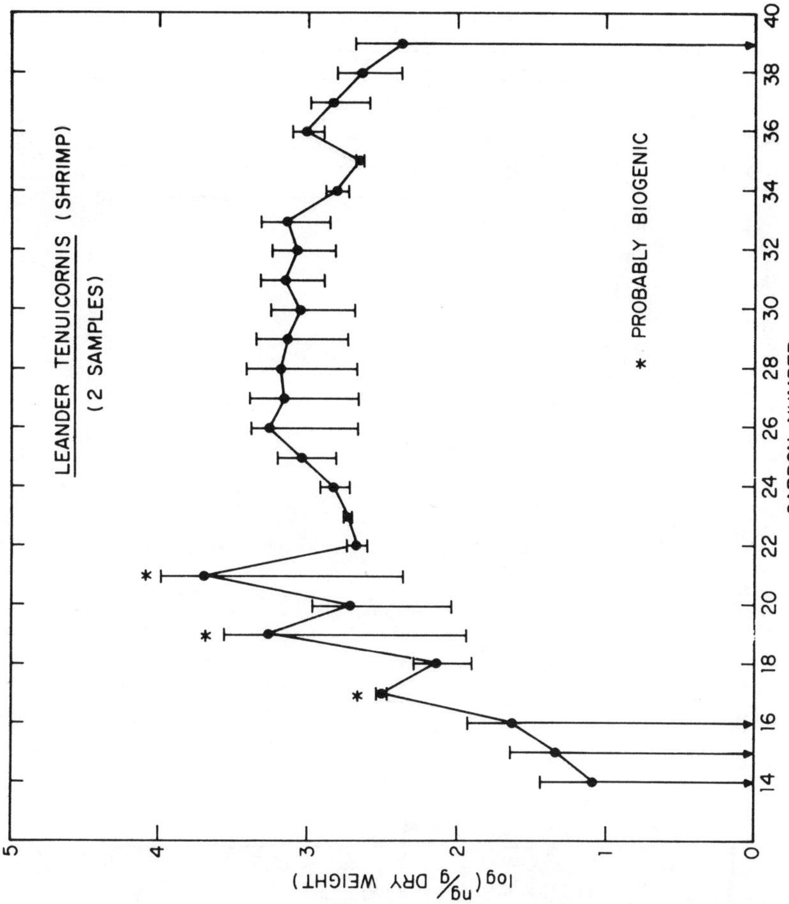

Figure 11-13. Hydrocarbon Content of *Leander tenuicornis.*

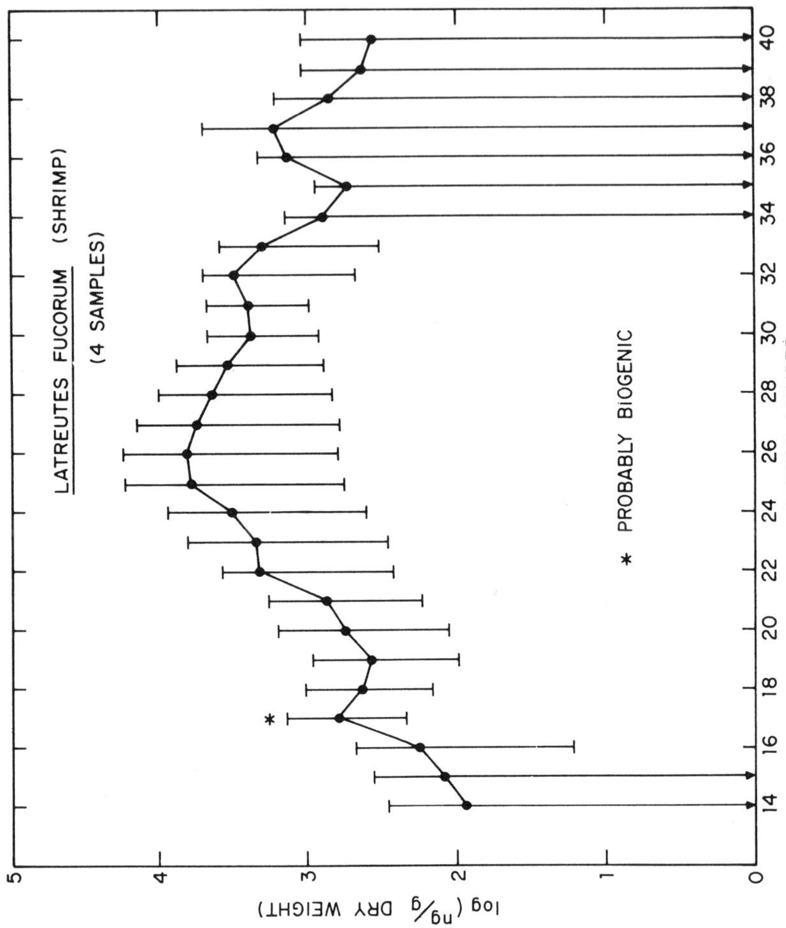

Figure 11-14. Hydrocarbon Content of *Latreutes fucorum.*

Table 11-2
Hydrocarbon Levels (μg/g Dry Weight) in Various Members of the Pelagic Sargassum Community

Organism	*Dry/Wet Weight Ratio*	*Number of Samples*	*Biogenic*	*Petroleum*	
Sargassum natans algae	0.23	37(12)	1.4(7.7)	22.8(10.6)	16(1.4)
Sargassum fluitans algae	0.17	10(1)	0.8(69.4)	11.1(13.2)	14(0.2)
Planes minutus crab	0.22	5(1)	0.6(0.0)	15.6(49.5)	26(∞)
Portunus sayi crab	0.19	3(1)	0.8(0.0)	490(177)	613(∞)
Snygnathus pelagicus pipefish	0.27	2(1)	35(1)	3230(32.6)	93(31)
Leander tenuicornis shrimp	0.14	2(1)	7.2(1.1)	69.3(20.4)	45(18)
Histrio histrio fish	0.25	–(1)	–(0.2)	–(0.5)	–(31)
Canthidermis sp. fish	0.25	–(1)	–(0.3)	–(1.0)	–(21)
Litiopa melanostoma[a] snail	0.04	4(–)	7.5(–)	190(–)	25(–)
Scyllaea pelagica nudibranch	0.03	1(–)	40.5(–)	524(–)	13(–)
Latreutes fucorum shrimp	0.11	4(–)	0.2(–)	132(–)	133(–)
Carangidae sp. fish	0.18	1(–)	3.5(–)	75(–)	21(–)
Mixed Fauna	0.16	3(–)	2.3(–)	77(–)	34(–)

NOTE: The present data is compared to the results reported by Burns and Teal (1973) given in parentheses.

[a]Shell-less weight.

to ingest tar, and tar lumps have been found in fecal pellets of *Litiopa* (Figure 11-15). However, although our samples of *Planes* were not gutted, they did not show particularly high hydrocarbon levels.

The amounts of biogenic hydrocarbons in the two *Sargassum* species did not follow any clear pattern, except that the oldest portions of the plants were found to contain the highest abundance of the natural *n*-alkanes and new growth the least. Only one sample out of ten of *Sargassum fluitans* showed the high biogenic hydrocarbon levels reported by Burns and Teal (1973).

Our observations confirm the earlier reports that the *Sargassum* community

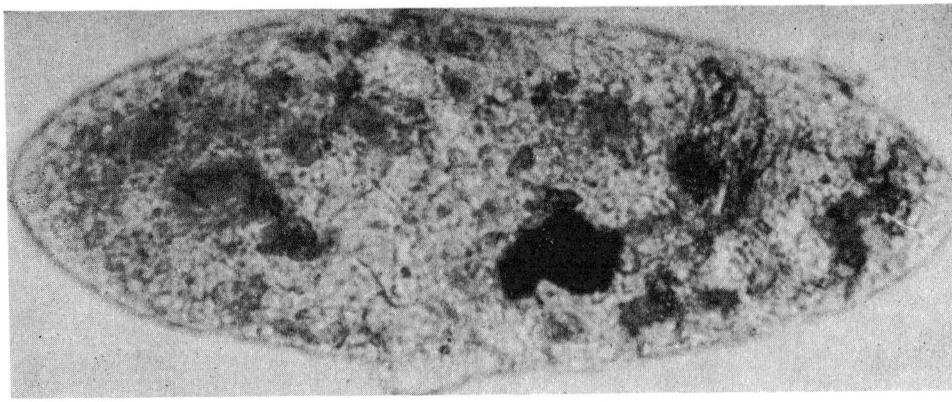

Figure 11-15. Fecal pellet from *Litiopa melanostoma* showing black tar lump inclusion.

is measurably contaminated by petroleum. Although the hydrocarbons may be accumulated in part through feeding (some top predators certainly had high hydrocarbon content), there does not appear to be any consistent magnification along the food web (Figure 11-16). No correlation was found between natural

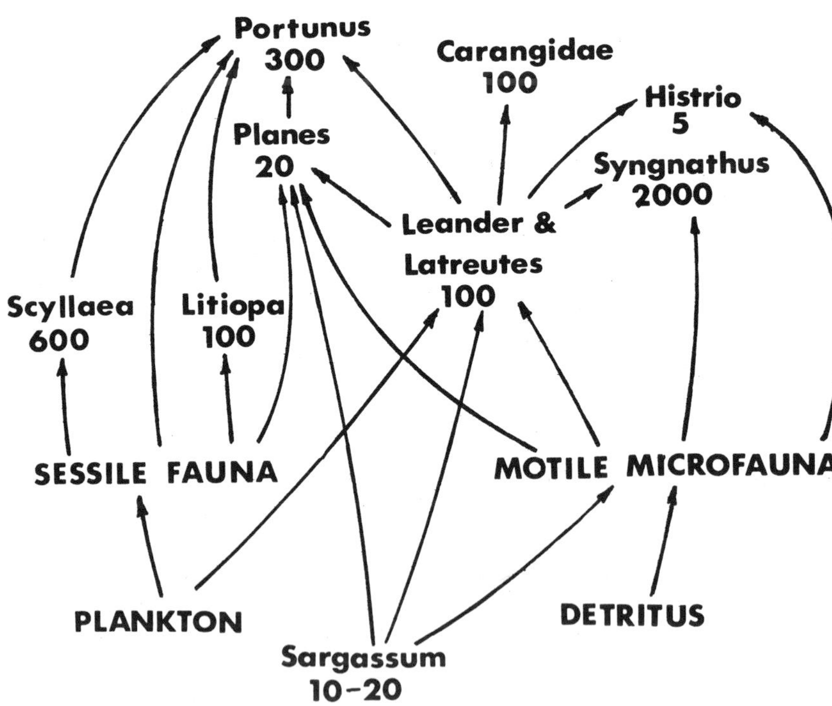

Figure 11-16. Simplified food web with approximate total hydrocarbon concentrations noted.

and biogenic hydrocarbon content, which might be expected if both were accumulated through the food web.

It seems likely that Burns and Teal are correct in suggesting that petroleum hydrocarbons are accumulated primarily across respiratory surfaces. Accumulation via the gut also seems likely for those species which ingest tar lumps. The levels of hydrocarbons in the different species must then be explained by differing body structure, age, and especially the physiological ability to store compounds found in petroleum.

Acknowledgments

This work was supported by the National Science Foundation, International Decade of Ocean Exploration, under Grant No. IDO 72-06411, and a grant from the Zemurray Foundation. The assistance of the Captain and crew of the R/V Panulirus, and various volunteers and students, particularly Barbara M. Baum, A. Barr Dolan, and Thomas D. Sleeter, is gratefully acknowledged.

References

Blumer, M., M. M. Mullin and D. W. Thomas (1964). Pristane in the marine environment. *Helgol. Wiss. Meeresunters, 10,* 187-201.

Blumer, M., R. R. L. Guillard and T. Chase (1971). Hydrocarbons of marine phytoplankton. *Marine Biology, 8,* 183-189.

Blumer, M. and J. Sass (1972). The West Falmouth Oil Spill II. Chemistry, *Woods Hole Oceanographic Institution Tech. Rept.* 72-19.

Burns, K. A. and J. M. Teal (1973). Hydrocarbons in the pelagic *Sargassum* community. *Deep-Sea Research 20,* 207-211.

Butler, J. N., B. F. Morris and J. Sass (1973). Pelagic tar from Bermuda and the Sargasso Sea. Special Publication No. 10, Bermuda Biological Station for Research.

Clark, R. C. (1966). *Saturated Hydrocarbons in Marine Plants and Sediments.* M.S. Thesis, MIT, Cambridge, Mass.

Clark, R. C. and M. Blumer (1967). Distribution of *n*-paraffins in marine organisms and sediment. *Limnology and Oceanography, 12,* 79-87.

Clark, R. C. and J. S. Finley (1974). Analytical techniques for isolating and quantifying petroleum paraffin hydrocarbons in marine organisms. *Marine Pollution Monitoring (Petroleum),* NBS Special Publication No. 409, pp. 209-212.

Corner, E. D. S. (1975). Proc. Royal Soc. B 189: 391 ff.

Corner, E. D. S., R. P. Harris, K. J. Whittle and P. R. Mackie (1975). Seminar Society of Experimental Biology, Liverpool.

Conover, R. J. (1971). Some relations between zooplankton and Bunker C oil in Chedabucto Bay following the wreck of the tanker *Arrow*. *J. Fish. Res. Board Can., 28,* 1327-1330.

Davis, J. B. (1968). Paraffin hydrocarbons in the sulfate-reducing bacterium *Desulfovibrio desulfuricans. Chem. Geol. 3,* 155-160.

Gordon, D. C. and P. D. Keizer (1974). Hydrocarbon concentrations in seawater along the Halifax-Bermuda section: lessons learned regarding sampling and some results. *Marine Pollution Monitoring (Petroleum),* NBS Special Publication No. 409, pp. 113-115.

Grice, G. D., G. R. Harvey, V. T. Bowen, and R. H. Backus (1972). The collection and preservation of open marine organisms for pollutant analysis. *Bul. Env. Contam. Toxicol., 7,* 125-132.

Keizer, P. D. and D. C. Gordon (1973). Detection of trace amounts of oil in seawater by fluorescence spectroscopy. *J. Fish. Res. Board Can., 30,* 1039-1046.

Lee, R. F., R. Sauerheber and A. A. Benson (1972a). Petroleum hydrocarbons: uptake and discharge by the marine mussel, *Mytilus edulis. Science, 177,* 344-346.

Lee, R. F., R. Sauerheber and G. H. Dobbs (1972b). Uptake, metabolism and discharge of polycyclic aromatic hydrocarbons by marine fish. *Marine Biology, 17,* 201-208.

McAuliffe, C. D. (1974). Determination of C_1-C_{10} hydrocarbons in water. In *Marine Pollution Monitoring (Petroleum),* NBS Special Publication No. 409, pp. 121-125.

Morris, B. F. (1971). Petroleum: Tar quantities floating in the Northwestern Atlantic taken with a new quantitative newston net. *Science, 173,* 430-432.

Morris, B. F. and D. Mogelberg (1973). Identification manual to the pelagic *Sargassum* fauna. Special Publication No. 11, Bermuda Biological Station. 63 pp. [This publication contains taxonomic information on 101 indigenous *Sargassum* species and a bibliography of 131 references to literature on the *Sargassum* community.]

Morris, B. F., J. N. Butler, J. Cadwallader and T. D. Sleeter (1976). Transfer of particulate hydrocarbon material from the ocean surface to the water column. In this volume.

Morris, B. F., J. Cadwallader and J. N. Butler (1976). Hydrocarbons in the *Sargassum* community (in preparation).

Morris, B. F. and J. Geiselman (1976). The food web of the *Sargassum* community (in preparation).

NAS (National Academy of Sciences) (1975). Petroleum in the Marine Environment. Washington, D. C., 107 pp.

Parker, C. A. (1971). The effect of some chemical and biological factors on the degradation of crude oil at sea. In *Water Pollution by Oil* (P. Hepple, ed.), Institute of Petroleum, London.

Stegeman, J. J. and J. M. Teal (1973). Accumulation, release, and retention of petroleum hydrocarbons by the oyster *Crassostrea virginica. Marine Biology, 22,* 37–44.

Wade, T. L. and J. G. Quinn (1975). Hydrocarbons in the Sargasso Sea surface microlayer. *Marine Pollution Bulletin, 6,* 54-57.

Zafiriou, O., M. Blumer and J. Myers (1972). Correlation of oils and oil products by gas chromatography. Woods Hole Oceanographic Institution Report 72-55.

APPENDIX VII

THE FOOD WEB OF THE *SARGASSUM* COMMUNITY

Joy Geiselman*

*Based on an unpublished manuscript (September 1973). References are given in Section 3 of the main text.

INTRODUCTION

It is the purpose of this paper to discuss the food web of the *Sargassum* community, for its trophic relations ultimately determine the transfer of energy and materials, including persistent pollutants derived from petroleum and related chemicals, within this oceanic ecosystem.

Sargassum clumps were collected in January 1972 and June-Aug. 1973 off-shore from Bermuda using a dip-net. The alga and its associated fauna were maintained in the laboratory at 22-25°C in aerated aquaria. Feeding behavior in the natural community was observed directly in these aquaria and in small finger bowls, both with and without the aid of a dissecting microscope. Food was introduced for qualitative and quantitative feeding experiments. Examination of gut contents and feces provided information for trophic categorization.

RESULTS AND DISCUSSION

Observations are summarized in Table I and are discussed in detail below. A model food web of the *Sargassum* community is presented in Figure 1.

The food sources of the *Sargassum* community include the *Sargassum* itself, the diatoms and other small epiphytic green, blue-green and red algae, the detritus deposited on the algae, the plankton and organic materials suspended in the surrounding water, and the sessile and motile animals living on the *Sargassum*. To discuss the feeding relationships, the organisms have been divided into the following trophic groups:

I. Primary producers and the detritus food chain

II. Herbivores

III. Omnivores
- Suspension feeders
- Detritus feeders
- Scavengers

IV. Carnivores

I. Primary producers and the detritus food chain:

The primary producers of the community are the *Sargassum* and its autotrophic epiphytes, e.g., the numerous diatoms and blue-green and red algae such as *Rivularia* and *Ceramium*.

The Sargasso Sea is an area of relatively low nutrient concentration and low primary productivity; iron, nitrogen, and phosphorous are limiting factors (Menzel and Ryther, 1960). Howard and Menzies (1969) found that net productivity of *Sargassum* in the Sargasso Sea ranged from 1.3 - 1.6 mg O_2/g dry weight/hr or 0.32 mg C/m^2/day, contributing as much as 60% to the total primary production in the upper meter of water. In addition, phytoplankton productivity is apparently higher in *Sargassum* patches than in the surrounding water (Blankley, unpublished data, cited by Smith, Burns and Carpenter, 1973). Thus, the floating *Sargassum* patch is an area of relatively high production compared to the surrounding water.

Culliney (1970) demonstrated that reactive phosphate concentration from water in a floating *Sargassum* patch was two to three times greater than that in the surrounding water, apparently from exudates of the *Sargassum* or from the excretory products of the animals. Carpenter (1970) suggested that the relatively high density of epiphytic diatoms on *Sargassum* may result from this enrichment of the immediately surrounding water.

The extensive surface area presented by *Sargassum* with its detritus covering is conducive to the adsorption of dissolved macromolecular compounds and to biochemical processes and acts to provide a favorable environment in the open ocean. Khailov and Finenko (1970) showed that organic macromolecular compounds dissolved in seawater adsorb on particulate matter populated by microorganisms and are important in the marine food chain. Much particulate detritus can be seen over the surface of *Sargassum*, resulting from decomposition of the alga and its fauna, as well as from fecal material.

Phaeophytes such as *Sargassum* exude phenolic compounds which combine with proteinaceous and carbohydrate material in seawater to produce organic aggregates (Sieburth and Jensen, 1969) which may be as much as 39% of the gross production (Khailov and Burlakova, 1969; Brylinsky, 1977). Species of *Fucus* are known to produce antibiotic materials during the growth period (Pratt *et al*., 1951; Sieburth and Conover, 1965) *Sargassum* is thus producing organic substances as well as serving as a collection site (Smith, Burns and Carpenter, 1973).

The Langmuir circulation vortices which produce the long parallel rows of *Sargassum* also collect foam, particulate matter, and organic aggregates. In such a condition of concentrated dissolved organic matter (DOM) and detritus, the inclusion of DOM into the *Sargassum* food web through bacterial activity on the detritus and utilization of detritus by suspension and detritus feeders may be significant:

Sargassum -> DOM -> Detritus -> Suspension and detritus feeders

II. Herbivores

Many animals feed on the *Sargassum* and its algal epiphytes but also utilize other food sources and are best classified in the next group, the omnivores. The gammaridean amphipods, the isopods, the polychaete *Platynereis dumerillii*, and the crabs *Planes minutus* and *Portunus sayi* all consume much *Sargassum*.

Cyclopoid copepods are generally believed to be herbivores (Green, 1963) and were often observed on *Sargassum* blades where the eipdermal layer had been disrupted, as well as on bushy blue-green algae such as *Rivularia*.

III. Omnivores

Omnivorers feed on a variety of living and dead plant and animal material in the community and include the suspension feeders, the detritus feeders, and the scavengers. As described in Section I, the bacteria are an important part of the detritus food chain. They help mobilize organic material and thus link the DOM and particulate matter with the suspension and detritus feeders.

The suspension feeders of the community are the bryozoans *Membranipora*, *Aetea*, and *Thalamoporella*, the polychaete *Spirorbis*, the barnacle *Lepas*, and the tunicate *Diplosoma*. *Membranipora* quite commonly covers extensive areas of the *Sargassum* stipes, blades and bladders, its lophophores extending to filter suspended matter. *Spirorbis* settles on top of the calcareous skeletal remains of the bryozoans as well as over the algal surfaces and tends to be gregarious. Its ciliary gills extend and sort particles by size without regard to food value (Jørgensen, 1966). *Lepas* is also described as a non-selective filter-feeder as is the tunicate *Diplosoma*, both of which are more sporadic in their occurrence on *Sargassum* than the common *Membranipora* and *Spirorbis*.

Platynereis dumerillii lives in clear chitinous tubes which it spins between *Sargassum* blades and stipes, often among bryozoan and hydroid colonies. Fragments of *Membranipora* skeleton and *Spirobis* tubes are often present at the opening of these tubes. Small polychaetes often crawl about over the weed surface, especially among *Membranipora* and *Zanclea* colonies; once they build a tube and attain a larger size, they remain in or near their tubes, extending only to feed or when disturbed. *Platynereis* was observed feeding on *Sargassum*, *Ceramium*, *Membranipora*, and hydroids, especially *Zanclea* and *Clytia*. *Sargassum* seems to be a major part of its diet; its jaws often bit off large chunks from the stipes and within a week, it consumed several blades surrounding its tube. The polychaete has been described as an herbivore upon algae (Pettibone, 1963) but it was observed to feed on much animal material as well. It often consumed whole hydranths of *Clytia*, *Zanclea*, *Aglaeophenia* and other available hydroids and also fed upon the remaining stolons. *Membranipora* skeletons were usually empty near *Platynereis* tubes, and the polychaete was also observed to uproot the skeletons, presumably in search of food. The polychaetes seemed to feed primarily upon the algae and sessile fauna, ignoring nearby amphipods, flatworms, and other small motile fauna. It did capture and consume a pycnogonid, and fecal examination did reveal a few copepod remains; however, as food sources, these are probably not significant.

The gastropod *Litiopa* is an omnivore which feeds on diatoms, detritus, *Sargassum*, bryozoans and hydroid remains. It was observed browsing along *Membranipora* colonies, uprooting the skeleton and sucking out the lophophores; hydroid stolons and decomposing hydranths were similarly consumed. The snail grazes on the epithelial layer of *Sargassum*, apparently consuming the epiphytic diatoms and particulate detritus as it leaves a trail on the blades. Microscopic observation of fecal pellets revealed many diatoms, black particles believed to be tar, *Membranipora* opercula and skeletal fragments, and several copepod appendages.

Harpacticoid copepods continuously crawl over the *Sargassum* and are most commonly found among old bryozoan skeletons. They appear to be inhabiting these protected areas which are a rich source of particulate detritus. Harpacticoid copepods are believed to be primarily detrital feeders (Green, 1963).

The gammaridean amphipods *Sunampithoë* and *Biancolina* consume much *Sargassum*. They were often observed living inside the air bladders and also in depressions in the thicker stipes, having chewed their way into these places; they also build mucous tubes among the blades and bladders. Whole bladders and blades wre consumed by a single amphipod in several days. The amphipods usually stay in or near the tube, extending short distances to feed upon the algae, particulate detritus, or their molted exoskeletons. They made no attempts to capture nearby copepods. Some species of *Amphitol* feed mainly on algae and detritus but under unfavorable conditions become predacious on small crustaceans (Skutch, 1926).

Hemiaegina minuta, a caprellid amphipod, is usually found clinging to hydroids and snaps at their surfaces with its large gnathopods. It appears that diatoms and detritus make up the bulk of its diet; the caprellid was never observed to attack the living hydroid zooids or to capture nearby copepods. Saunders (1966) found that caprellids feed on sessile diatoms, some small crustacea, and some detritus clinging to hydroids; whether they consume protozoa and what quantity of animal vs. plant food was questioned. Examination of gut contents of *Hemiaegina* revealed several diatoms and occasionally some black particulate matter, believed to be tar bits.

The isopods *Bagatus* and *Idotea* feed mainly upon *Sargassum*, detritus, and decomposing hydroid colonies. Naylor (1955) described most *Idotea* species as omnivores, with seaweeds providing the bulk of available food and some animal remains and living animals being eaten.

The shrimps *Leander*, *Latreutes*, and *Hippolyte* are omnivorous scavengers of the *Sargassum* community. They were observed to capture and devour *Platynereis*, amphipods, isopods, and flatworms. Using their chelae, they picked at *Lepas*, *Membranipora* and hydroid colonies, probably consuming many small crustaceans (both living and dead) and detritus as they grazed.

However, the tubeworm Spirorbis and snail Litiopa appeared impervious to their attack. When other food was not available, the shrimps ate Sargassum, but the alga does not constitute a major part of their diet. Gut contents included hydroid hydranths and copepod and other crustacean remains. The shrimps occasionally attacked anemones and nudibranchs, but quickly ceased as they were obviously stung by nematocysts.

Planes minutus and Portunus sayi are particularly voracious, and although scavengers, they are the top carnivores of the community. Portunus is larger in size than Planes and sometimes will attack and devour the smaller crab. In addition, Portunus was observed to feed upon Spirorbis, Litiopa, Anemonia, and nudibranchs, and the pipefish Syngnathus. These were ignored or unable to be captured by Planes, presumably due to impervious shells, nematocysts or speed of prey. Planes, Platynereis, amphipods, isopods, flatworms, and Histrio histrio were caught and eaten by the crabs. Hydroids, Lepas and Membranipora colonies were torn, uprooted and consumed as were Sargassum blades and bladders; tar lumps adhering to the algae were also eaten. When food supply diminished, the crabs became cannibalistic. They ate many dead animal remains, particularly those of shrimps. Live shrimps were never observed to be captured by the crabs; the crabs and shrimps occupied their own territories and generally did not disturb each other. Examination of gut contents from both crab species revealed small crustacean limbs, shrimp carapaces, hydroid hydranths, Lepas shells, Membranipora skeletal fragments, and black tar bits.

IV. Carnivores

Hydroids are chiefly carnivorous, ingesting any small animals of appropriate size that come into contact with the tentacles. The abundance of hydroid colonies in the community may be explained by the plentiful supply of small crustacea, worms, eggs and larvae in the surrounding water.

Anemonia sargassensis is a motile coelenterate, utilizing its tentacles and mucous masses to catch copepods, small worms, ciliates, and flagellates from the Sargassum surface and plankton. Larger motile animals such as pycnogonids were often caught but usually escaped before being consumed.

The nudibranchs are carnivores preying largely upon hydroids and anemones. Scyllaea pelagica, Doto pygmaea, and Spurilla neapolitana devoured many Anemonia clinging to Sargassum. As they browsed over the weed, the nudibranchs sucked out the living contents from hydroid colonies, leaving only the perisarc covering of hydranths and stolons. They did not appear selective in their choice of hydroids; any present on the blades were consumed including Clytia, Aglaeophenia, Plumularia, Obelia, and Zanclea. Miller (1961) reported that Doto feeds on over twenty species of hydroids, although some nudibranchs tend to be selective to either gymnoblast or to calyptoblast hydroids.

Membranipora was also observed to have been eaten, but only when hydroids and anemones were not present. Feces of Scyllaea contained bryozoan opercula and skeletal fragments as well as hydroid remains. The nudibranchs did not appear to graze on the Sargassum itself or to attempt capture of small motile animals. They may consume diatoms and detritus.

The flatworms Amphiscolops, Acerotisa, and Gnescioceros glide very rapidly over the Sargassum and are most commonly found among bryozoan colonies. Amphiscolops was observed to feed mainly upon copepods, capturing them by rapidly extending its body and grasping the prey with its curved anterior margin. The copepod is engulfed in an amoeboid fashion and is visible through the body of the flatworm. The flatworms also spend long periods of time attached to Membranipora skeletons. Gnesciosceros appeared to feed on the bryozoan by sucking out lophophores from the skeleton but it may also have been eating small protozoa, crustaceans or diatoms (Jennings, 1957).

Anoplodactylus petiolatus was found crawling over Membranipora and hydroid colonies, picking at their surfaces. They are generally believed to be carnivorous, probably feeding on these sessile fauna.

Syngnathus pelagicus and Histrio histrio feed on the small motile animals living on Sargassum as well as on the surrounding plankton. Guts of both fish species contained shrimp, polychaete setae (Platynereis) and copepods. Adams (1960) found euphausiids, shrimps, amphipods, fish larvae, and medusae in the guts of juvenile Histrio. The fish were observed to hover motionless among the Sargassum blades, then rapidly attack and swallow a nearby animal. Latreutes, Leander, Hemiaegina and Sunampithoë were all prey. Crabs, except small juvenile stages, were never attacked; neither were Scyllaea, Anemonia, Litiopa, or the sessile fauna (hydroids, bryozoans, tube worms). Histrio attempted attacks at Syngnathus and is also cannibalistic.

Pollutants in the Sargassum food web

In the act of feeding, organisms may take up pollutants from their environment and transfer them throughout the food web. Hydrocarbons from oil pollution have been found in the lipids of marine animals (Blumer, Souza and Sass, 1970). Petroleum hydrocarbons may enter the Sargassum food web by the direct consumption of tar lumps or by uptake of the soluble components of oil either across respiratory surfaces or on the surfaces of particulate matter. Portunus and Planes were observed to eat tar adhering to the Sargassum surface and black lumps believed to be tar were found in the guts of crabs and amphipods. Litiopa fecal pellets also contained black particles; these may have been taken up from the surface of the weed as the snail grazed.

A more important source of contamination may be through aggregation of soluble petroleum hydrocarbon molecules on the surface of particulate matter which serves as a food source for many animals of the community, similar to the aggregation of other molecules on detritus as previously discussed:

petroleum -> dissolved/emulsified -> detritus -> filter feeders, detritus feeders

Langmuir circulation may be acting to concentrate these contaminants in the windrows of *Sargassum*. The abundance of filter feeding organisms (bryozoans, tubeworms, barnacles, tunicates) and other animals utilizing particulate matter as a food source (hydroids, anemones, copepods, amphipods and other detritus-feeders) indicate the significance of this source of contamination. In addition, these animals constitute the main food source for the predators of the community, which could lead to concentration of pollutants in the higher trophic levels. Whether bacteria on the particulate detritus are metabolizing the aggregating contaminants is an important question and should be pursued.

The primary producers of the community may also take up petroleum hydrocarbons from the surrounding water. Burns and Teal (1973) found that both species of pelagic *Sargassum* showed petroleum contamination, and Nuzzi (1973) demonstrated that water soluble components of oil were detrimental to phytoplankton growth. Since many animals consume the *Sargassum*, it, too, is acting as a concentrator, and its possible reduced production due to the toxicity of the petroleum hydrocarbons is a potential danger to the community. The inhibition of phytoplankton populations is, of course, of major importance since the growth of all marine fauna is ultimately dependent upon the productivity of the phytoplankton.

Burns and Teal (1973) found that all the *Sargassum* animals they analyzed showed contamination by petroleum hydrocarbons (Table 2). They concluded that neither the total amount of petroleum hydrocarbons nor the ratio of petroleum to biogenic compounds could be correlated with the supposed positions of the animals in the food chain. This conclusion depended on the observation that the crabs showed a greater concentration of hydrocarbons than the fish, and they assumed that fish were higher in the food chain than crabs. However, the present study has shown that the crabs prey on fish, and also consume tar lumps directly. Hence, the higher concentration of petroleum hydrocarbons would be expected in the crabs, as observed.

Burns and Teal also suggested that since the ratio of petroleum to biogenic hydrocarbons was so high in the animals, that the main route of entry is directly across the gills and other surfaces. Although entry of pollutant hydrocarbons across respiratory surfaces is most likely, particularly in animals not directly consuming tar particles, entry via the food web appears also to be significant.

Table I. Summary of Food Sources

Organism	Observed Food Sources	Gut and Fecal Contents
Anemonia sargassensis (anemone)	copepods, flatworms, ciliates, flagellates symbionts?	*Acartia* copepod
Amphiscolops sargassi *gnescioceros sargassicola* *Acerotisa notulata* (flatworms)	copepods bryozoans (?)	copepods diatoms
Platynereis dumerillii (polychaete worm)	*Sargassum*, *Ceramium*, *Clytia*, *Aglaeophenia*, *Zanclea* and other hydroids *Membranipora*, *Spirorbis* (?)	*Sargassum*, hydroid hydranths, *Membranipora* skeletal remains, copepod appendages
Bagatus minutus (isopod)	*Sargassum*, detritus, hydroid remains, epiphytic algae	
Sunampithoë pelagica	*Sargassum*, detritus epiphytic algae	diatoms, epiphytic algae
Biancolina sp.	*Sargassum*, epiphytic algae	diatoms, epiphytic algae
Hemiaegina minuta (amphipods)	*Sargassum*, detritus, hydroids epiphytic algae	diatoms, tar bits
Leander tenuicornis *Latreutes fucorum* *Hippolyte coerulescens* (shrimps)	Platynereis, amphipods, isopods, flatworms, *Lepas*, hydroids? Membranipora, dead animals *Sargassum*	hydroid hydranths, copepod remains epiphytic algae, diatoms
Planes minutus (crab)	Hydroids, *Platynereis*, *Lepas*, amphipods, *Membranipora*, flatworms, *Sargassum*, dead bodies, *Histrio*, isopods?	shrimp carapaces, hydroid hydranths, *membranipora* skeletal fragments, *Lepas* shells, tar bits, *Sargassum*
Portunus sayi (crab)	Same as *Planes minutus* above plus *Scyllaea*, *Anemonia*, *Spirorbis*, *Litiopa*, *Syngnathus*, plants	same as *Planes minutus* above
Anoplodactylus petiolatus	*Clytia*, *Zanclea*, *Membranipora*	

Table I. Summary of Food Sources (cont.)

Organism	Observed Food Sources	Gut and Fecal Contents
Litiopa melanostoma (gastropod snail)	*Membranipora*, hydroid remains, detritus, *Sargassum*	diatoms, copepod legs, *Membranipora* remains, tar
Scyllaea pelagica *Spurilla neapolitana* *Doto pygmaea* (nudibranchs)	*Anemonia*, *Membranipora*, *Clytia*, *Plumularia*, *Obelia*, *Zanclea*, and other hydroids	diatoms, *Membranipora*, remains, hydroid hydranths
Histrio histrio *Sygnathus pelagicus* (fish)	*Leander*, *Latreutes*, amphipods, *Platynereis*, flatworms	shrimps, copepods, *Platynereis* setae
Copepods Bryozoa *Spirorbis* sp. *Lepas* sp.	?	

Figure 1. Food Web

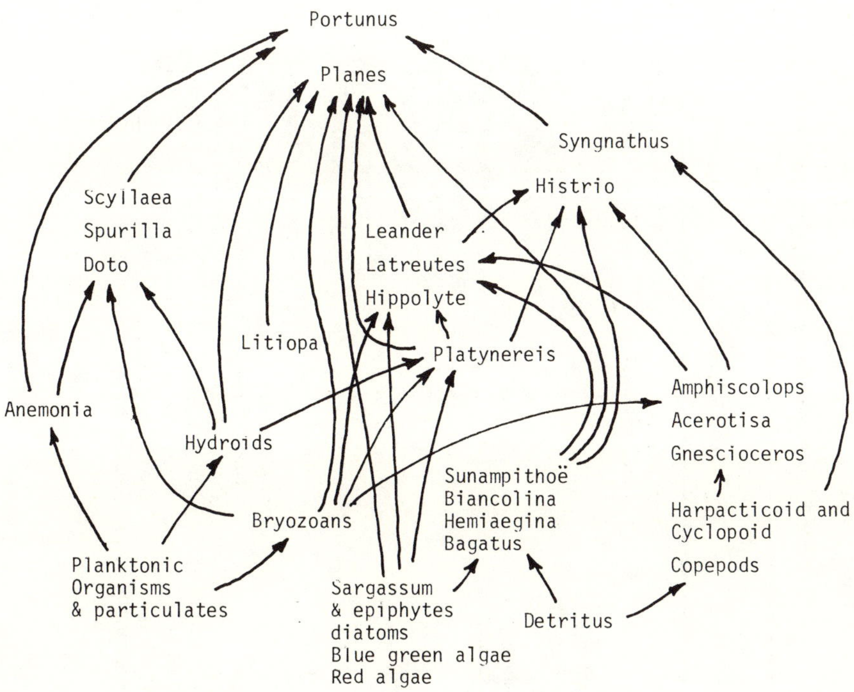

*Table 2. Hydrocarbons found in Sargasso plants and animals. Values all in μg/g wet weight (PPM).**

Sample	Description	Position collected	Natural	Petroleum n-Alkane	Petroleum Unresolved	Petroleum/natural
1. (25·3 g)	*S. natans*	31°59′N 63°27·5′W	1·38	0·36	4·72	3·7
2. (25·6 g)	(narrow blade)	31°29′N 62°23′W	0·55	0·16	1·71	3·4
3a. (26·4 g)	,, ,,	30°22′N 60°46′W	0·58	0·18	1·59	3·0
b. (Pentane washings)	,, ,,		(0·03)	(0·07)	(0·21)	(9·3)
4. (40·4 g) (new fronds)	,, ,,	31°43′N 62°35′W	0·51	0·06	1·21	2·5
5a. (26·0 g)	,, ,,	30°22′N 60°46′W	1·84	0·36	3·57	2·1
b. (Pentane washings)	,, ,,		(0·06)	(0·09)	(0·97)	(17)
6. (46·4 g) (old fronds)	,, ,,	31°43′N 62°35′W	0·68	0·05	1·28	1·9
7. (26·4 g)	,, ,,	31°59′N 63°27·5′W	0·78	0·33	0·99	1·7
8. (32·7 g)	,, ,,	31°43′N 62°35′W	2·11	0·23	2·71	1·5
9a. (25·6 g)	,, ,,	30°22′N 60°46′W	2·36	0·30	2·68	1·3
b. (Pentane washings)	,, ,,		(0·06)	(0·24)	(1·62)	(31)
10. (41·7 g)	,, ,	31°59′N 63°27·5′W	4·28	0·44	3·16	0·84
11a. (25·9 g)	,, ,,	30°22′N 60°46′W	2·32	0·12	1·69	0·77
b. (Pentane washings)	,, ,,		(0·50)	(0·08)	(0·40)	(0·96)
12. (37·2 g)	,, ,,	35°19′N 63°56′W	3·73	0·15	1·03	0·32
13. (34·8 g)	(wide blade)	35°19′N 63°56′W	11·8	0·19	2·05	0·19
14. 4 crabs (2·94 g)	*Planes minuta*	31°59′N 63°27·5′W	0·00	0·28	10·60	∞
15. 4 crabs (1·18 g)	*Portunus sayi*	35°19′N 63°56′W	0·00	8·22	25·48	∞
16. 18 *Sargassum* fish (8·76 g)	*Histrio histrio*	35°19′N 63°56′W	0·05	0·13	1·45	31
17. 13 pipefish (2·44 g)	*Syngnathus pelagicus*	,, ,,	0·27	1·41	7·39	31
18. 1 triggerfish (7·42 g)	*Canthidermis sp.*	,, ,,	0·08	0·26	1·41	21
19. 53 *Sargassum* shrimp (7·35 g)	*Leander tenuiformis*	,, ,,	0·16	0·51	2·35	18

(From Burns and Teal, 1973)

APPENDIX VIII

HYDROCARBONS IN SOME FAUNA OF THE *SARGASSUM* COMMUNITY*

Dorset Trapnell
Anthony H. Knap
James N. Butler

*Based on an unpublished (Aug. 1978) manuscript. References are given in Section 3 of the main text.

INTRODUCTION

The *Sargassum* community, consisting of floating clumps of pelagic *Sargassum natans* or *Sargassum fluitans* together with about 15 common and 85 uncommon native species of macrofauna, is found throughout the Sargasso Sea (Morris and Mogelberg, 1973). This region of the North Atlantic Ocean is also heavily contaminated with pelagic tar, which results primarily from ballast and tank-washing operations of the Europe-Persian Gulf tanker traffic (Morris, 1971; Butler *et al*., 1973; NAS, 1975).

Previous studies (Burns and Teal, 1973; Morris *et al*., 1976) of hydrocarbons in the pelagic *Sargassum* community have been primarily concerned with the analysis of whole organisms, although Burns and Teal removed the guts of their animals before extracting the hydrocarbons. In addition to the alga, four animals were analyzed by both teams: (1) the crab *Planes minutus*, (2) the crab *Portunus sayi*, (3) the pipefish *Syngnathus pelagicus*, and (4) the shrimp *Leander tenuicornis*. Both teams agreed that petroleum hydrocarbons dominated in the crabs, with about 30 μg/g dry weight in *Planes* and 300 μg/g dry weight in *Portunus*. Clearly defined biogenic hydrocarbons were less than 1 μg/g. *Leander* showed a total hydrocarbon content in the same range as *Planes*, 30 μg/g, but more of this was biogenic (5 μg/g). Variability within a set of replicate samples (Morris *et al*.) is at least a factor of two, and although Burns and Teal reported no replicates for the animals, their replicates for *Sargassum natans* showed similar variability.

The largest discrepancy between the results of these two studies was with *Syngnathus pelagicus*. Burns and Teal reported 1 μg/g biogenic and 30 μg/g petroleum hydrocarbons: Morris *et al*. reported 35 μg/g biogenic and 3200 μg/g petroleum. This appears to be far beyond the normal variability of sampling and analysis and calls for an explanation. Examination of the Burns and Teal gas chromatogram reveals the expected biogenic C_{17} normal alkane plus other compounds eluting at approximately the same retention volume. The petroleum hydrocarbons consisted of C_{20} to C_{28} normal alkanes plus a small unresolved envelope with retention index between C_{16} and C_{23}. In the sample of 18 *Syngnathus* totalling 1.66 g dry weight analyzed by Morris *et al*., the unresolved envelope was too small to be visible on the same scale as the normal paraffins and biogenic olefins. Normal paraffins from C_{20} to C_{37} were present in a regular envelope typical of a paraffinic crude oil residue, with the maximum at C_{25} being 500 μg/g.

One obvious hypothesis to explain this result was that at least some of the fish had ingested paraffinic petroleum particles, which remained in their gut when the fish were washed prior to extraction. However, at the time no separate analysis of the gut contents was made.

This work was an attempt to illuminate the question of

whether petroleum hydrocarbons in the *Sargassum* fauna reside primarily in their guts or whether substantial uptake of identifiable petroleum hydrocarbons into the other tissues can occur.

Because of the necessity of dissecting the organisms in sufficient quantity to make a satisfactory extraction and analysis, without contamination by hydrocarbons from the laboratory, only the largest and most abundant animals were chosen for the study. These were the crabs *Planes minutus* and *Portunus sayi*, the pipefish *Syngnathus pelagicus*, and mixed shrimp (mostly *Leander tenuicornis*). Dissections were as simple as possible. The guts and gills of the crabs were removed: the guts of the pipefish were removed; the shrimp were not easily dissected and were analyzed in their entirety.

MATERIALS AND METHODS

All glassware was cleaned with detergent, then with dichromate-sulphuric acid and rinsed with tap water. The glassware was oven-dried and rinsed with distilled pentane just before use.

Reagents

All water used in the analyses was extracted by shaking 50 ml of methylene chloride with every liter of deionised water in a separatory funnel. After separation the bottom layer of methylene chloride was discarded. Fresh methylene chloride was added in the same proportion and the procedure repeated. After the extraction, the remaining traces of methylene chloride were distilled off and then the water was distilled. Approximately the first 100 ml of distillate were discarded. The last 200 ml of water remaining in the distillation flask were also discarded.

All the solvents were distilled once before use. A 20% benzene - 80% methanol solution was used for the hydrocarbon extractions and column chromatography. A 20% benzene - 80% pentane solution was used for the column chromatography.

Sodium chloride, sodium sulphate, alumina, and silica gel were cleaned by Soxhlet extraction with benzene-methanol for at least 48 hours, and air-dried for 24 hours in the fume hood. Sodium chloride and sodium sulphate were oven-dried at 110°C for another 24 hours.

Sodium sulphate was activated by heating at 230°C for 24 hours. Alumina was activated by heating at 250°C for 24 hours, then partially deactivated by the addition of 6% (by weight) of water after it had cooled to room temperature. Silica gel was activated at 120°C for 24 hours, cooled to room temperature, and partially deactivated with 5% (by weight) of water. Reagents were stored in sealed containers for at least a day before use.

The pre-folded filter papers and cellulose extraction thimbles were cleaned in a Soxhlet apparatus with benzene-methanol solution for 48 hours, air dried and placed in a dessicator for several more days before use.

Column Preparation

Alumina-silica chromatography columns were packed by the following method: A glass wool plug was inserted into a 50 ml glass column. Approximately 30 ml of pentane were added and all air bubbles in the glass wool were allowed to escape. Then 14 cm^3 of silica gel were poured very gently into the column, and the column was tapped until it appeared that all the air bubbles had been removed. The silica gel was allowed to sit in pentane for several hours, and any more air bubbles that formed were also removed. Another 8 cm of alumina were poured into the column very slowly to avoid trapping any air bubbles. Once all the air bubbles had escaped, fresh pentane was passed through the column. The column was rinsed in this way several times. The column was never allowed to run dry.

Collection

During the months of July, August and September 1977 pelagic *Sargassum* was collected by dip-netting. All samples were collected at least one mile off the South Shore of Bermuda. To prevent shipboard contamination, everything that came in contact with the samples was washed with hexane. The *Sargassum* was kept in uncontaminated seawater until ashore where the organisms were removed. The animals were sorted and frozen in hexane-cleaned glass jars.

Dissection

Precautions were taken against contamination by cleaning all the instruments in dichromate-sulphuric acid, allowing them to air dry and rinsing them with pentane. Pentane-rinsed rubber gloves were worn during the dissections. The dissection instruments were cleaned in pentane between each dissection. When dissecting, the thawed specimens were placed in small glass dishes and submerged in water.

During the dissection of a crab, the body is initially placed in a dish with the dorsal surface up. With a size 15 surgical blade, the posterior hinge of the carapace was entered and the muscles attached to the carapace were severed. The carapace was lifted exposing the stomach, most of the intestinal tract and surrounding organs. Using either scissors or a scalpel, the stomach was detached from the mouth region. The tissues and organs surrounding the stomach and intestinal tract were removed so that the stomach and intestinal tract could be lifted from the body easily. The crab was gently turned over so that the ventral surface was up. The telson was lifted and pulled away from the body so that the intestinal tract was straightened out. Using blunt forceps, the colon and rectum were pushed through the aperture so that the entire alimentary canal coiled up in the dorsal area. The crab was turned over again so that the dorsal surface was up again. With forceps the stomach was clasped and the entire alimentary canal removed from the body.

With the crab dissections involving the removal of the gills, the same procedure was used for separating the carapace from the rest of the body. Each gill filament was detached separately by gripping the filament at its base and pulling gently.

The gills, alimentary canals and remaining body tissues were stored separately in glass jars sealed with aluminum-lined lids. When the remaining body tissues were transferred to a container, the water in which the specimen was dissected was also transferred and later analysed with the body tissues. All tissues were frozen before they were extracted.

Prior to the dissection of a pipefish, the body was immersed in water with the ventral surface up. The body was held firmly at the posterior end with forceps while a shallow incision with a surgical blade was made beginning at the anus and ending in the head region. Scissors were used to complete the incision to the mouth. Using blunt forceps, the rectum region was lifted from the body and then the rest of the alimentary canal was very gently lifted out intact. The alimentary canals were stored separately from the rest of the body tissues.

Extraction

The extraction apparatus consisted of a 250 ml or 300 ml round-bottomed flask two thirds full of benzene-methanol solution, resting on an insulated heating mantle. A Soxhlet extractor body was fitted to the flask. About five clean Raschig rings were placed in the bottom of the extractor body to ensure proper drainage. A condenser was then fitted to the extractor body. All the joints were ground glass joints.

The thawed tissues were placed in a pre-weighed cellulose thimble which was then placed in the Soxhlet system. Heat was applied to the flask until the solvent started to boil ($80^{\circ}C$). The extraction was conducted until the solvent in the extractor body was completely colourless (24 to 48 hours).

The extracted tissues and thimble were air dried, placed in a $70^{\circ}C$ oven for about two days and allowed to cool in a dessicator. The thimble and tissues were weighed together.

The solvent remaining in the extractor body was discarded. The solvent in the flask containing the extracted hydrocarbons was filtered into a 250 ml glass separatory funnel, using a pre-weighed filter paper. The flask was rinsed several times with fresh pentane, which was also filtered. The sides of the filter paper were repeatedly washed with fresh pentane to ensure that all the hydrocarbons were carried into the separatory funnel. The filter paper was air dried, placed in a desiccator for about three days and weighed.

Approximately 50 ml of fresh pentane were added to the solution in the separatory funnel, and shaken vigorously for about five minutes. The layers were allowed to separate, and the bottom (benzene-methanol) layer was run into another 250 ml separatory funnel. Approximately 75 ml of fresh pentane were added to this second funnel, and shaken, and the bottom solvent layer run into a third separatory funnel. Another 75 ml of fresh pentane were added to the second funnel and the procedure repeated. After these separations, the benzene-methanol layer was discarded. The pentane portions in the second and third separatory funnels were combined with the first.

The combined pentane extracts were washed with about 50 ml of sodium chloride-saturated water in the first funnel, and the salt-water layer discarded.

The remaining pentane layer was decanted into a volumetric flask dried over anhydrous sodium sulphate, and filtered into a round-bottomed flask. Fresh pentane was used to wash the volumetric flask. This wash pentane was also filtered into the round-bottomed flask. The filter paper and residue were repeatedly rinsed with fresh pentane. The round-bottomed flask was placed on a rotary evaporator and the sample evaporated to dryness.

The sample was redissolved in about 10 ml of pentane and run through a previously prepared alumina-silica column (see above). The flask was rinsed several more times with small portions of pentane which were also run through the column. A total of about 75 ml of pentane containing the dissolved saturated hydrocarbons were collected in a pre-weighed 100 ml round-bottomed flask. The saturated hydrocarbon fraction was evaporated to dryness on a rotary evaporator. Dry nitrogen was passed over the sample for about ten minutes to drive off any remaining solvent vapours. The flask was kept in a desiccator for several days and then weighed.

For gas chromatography, the saturated hydrocarbon sample was redissolved in about 1 ml of pentane and transferred into a small glass vial. Another 1 ml of pentane was used to wash out the flask and added to the vial. This was repeated several times to ensure that the entire sample had been transferred. The sample in the vial was allowed to evaporate to dryness under a fume cupboard. The labelled vial was tightly sealed with an aluminum-lined lid and frozen until analyzed in the gas chromatograph.

After the pentane fraction had run through, about 70-80 ml

of benzene-pentane solution were passed through the column to elute the aromatics. The procedure was identical to that described above for the saturated hydrocarbon fraction.

Finally, approximately 70-80 ml of fresh benzene-methanol solution were passed through the column. The eluted material containing the dissolved nitrogen, sulphur, oxygen containing compounds was collected in a pre-weighed 100 ml round-bottomed flask. The procedure then followed that of the saturated hydrocarbon fraction.

RESULTS

The gravimetric results are given in Table 1. Because of the humid conditions in Bermuda, errors of several milligrams are common even with considerable precautions to assure sample container dryness, and thus samples smaller than 1 mg often appeared to have negative weight. Accurate quantitation of the fractions was made from the gas chromatograms, and these results are included as a second column for those analyses which were made.

Gas chromatograms of the pentane-extractable (aliphatic) portion of the hydrocarbons are given in Figures 1 through 14, and a blank sample carried through the entire extraction procedure is shown in Figure 15. These chromatograms exhibit considerable structure, which will be discussed below; the aromatic analysis, while useful for quantitation of the fraction, did not show easily recognizable patterns because of the large number of overlapping peaks.

DISCUSSION

Leander tenuicornis (Fig. 1) showed the same major peaks as found by Morris *et al*. (1976), at or near retention index corresponding to normal alkanes with carbon numbers 19, 21, 28, 35, and 37. These last two peaks are definitely not normal alkanes, but the first three may be. Their precise identity should be confirmed by mass spectrometry.

The tissues of female crabs (mixed species) with guts and gills removed (Fig. 3) showed the same peaks as *Leander*, with an additional peak near 21. The mixed-species, mixed sex crabs (Fig. 4) with guts and gills removed, on the other hand, showed only the three peaks at 28, 35, and 37; as did the tissues of male crabs with guts and gills removed (Fig. 5). The hydrocarbons in all these samples appear to be predominately biogenic, although the unresolved envelope and the series of small normal paraffin peaks are indicative of at least an equal amount of petroleum hydrocarbons.

When the crabs were separated by species (and the technique had been more fully developed) a high similarity was found between the hydrocarbon contents of *Planes* and *Portunus*. The gills of *Planes* (Fig. 12) and of *Portunus* (Fig. 11) both show a large unresolved complex mixture and small normal alkane series characteristic of petroleum, with only a single C_{28} peak that is an obvious biogenic component. It seems as if the gills of the crabs were entrapping particles of pelagic tar, from which the paraffins had been largely degraded (pelagic tar typical of the Bermuda region is more paraffinic than these chromatograms). Whether this elimination of paraffins was a function of the gills and their associated microbial community or whether the sample analyzed happened to come from a region polluted by a low-paraffin crude oil residue, cannot be established firmly. A second sample (Fig. 7) of *Planes* gills was too small to interpret; the guts and gills of female mixed-species crabs (Fig. 6) and male crabs (Fig. 2) are more paraffinic, but for some unexplained reason lack the C_{28} peak found in Figs. 11 and 12.

The *Planes* (Fig. 13) and *Portunus* (Fig. 14) tissues with gills removed show little that is obviously biogenic, and much more paraffin compared to the unresolved complex mixture, than was found in any of the gill samples. These chromatograms might easily be mistaken for a typical tar lump sample.

The *Syngnathus* sample did not confirm the results of Morris *et al.*, although both the digestive tracts (Fig. 9) and the tissues without digestive tracts (Fig. 8) showed an unresolved complex mixture and paraffins. The digestive tracts were similar to the crab gills (Figs. 11 and 12) in that there was a distinct C_{28} peak; but the structure was more complicated. The tissues with guts removed showed distinct peaks near 30, 32, 34, 36, which would appear to be biogenic because of the even-carbon preference, but such a series was not evident in the chromatogram of Morris *et al.*

CONCLUSIONS

This study has confirmed that the hydrocarbons of animals from the pelagic *Sargassum* community are predominantly of petroleum origin, and that these hydrocarbons are indeed distributed differently in the different tissues of the crabs and pipefish. However, the hypothesis that most of the petroleum would be found in the guts and gills was *not* confirmed. Petroleum hydrocarbons appear to be partitioned in a complex but reproducible way. In particular, samples of two different species of crab, *Planes minutus* and *Portunus sayi*, showed similar gill composition and extremely similar composition of the tissues with gills removed.

ACKNOWLEDGEMENTS

The authors would like to thank John A. Barnes for advice and Jacqlyn McLohne for assistance. This work was supported by the National Science Foundation under grants No. GX32883 and OCE-77-18662, and by the Bermuda Government.

Table I. Weight of Extractable Hydrocarbon Fractions from *Sargassum* Fauna

Sample	GC Fig.	Tissues (g)	Extracted Residue (g)	Saturated Hydrocarbons (g) gravimetric	Aromatics (g) gravimetric	NSO's* (g) gravimetric
Whole shrimp	1	1.92617	0.07421	0.00244	0.01278	0.04223
Mixed crabs with guts & gills removed	4	1.61450	0.09223	<0.001	0.01418	0.03976
Female crabs with guts & gills removed	3	0.75425	0.55254	<0.001	0.00720	0.01027
Guts & gills of female crabs	6	0.00594	0.05372	0.00563	<0.001	0.00488
Male crabs with guts & gills removed	5	0.25466	0.02295	<0.001	<0.001	<0.001
Guts & gills of male crabs	2	0.00010	0.00720	<0.001	<0.001	0.00442
Planes minutus with gills removed	13	1.32274	0.08583	0.00168	0.00491	0.01989
Gills of *Planes minutus*	7	<0.001	0.21591	0.00071	<0.001	0.00544
Gills of *Planes minutus*	12	0.00009	0.00713	<0.001	0.00393	0.01016
Portunus sp. with gills removed	14	0.64602	0.06582	<0.001	0.01104	0.00817
Gills of *Portunus sp.*	11	<0.001	0.00450	0.00200	0.00710	0.00118

Table I. Weight of Extractable Hydrocarbon Fractions from *Sargassum* Fauna (cont.)

Sample	GC Fig.	Tissues (g)	Extracted Residue (g)	Saturated Hydrocarbons (g) gravimetric	Aromatics (g) gravimetric	NSO's* (g) gravimetric
Syngnathus pelagicus with guts removed	8	0.79588	0.07087	0.00325	0.00196	0.02250
Guts of *Syngnathus pelagicus*	9	<0.001	0.00711	<0.001	<0.001	0.00573
Blank	15					

*NSO's are nitrogen, sulphur, oxygen-containing compounds

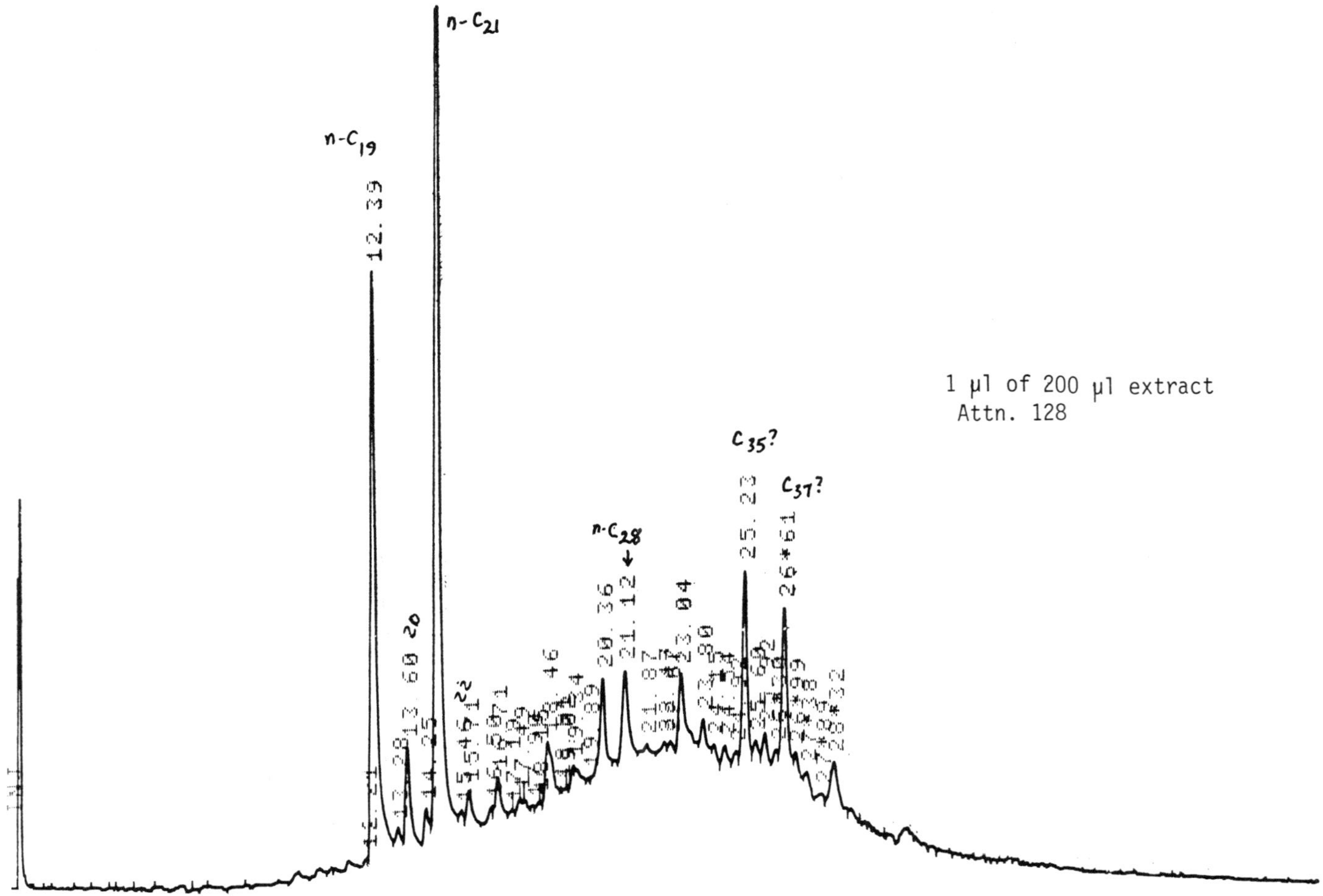

Fig. 1. Gas chromatogram of extract from whole shrimp (Leander tenuicornis) 1 ul of 200 ul total extract.

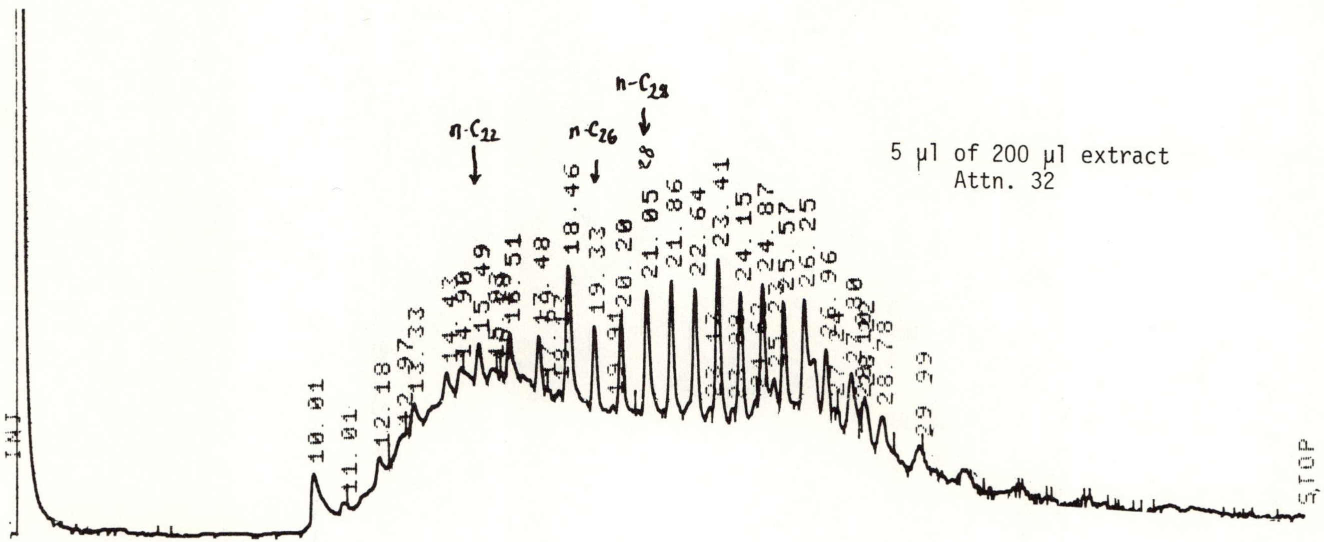

Fig. 2. Gas chromatogram of extract from guts and gills of male crabs (mixed species).

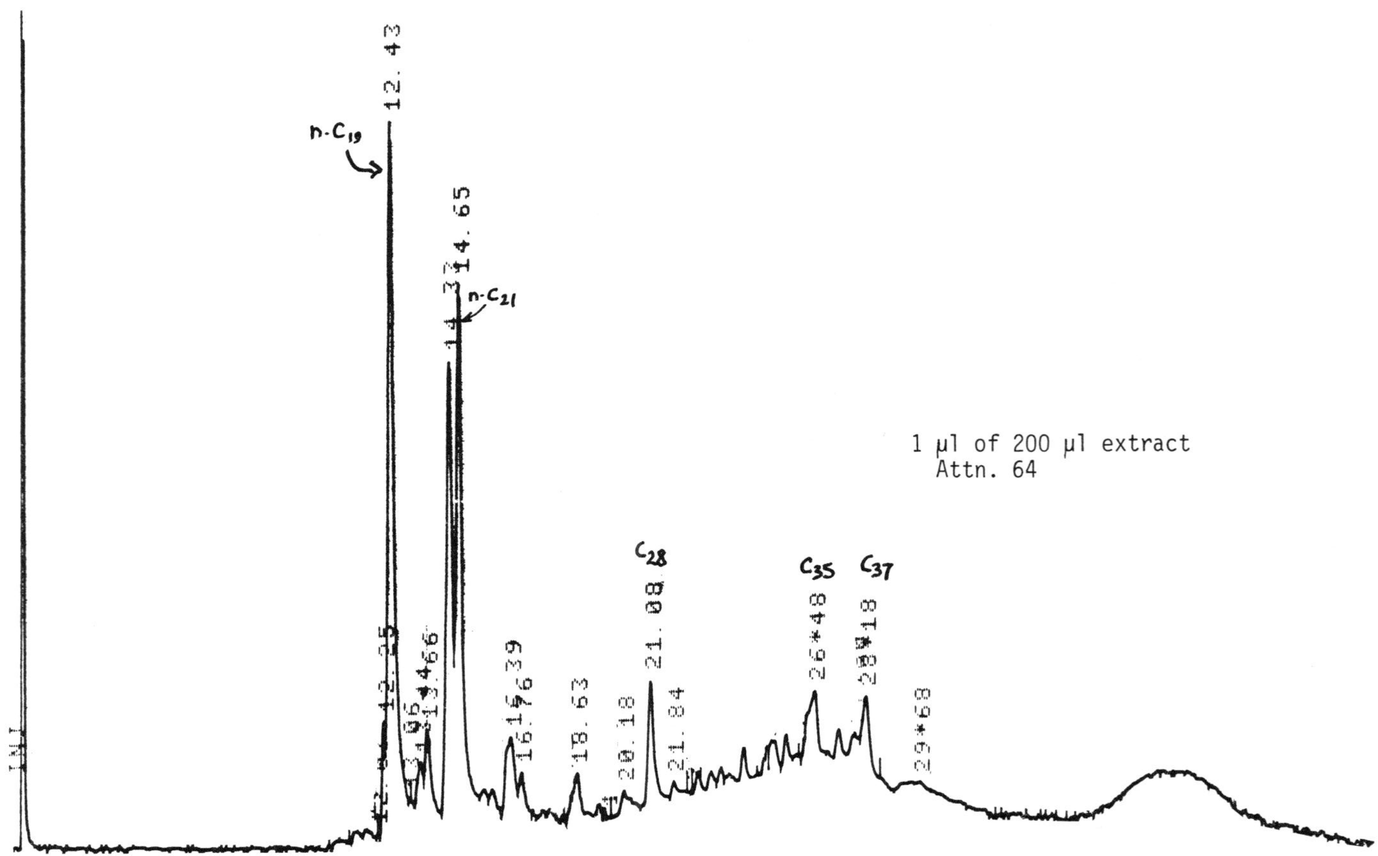

Fig. 3a. Gas chromatogram of extract from tissues of female crabs (mixed) with guts and gills removed.

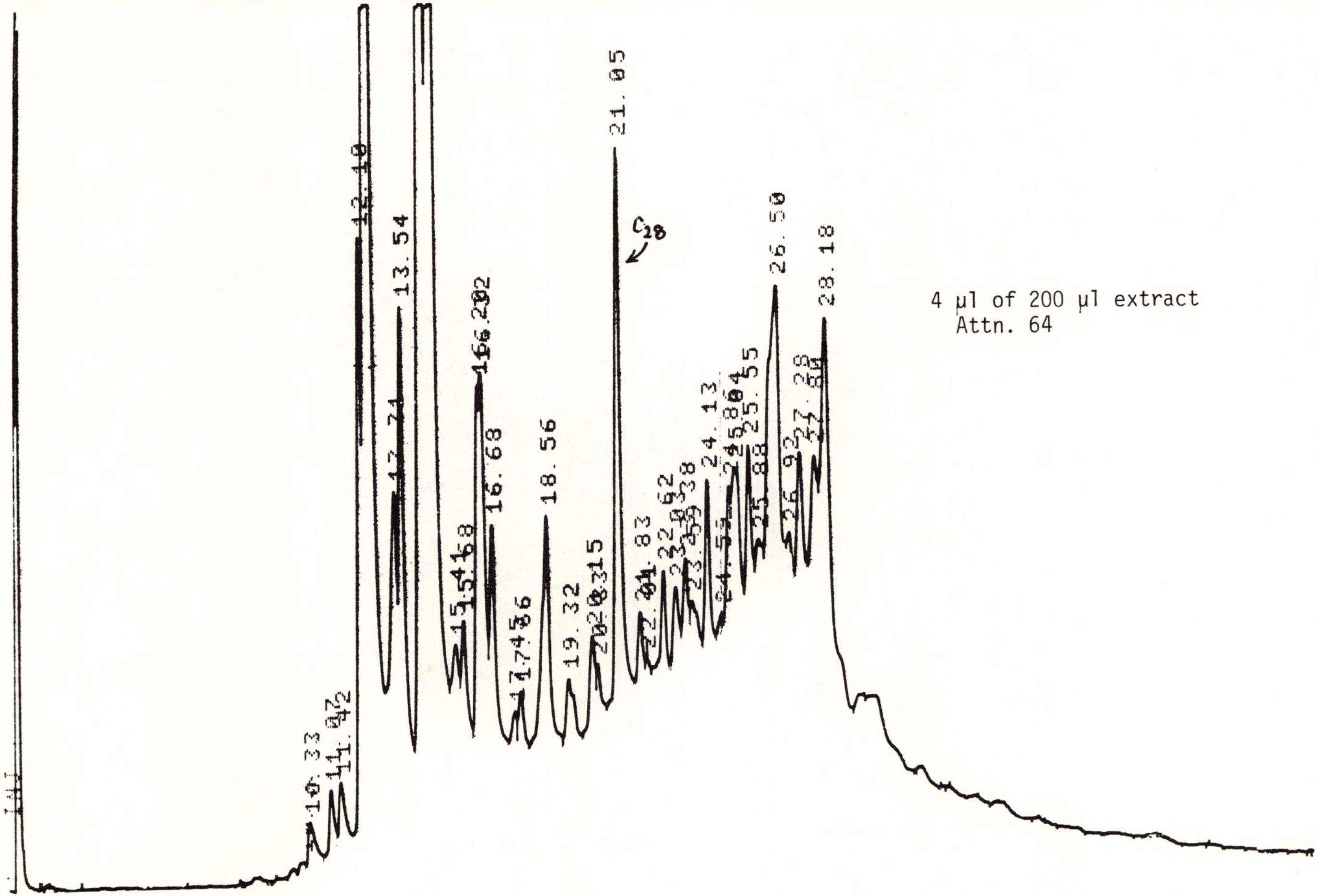

Fig. 3b. Larger injection of same extract as 3a.

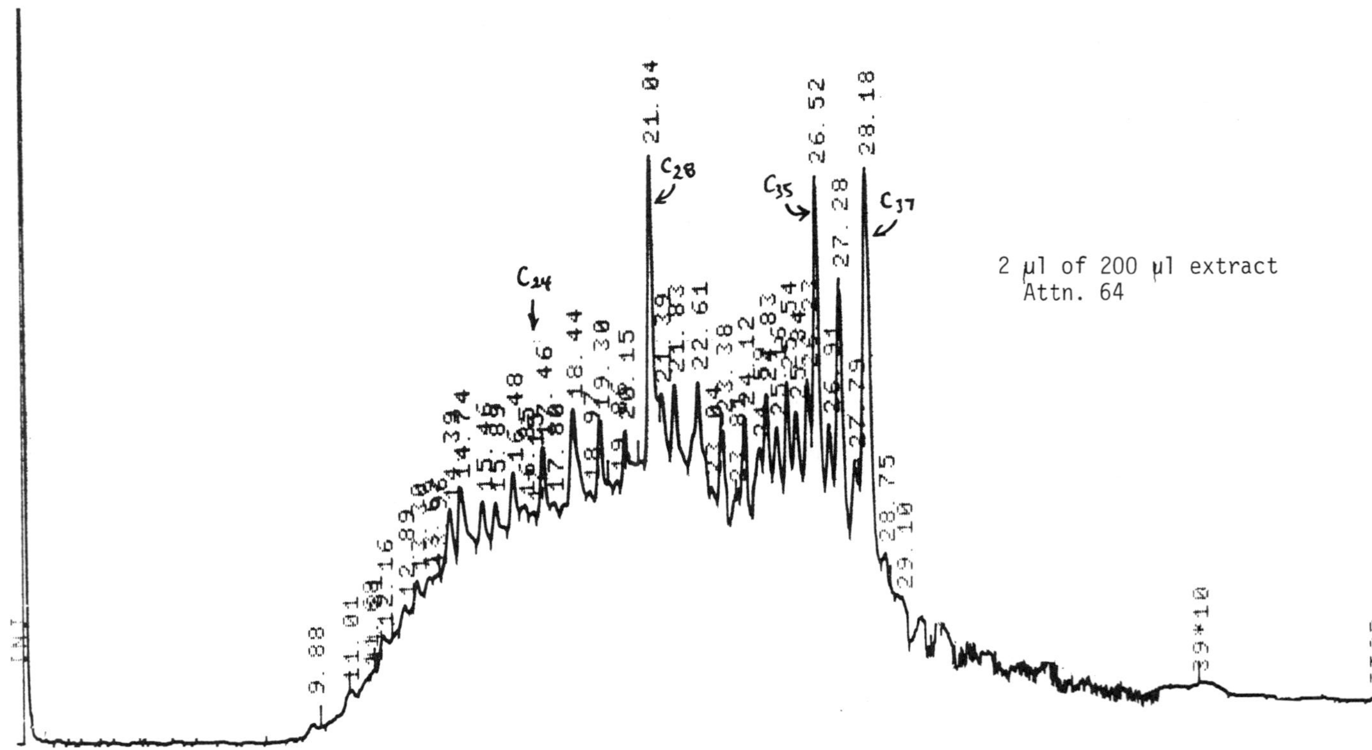

Fig. 4. Gas chromatogram of extract from crab tissues (mixed species, mixed sex) with guts and gills removed.

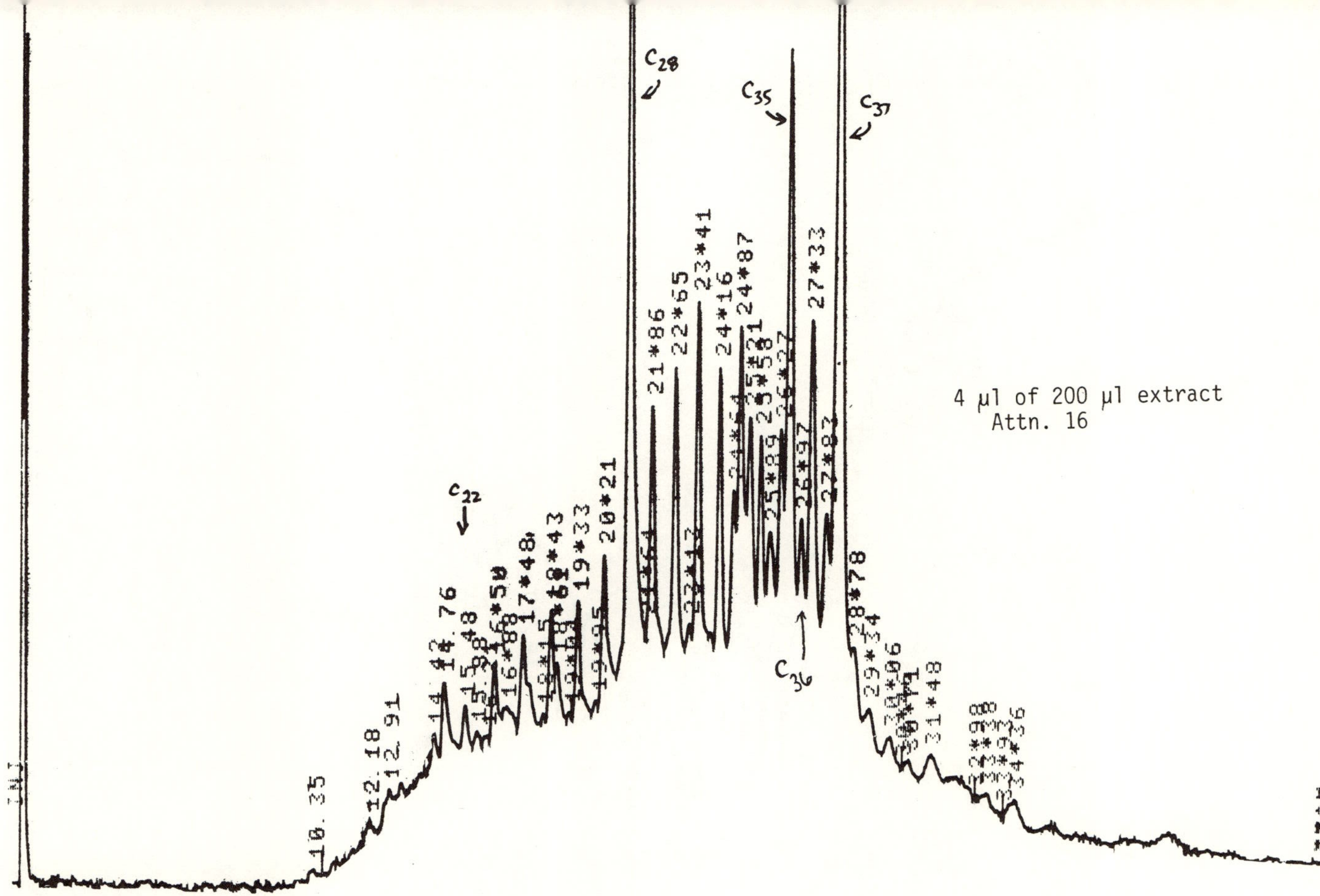

Fig. 5. Gas chromatogram of extract from male crabs with guts and gills removed.

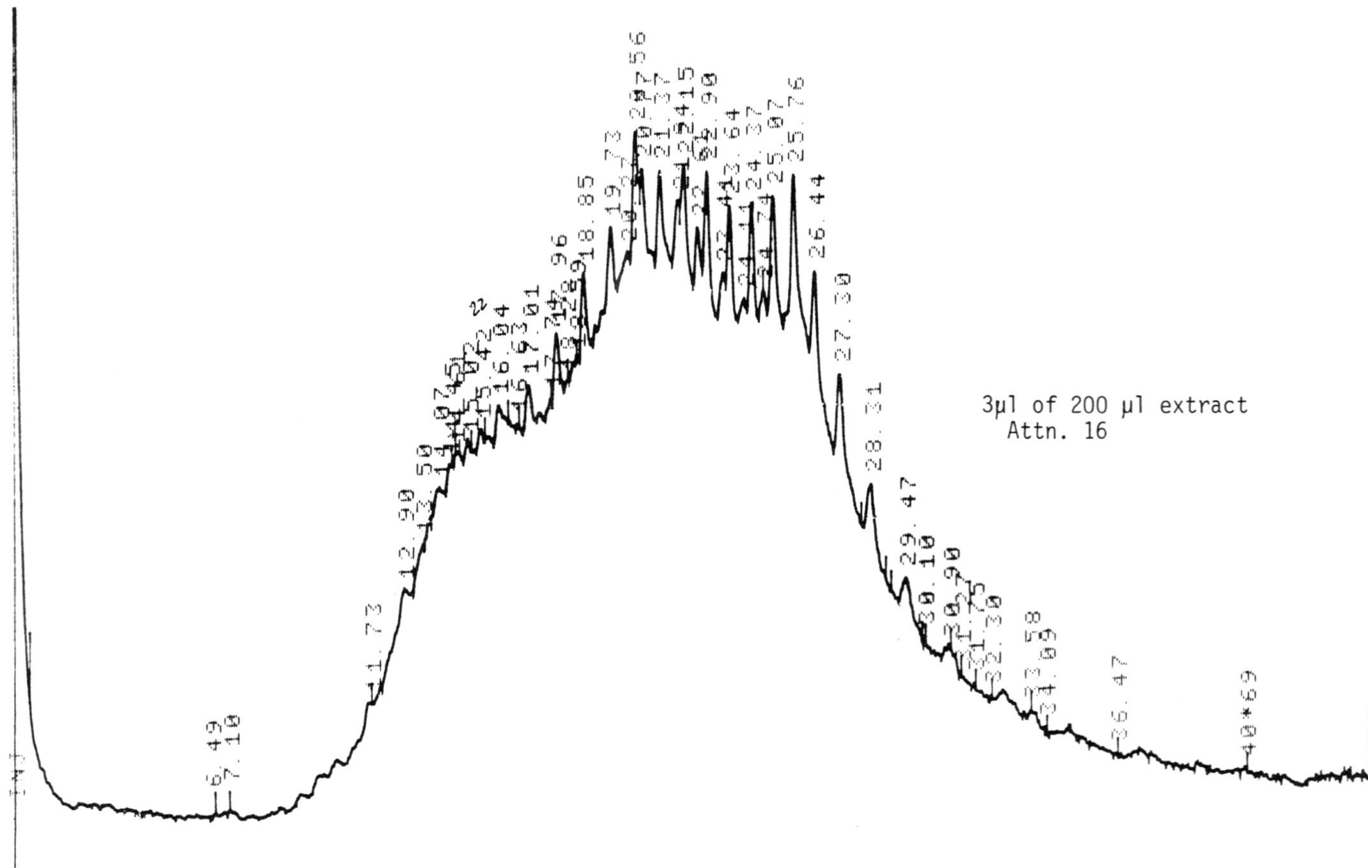

Fig. 6. Gas chromatogram of extract from female crabs (mixed species). Guts and gills only.

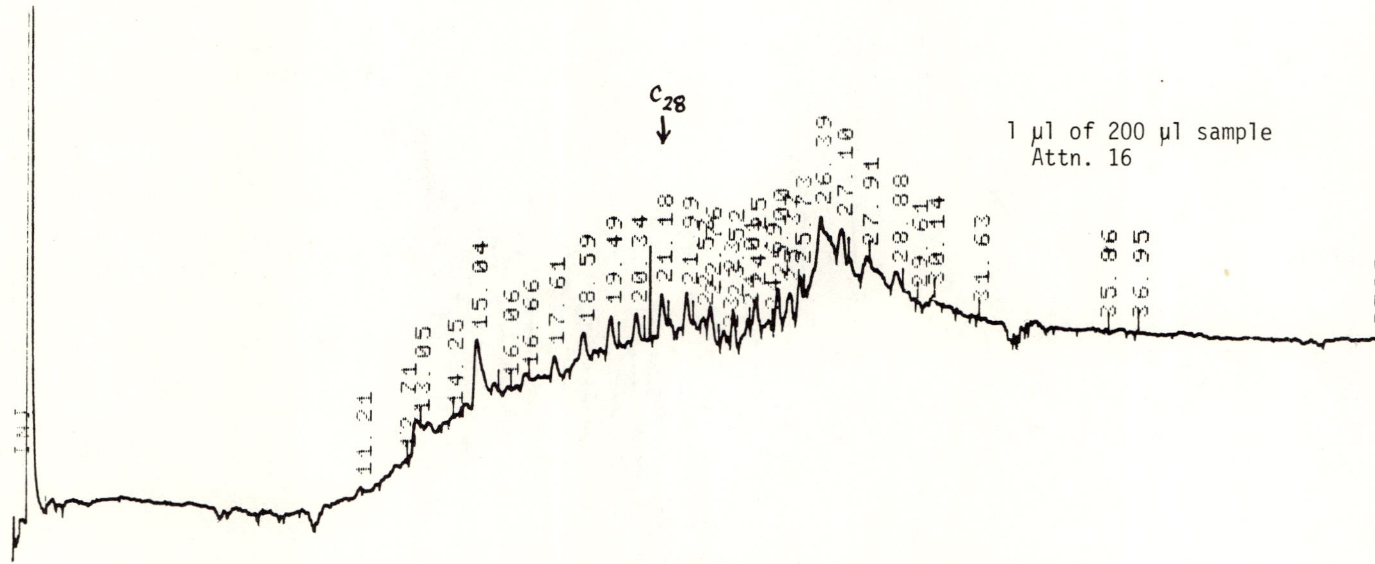

Fig. 7. Gas chromatogram of extract from gills of crab (<u>Planes minutus</u>).

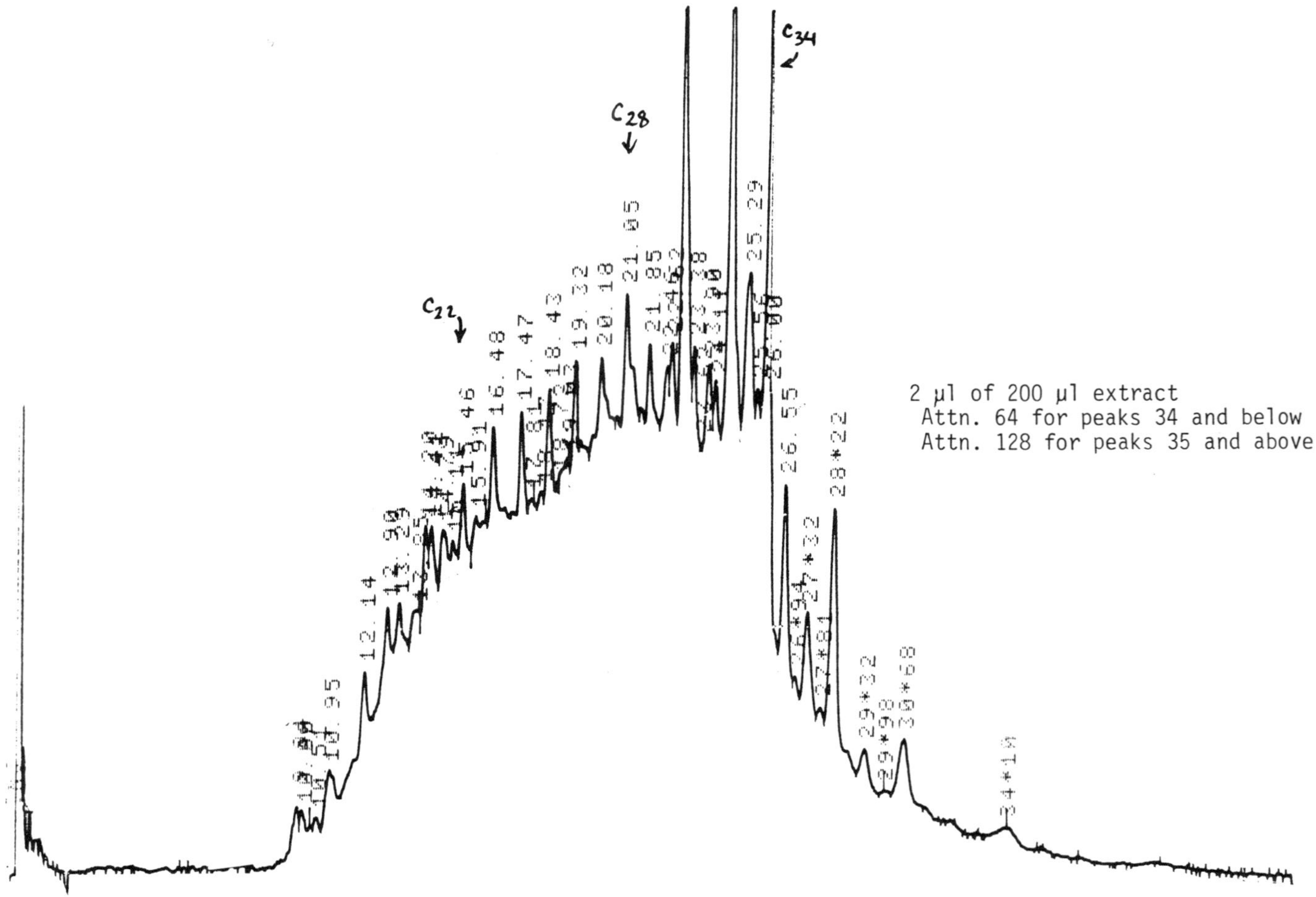

Fig. 8. Gas chromatogram of extract from pipefish (Syngnathus pelagicus) with guts removed.

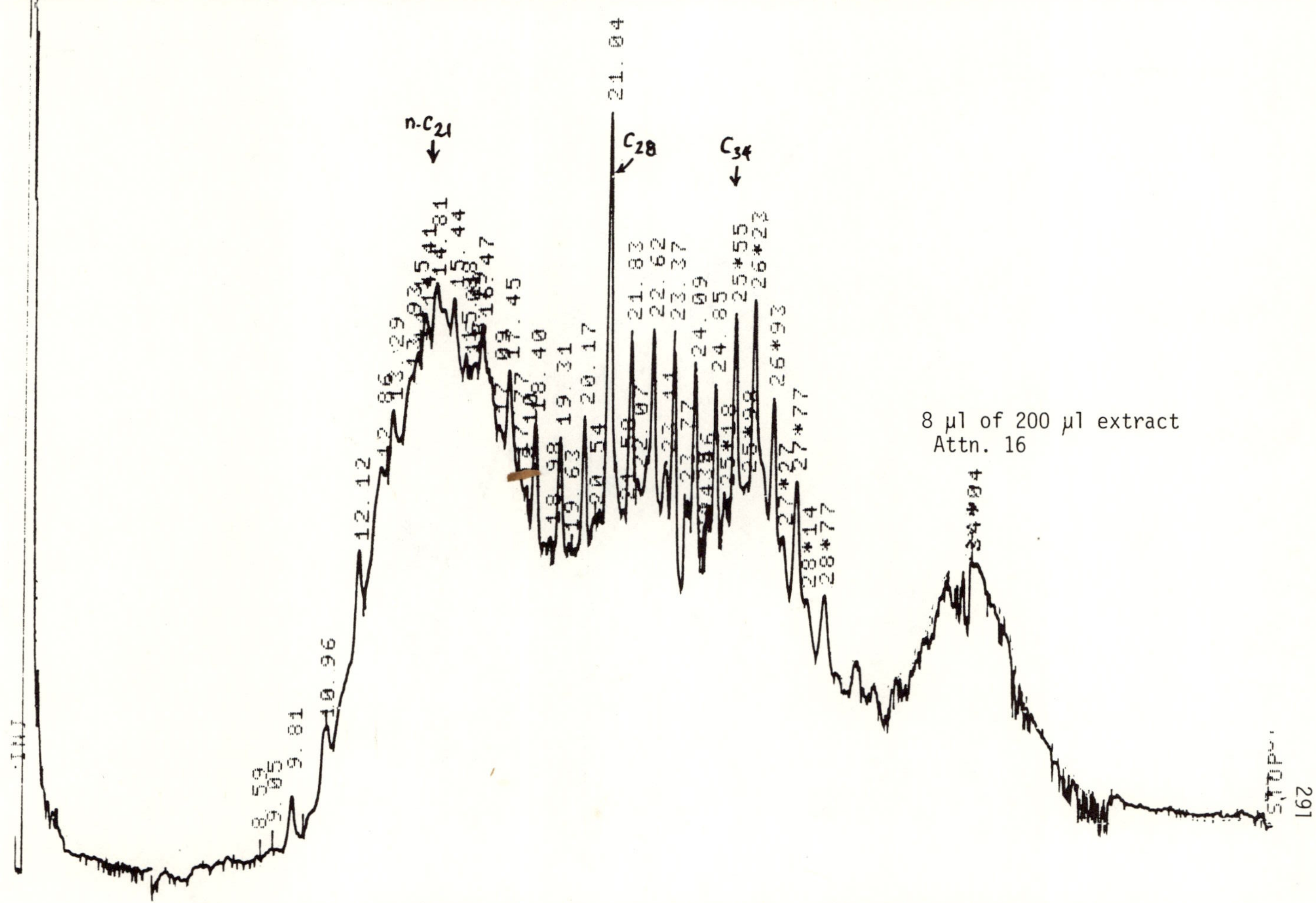

Fig. 9. Gas chromatogram of extract from digestive tracts of <u>Syngnathus pelagicus</u>.

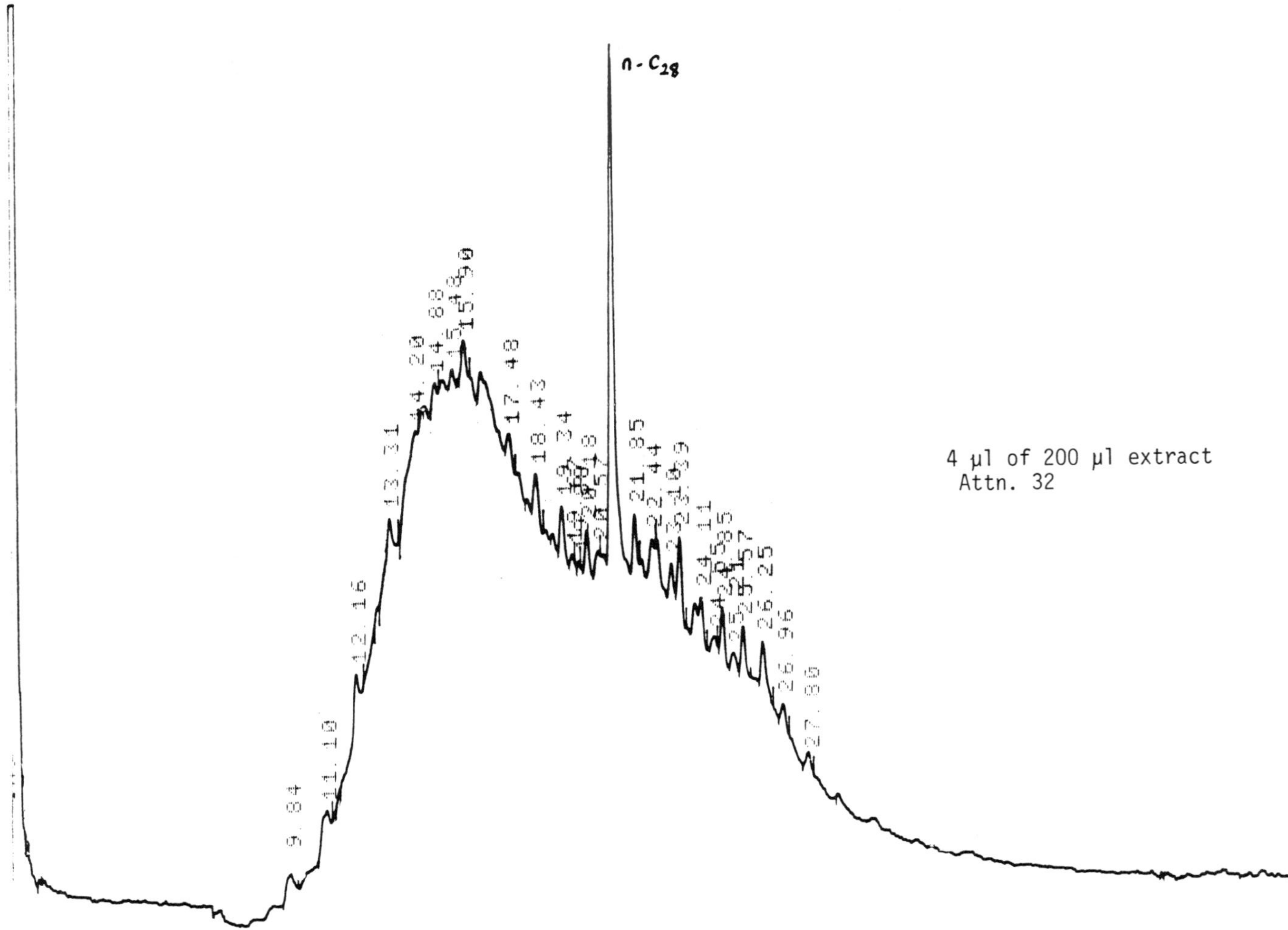

Fig. 11. Gas chromatogram of extract from gills of crab (<u>Portunus sayi</u>).

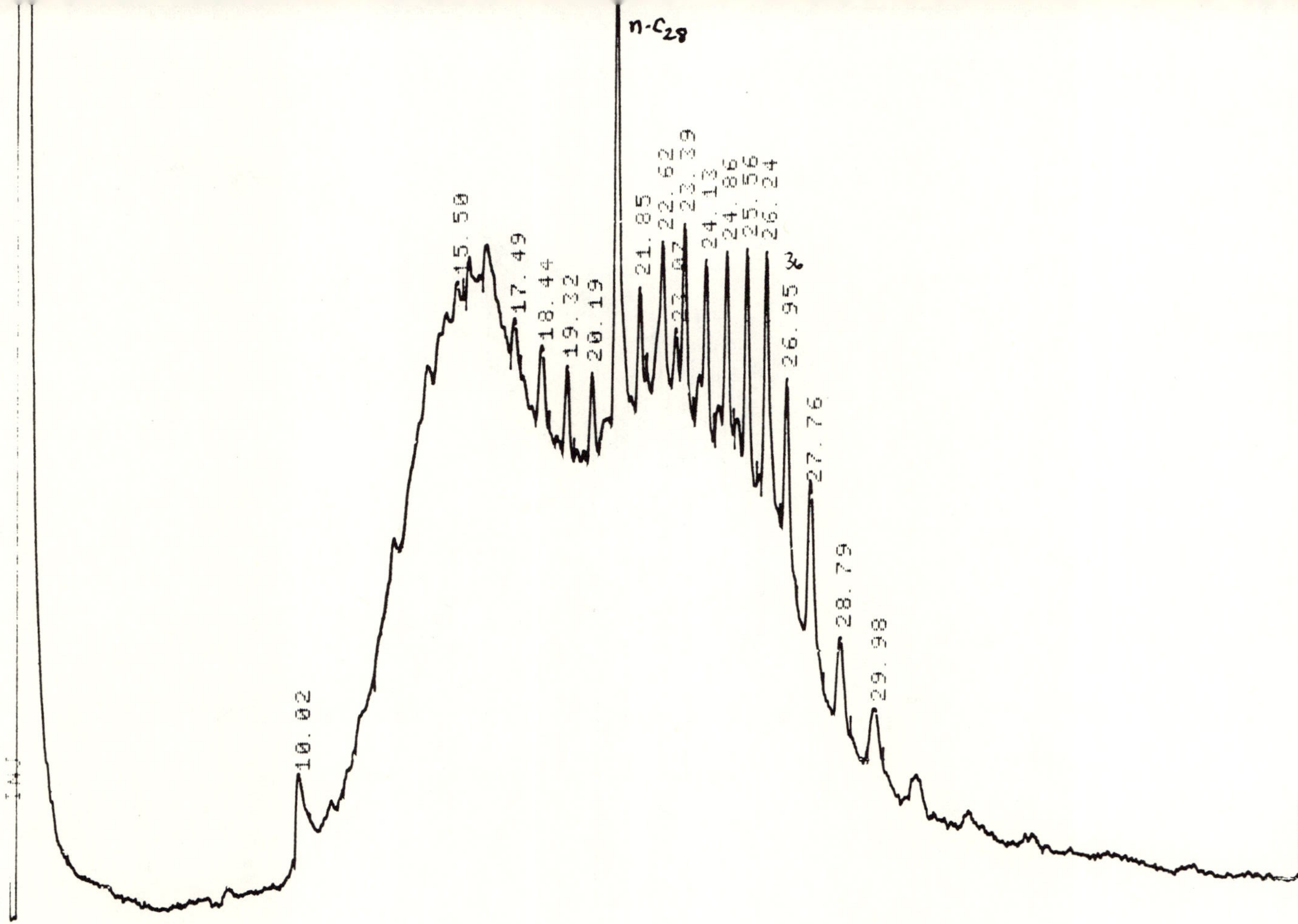

Fig. 12. Gas chromatogram of extract from gills of crab (<u>Planes minutus</u>).

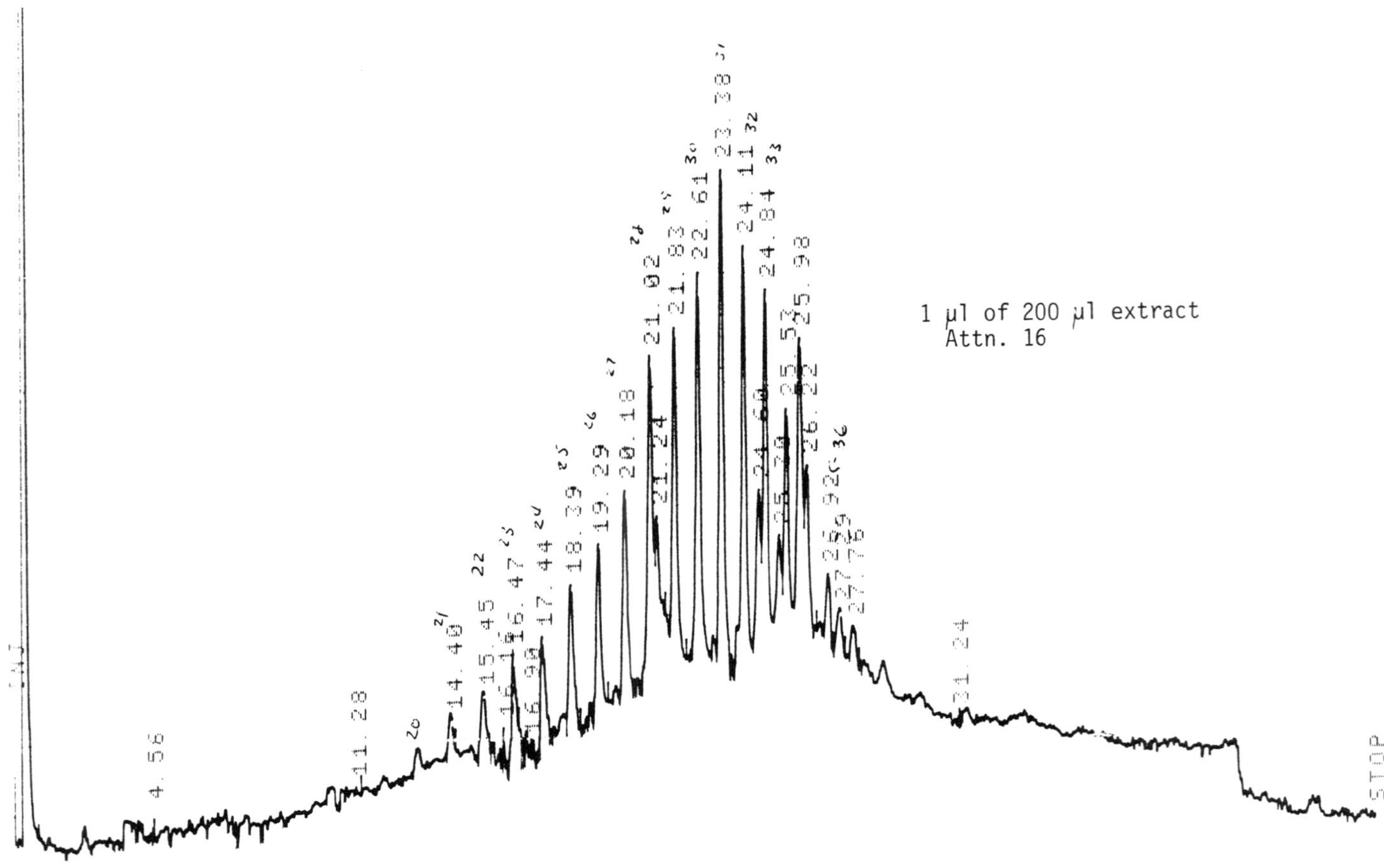

Fig. 13. Gas chromatogram of extract from tissues of crab (<u>Planes minutus</u>) with gills removed.

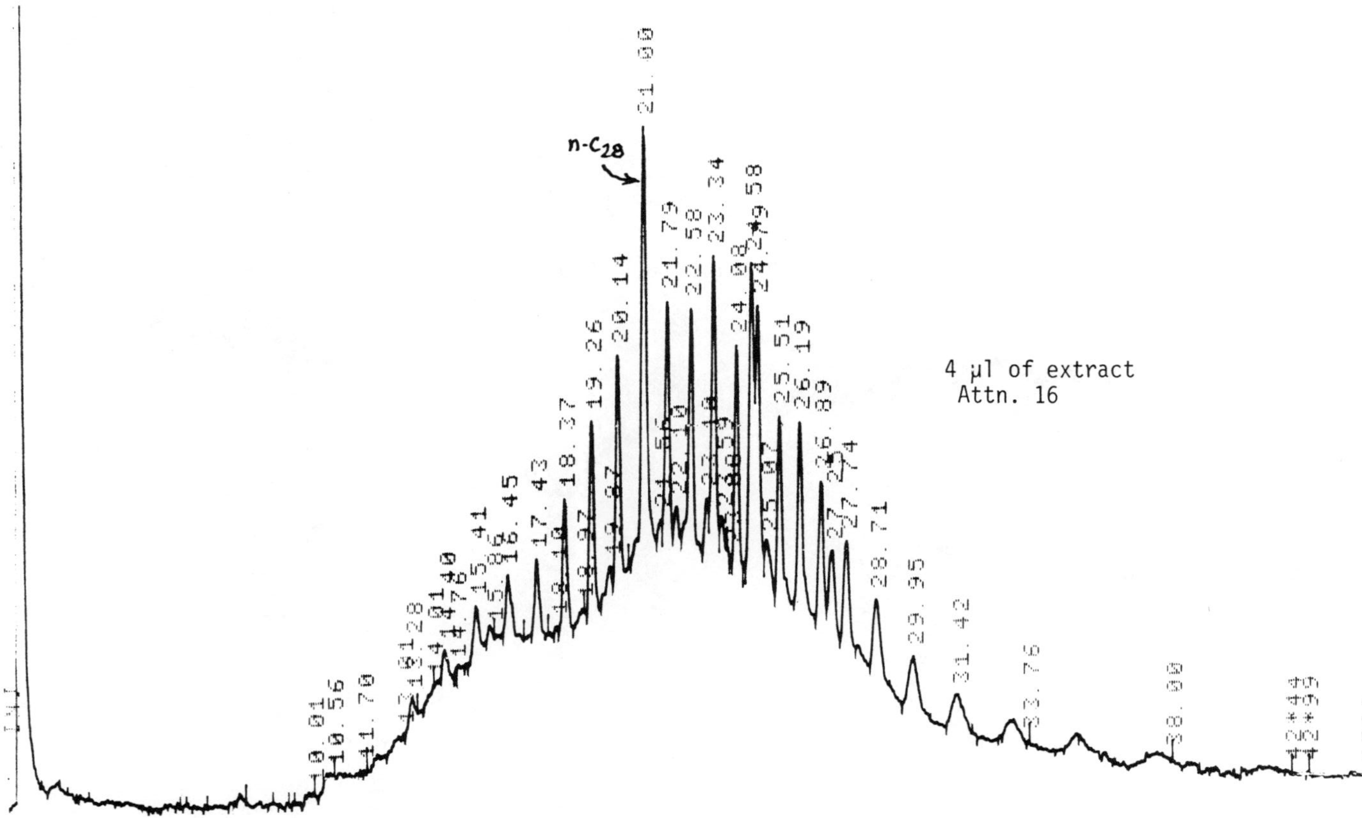

Fig. 14. Gas chromatogram of extract from <u>Portunus sayi</u> tissues with gills removed.

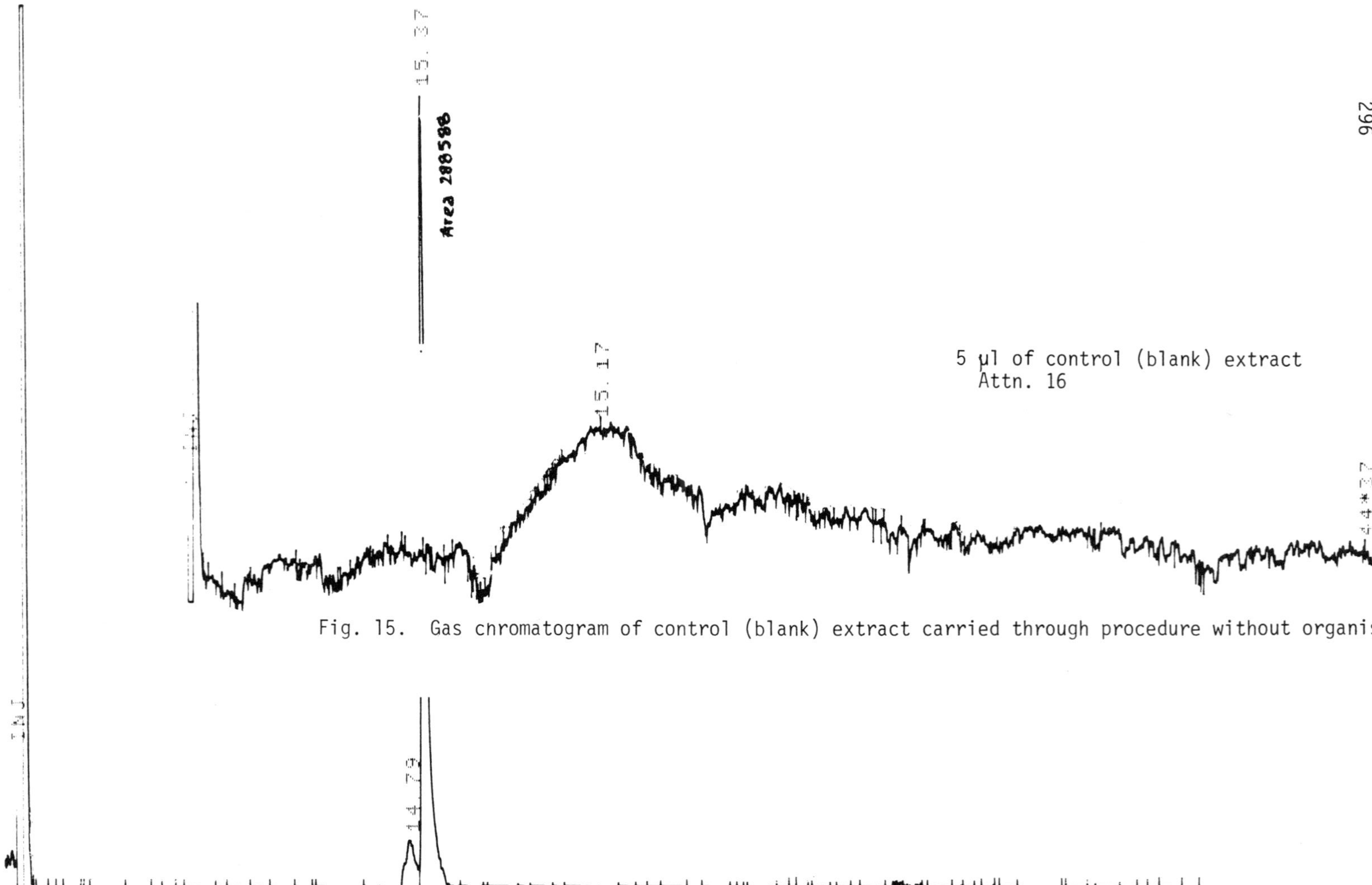

Fig. 15. Gas chromatogram of control (blank) extract carried through procedure without organisms.